The
HISTORY
of
MODERN
MATHEMATICS

Volume I:
IDEAS AND THEIR RECEPTION

Photograph of Participants, Symposium on the History of Modern Mathematics,
Vassar College, Poughkeepsie, New York, June 20–24, 1989.
Photo copyright © 1989 by I. Grattan-Guinness.

The
HISTORY
of
MODERN
MATHEMATICS

Volume I:
IDEAS AND THEIR RECEPTION

Proceedings of the Symposium on the
History of Modern Mathematics
Vassar College, Poughkeepsie, New York
June 20–24, 1988

Edited by

DAVID E. ROWE

JOHN MCCLEARY

ACADEMIC PRESS
Harcourt Brace Jovanovich, Publishers
Boston San Diego New York
Berkeley London Sydney
Tokyo Toronto

ACADEMIC PRESS, INC
San Diego, California 92101

United Kingdom Edition published by
ACADEMIC PRESS LIMITED
24-28 Oval Road, London NW1 7DX

Library of Congress Cataloging-in-Publication Data

Symposium on the History of Modern Mathematics (1988 : Vassar College)
    The history of modern mathematics : proceedings of the Symposium
on the History of Modern Mathematics, Vassar College, Poughkeepsie,
New York, 20–24 June 1988 / edited by David E. Rowe and John
McCleary.
        p.   cm.
    Includes bibliographical references.
    Contents: v. 1. Ideas and their reception — v. 2. Institutions
and applications.
    ISBN 0-12-599661-6 (v. 1 : alk. paper). — ISBN 0-12-599662-4 (v.
2 : alk. paper)
    1. Mathematics — History — Congresses. I. Rowe, David E., Date–
. II. McCleary, John, Date– . III. Title.
QA21.S98   1988
510'.9—dc20   89-17766
                                                 CIP

PRINTED IN THE UNITED STATES OF AMERICA
90 91 92 93 94   9 8 7 6 5 4 3

Contents

Table of Contents for Volume II   vii
Contributors List   ix
Preface   xi

The Context of Reception

A Theory of Reception for the History of Mathematics   3
John McCleary

Riemann's *Habilitationsvortrag* and the Synthetic *A Priori* Status   17
of Geometry
Gregory Nowak

Foundations of Mathematics

Cantor's Views on the Foundations of Mathematics   49
Walter Purkert

Kronecker's Views on the Foundations of Mathematics   67
Harold M. Edwards

Toward a History of Cantor's Continuum Problem   79
Gregory H. Moore

National Styles in Algebra

British Synthetic vs. French Analytic Styles of Algebra   125
in the Early American Republic
Helena M. Pycior

Toward a History of Nineteenth-Century Invariant Theory   157
Karen Hunger Parshall

Geometry and the Emergence of Transformation Groups

The Early Geometrical Works of Sophus Lie and Felix Klein   209
David E. Rowe

Line Geometry, Differential Equations, and the Birth   275
of Lie's Theory of Groups
Thomas Hawkins

Projective and Algebraic Geometry

The Background to Gergonne's Treatment of Duality: 331
Spherical Trigonometry in the Late 18th Century
Karine Chemla

Algebraic Geometry in the Late Nineteenth Century   361
J. J. Gray

Abel's Theorem

Abel's Theorem   389
Roger Cooke

Number Theory

Heinrich Weber and the Emergence of Class Field Theory   425
Günther Frei

Notes on the Contributors   451

Contents for Volume II

Table of Contents for Volume I   vii
Contributors List   ix
Preface   xi

The Crossroads of Mathematics and Physics

Crystallographic Symmetry Concepts and Group Theory (1850-1880)   3
Erhard Scholz

Physics as a Constraint on Mathematical Research:
The Case of Potential Theory and Electrodynamics   29
Thomas Archibald

The Geometrization of Analytical Mechanics: A Pioneering   77
Contribution by J. Liouville (ca. 1850)
Jesper Lützen

Schouten, Levi-Civita, and the Emergence of Tensor Calculus   99
Dirk J. Struik

Applied Mathematics in the Early 19th-Century France

Modes and Manners of Applied Mathematics: 109
The Case of Mechanics
I. Grattan-Guinness

La Propagation des Ondes en Eau Profonde et ses Développements   129
Mathématiques: (Poisson, Cauchy 1815–1825)
Amy Dahan Dalmedico

Pure versus Applied Mathematics in Late 19th-Century Germany

Pure and Applied Mathematics in Divergent Institutional Settings   171
in Germany: The Role and Impact of Felix Klein
Gert Schubring

On the Contribution of Mathematical Societies to Promoting   223
Applications of Mathematics in Germany
Renate Tobies

Mathematics at the Berlin Technische Hochschule/Technische   251
Universität: Social, Institutional, and Scientific Aspects
Eberhard Knobloch

Applied Mathematics in the United States During World War II

Mathematicians at War: Warren Weaver and the Applied   287
Mathematics Panel, 1942–1945
Larry Owens

The Transformation of Numerical Analysis by the Computer: 307
An Example from the Work of John von Neumann
William Aspray

Notes on the Contributors   323

Contributors List

The number in the parentheses refers to the pages on which the author's contribution begins.

Karine Chemla (331)
REHSEIS–CNRS
Université Paris-Nord
Centre Scientifique et Polytechnique
Avenue J. B. Clément
93430 Villetaneuse, France

Roger Cooke (389)
Department of Mathematics
University of Vermont
Burlington, VT 05405

Harold M. Edwards (67)
Courant Institute of Mathematics
New York University
251 Mercer Street
New York, NY 10012

Günther Frei (425)
Université Laval
Faculté des Science
Quebec, Canada G1K 7P4

Jeremy Gray (361)
Faculty of Mathematics
Open University
Milton Keynes MK7 6AA, England

Thomas Hawkins (275)
Department of Mathematics
Boston University
111 Cunningham Street
Boston, MA 02215

John McCleary (3)
Department of Mathematics
Vassar College
Poughkeepsie, NY 12601

Gregory H. Moore (79)
Department of Mathematics
McMaster University
Hamilton, Ontario, Canada, TSH 719 B

Gregory Nowak (17)
History of Science Program
Dickinson Hall
Princeton University
Princeton, NJ 08544

Karen Parshall (157)
Department of Mathematics
University of Virginia
Charlottesville, VA 22903

Walter Purkert (49)
Sektion Mathematik
Karl-Marx-Universität
7010 Leipzig, German Democratic Republic

Helena Pycior (125)
Department of History
University of Wisconsin
Milwaukee, WI 53201

David E. Rowe (209)
Department of Mathematics
Pace University
Pleasantville, NY 10570

Preface

Within the past decade, considerable attention has been directed toward the task of studying and rewriting the history of mathematics during the 19th and early 20th centuries. While earlier generations of scholars expended most of their efforts researching the history of ancient mathematics or the mathematics of the Renaissance and Scientific Revolution, a growing number of contemporary historians have been drawn to the modern era, i.e., the period roughly spanned by the publication of Gauss's *Disquisitiones arithmeticae* in 1801 to the advent of high-speed electronic computers around 1950. This recent interest reflects a widespread recognition of the importance of an era in which mathematicians produced a vast corpus of rich new ideas that form the foundations of the subject as we know it today. It should not be overlooked, however, that this period was marked not only by revolutionary intellectual achievements but that it also witnessed the emergence of mathematics as an autonomous academic discipline. Over the course of the 19th century, mathematicians began to articulate new standards for teaching and research. The process began slowly, but by the last third of the century such issues were often debated, adopted, or promoted by various professional organizations. These mutually reinforcing intellectual and institutional developments profoundly affected the traditional role of mathematics within the exact sciences, providing it with a radically new orientation, one whose background has not received sufficient attention in the historical literature.

The twenty-four essays presented herein represent a sampling of recent efforts to redress this situation. They constitute the fruits of the Symposium on the History of Modern Mathematics held at Vassar College from June 20–24, 1988. Forty-five participants from twelve countries took part in this symposium which featured twenty-nine lectures covering a variety of topics from the era of Euler and Clairaut to the age of modern computers. In organizing this meeting, we sought to adopt a program that would accomplish three goals: (1) reflect the diverse approaches recently taken by historians of mathematics who have studied various aspects of this period; (2) stress certain characteristic themes and ideas that illustrate its development; and (3) emphasize issues and ideas that shed some light on relationships between mathematics and other scientific disciplines. By so doing, we hoped to give the symposium reasonable thematic unity while at the same time showcasing the wide spectrum of research that is now being done on the history of modern mathematics. Above all else, we wanted to avoid the pitfall of producing a hodgepodge collection of highly specialized articles. Although some of the essays below obviously presuppose a fairly sophisticated knowledge of and interest in the history of mathematics during this period, they are intended for the *general reader* who satisfies these prerequisites. These essays do not pretend to cover all major areas of research—indeed, the monumental achievements of 19th-century mathematicians in the fields of real and complex analysis receive only passing attention here—nor should they be regarded as the last word on the subjects they do address. In most cases, the authors set forth conclusions based on research that is still very much in progress. The standards by which we would like to see these studies measured, therefore, are those of cohesiveness, originality, significance, and the strength of the argumentation and the documentary evidence used to support it.

Volume I is concerned with the emergence and reception of major ideas in fields that range from foundations and set theory, algebra and invariant theory, and number theory to differential geometry, projective and algebraic geometry, line geometry, and transformation groups. The introductory essay by John McCleary raises some larger historiographical concerns regarding the reception of mathematical works *qua* literature. In one form or another these issues are echoed in many of the contributions to both volumes. For example, the complex process of assimilating and enlarging the fundamental achievements of Gauss and Riemann in differential geometry is examined from three different standpoints in the articles by Gregory Nowak, Jesper Lützen, and Dirk Struik. The essays by Walter Purkert and Harold Edwards, on the other hand, recreate the *Urwelt* of two opposing views on the foundations of mathematics—those of Cantor and Kronecker, respectively. By so doing, they reveal that both men have been badly misunderstood not only by their contemporaries but by those who have represented their respective philosophies ever since. Whereas Cantor's famous continuum hypothesis was regarded as somewhat akin to a "self-evident truth" (albeit one that required proof), Gregory Moore's analysis reveals the large number of hats it has worn and the key part it has played in the emergence of axiomatic set theory.

The reception of European educational traditions in the United States forms the larger background against which Helena Pycior describes early American algebra. By examining the British and French textbooks adopted at leading American educational institutions like Harvard and West Point, she offers some new insights about the dynamics that shaped this transmission process. Karen Parshall's essay, on the other hand, illustrates how difficult the transmission of foreign ideas can sometimes be, particularly when they are imbedded within an indigenous research tradition. J. J. Sylvester and Paul Gordan, two leading exponents of the British and German approaches to invariant theory, were both eager to understand the others work, particularly since both shared a strong desire to solve the finite basis problem. Yet despite their best efforts, neither was able to penetrate fully the other man's techniques.

Karine Chemla's essay documents the prehistory of the duality concept, which was vitally important for 19th-century projective geometry, by tracing its roots in 18th-century spherical trigonometry. The central role duality, which came into play fifty years after Gergonne called attention to it in the 1820s, is amply illustrated in the articles by Rowe, Hawkins, and Gray. David Rowe and Thomas Hawkins take up a theme that is also discussed at length in Erhard Scholz's paper, namely the work of Lie and Klein on line geometry and transformation groups during the years 1869-1873. Rowe's article stresses the background events that led up to Lie's discovery of the line-to-sphere transformation and the "Erlanger Programm," whereas Hawkins focuses his attention on the rich variety of ideas that contributed to Lie's creation of a new theory of continuous groups. Jeremy Gray's overview of major developments in algebraic geometry highlights, among other things, the fundamental significance of Abel's Theorem for the whole subject. Abel's Theorem was a major tool for two of the key players Gray considers, Alexander Brill and Max Noether. As mathematicians are sometimes wont to do, these two tried to reconstruct the source of Abel's original inspiration in their classic study *Die Entwicklung der Theorie der algebraischen Funktionen in älterer und neuerer Zeit*. Following in this grand tradition, Roger Cooke offers his own interpretation of what Abel was up to and where Brill and Noether went wrong. Volume I then concludes with a contribution to the history of number theory by Günther Frei who traces the roots of Heinrich Weber's class field theory back to Euler's ζ-function and the fundamental contributions of Dirichlet, Kronecker, and Dedekind to algebraic number theory.

Whereas Volume I is almost exclusively concerned with developments in pure mathematics, the emphasis in Volume II shifts to ideas that lie at the crossroads of mathematics and physics—crystallography and group structures, potential theory and electrodynamics, differential geometry and classical and relativistic mechanics. On the institutional front, particular attention has been focused on issues pertaining to the tensions and interplay between pure and applied mathematics as these arose in three different historical settings: early 19th-century France, Wilhelmian era Germany, and the United States during and after World War II. Erhard Scholz and Thomas Archibald offer two case studies of mathematical theories that emerged in symbiotic union with physical research. Scholz shows how crystallographers, in particular the French theorist August Bravais, were studying group structures decades before Camille Jordan, following in Bravais' footsteps, launched the abstract theory of geometrical groups. Archibald's article deals with the work of German mathematicians who sought to develop an approach to potential theory consistent with Wilhelm Weber's action-at-a-distance model for electrodynamics. This led Riemann, Carl Neumann, and others to study potentials propagated with finite velocity. As Maxwell's theory gradually displaced Weber's, however, this research program lost its initial impetus, and potential theory turned increasingly toward problems of a purely mathematical nature.

The interaction between mathematics and physics is also a major theme (though played in a minor key) in the articles by Jesper Lützen and Dirk Struik. Lützen's analysis of an unpublished paper of Liouville reveals the masterful command the Frenchman had of Gauss's intrinsic approach to surface theory. In this work Liouville derives the Hamilton-Jacobi equations of classical mechanics by treating the motion of a particle in a force field as equivalent to regarding the same particle in a force-free setting in which it moves along a geodesic curve on a certain surface. Struik's paper on the emergence of modern tensor analysis gives a first-hand account of the sudden interest this field attracted immediately after the publication of Einstein's theory of general relativity in 1915. As a final illustration of ideas that arose from a multiplicity of fields, William Aspray's essay analyzes the revolutionary impact of high-speed computers on numerical analysis by highlighting the solutions devised by John von Neumann to a variety of real-world problems.

That ubiquitous and despairingly simple question—what is applied mathematics?—crops up in a number of the essays in Volume II. Ivor Grattan-Guinness's essay shows how complex this matter had already become for Parisian mathematicians during the early decades of the 19th century. With respect to applications, he identifies two large camps of practitioners with diametrically opposite views: the theoreticians, whose prototype was Lagrange, and the engineer *savants*, typified by Lazare Carnot. Amy Dahan's paper examines how two of the theoreticians, namely Poisson and Cauchy, dealt with a problem in hydrodynamics, one which turned out to have important implications for the theory of Fourier series. Larry Owens' study reveals that a strikingly similar debate over applications took place more than 100 years later in the United States, but this time colored by the tense atmosphere of wartime. Owens identifies

Warren Weaver as the leading spokesman for "useful" applied mathematics, whereas Norbert Wiener and John von Neumann are representative examples of the theoretical applier. The debate over applications of mathematics in Wilhelmian Germany, the principal theme of Renate Tobies's article, was, if anything, even more electric owing to the institutional rift that separated the engineering schools from the universities. Her essay and the one by Gert Schubring analyze the central and oft-debated role of Felix Klein in reforming mathematics education in Germany. Both authors concur that the principal motivation behind Klein's numerous reform proposals was to ensure that mathematics instruction would not lose touch with the type of training demanded by modern engineering science. Finally, Eberhard Knobloch offers a very different approach to the study of the institutionalization of applied mathematics. His prosopographical analysis of the Berlin Technische Hochschule over the course of roughly 100 years may be regarded as a companion piece to the well-known volume by K.-R. Biermann, *Die Mathematik und ihre Dozenten an der Berliner Universität, 1810–1933.*

It goes without saying that an undertaking such as this one can only succeed through the joint effort, cooperation, and support of many people. We should note at the outset that this symposium was directly inspired by earlier meetings on the history of 19th-century mathematics held over the past eight years in Cambridge, Massachusetts, Siena, Italy, and Sandbjerg, Denmark. To the organizers of these earlier symposia, and particularly to Kirsti Andersen, go our gratitude for their efforts and inspiration. The immediate occasion for the Vassar meeting was to celebrate Walter Purkert's semester in the United States as a visiting professor at Pace University. Professor Purkert joins us in thanking Pace University, and particularly Dr. Martin Kotler, Chair of the Mathematics Department, Dr. Joseph E. Houle, Dean of Dyson College of Arts and Sciences, and Dr. Joseph M. Pastore, Jr., Provost of Pace University, for their generous support of this undertaking. We are pleased to acknowledge that major funding for the Vassar symposium was made possible through a grant from the National Science Foundation, and we are also grateful to the other organizations who helped to sponsor it: the History of Science Society, the International Commission on the History of Mathematics, and the International Union of the History and Philosophy of Science.

In editing these two volumes, we have been ably assisted by an editorial board consisting of: Joseph W. Dauben (Lehman College and the Graduate Center, CUNY), Jesper Lützen (Copenhagen University), Karen V. H. Parshall (University of Virginia), and Walter Purkert (Leipzig University). Our thanks to all four of them for their assistance, and particularly to Joseph Dauben for his advice and guidance in helping to organize the symposium and to Karen Parshall for going above and beyond the call of duty to help us edit the essays as swiftly as possible. We also wish to thank Klaus Peters of Academic Press for his enthusiastic support of this project and for his efforts in guiding it through some rough seas and safely into port.

Obviously, we could not publish these conference proceedings in a timely fashion unless the contributors themselves were willing to take time off from various other duties and projects in order to compose their articles. We were most fortunate that nearly everyone involved in this project held to the deadlines for submission of first drafts and revisions. The swift publication and general appearance of the volumes were made possible by the submission of articles as computer files and the magic of TeX, the typesetting program. We apologize for any errors that may have slipped through in the process of editing these volumes.

The highly nontrivial task of converting our authors' assorted manuscripts into the text that follows was greatly aided by the typing skills of MaryJo Santagate, Christina Mattos, Deborah Mock, and Anne Clarke. Our thanks to them, and to John Feroe for assisting us in producing several of the diagrams and tables that appear in these volumes. We greatly appreciate the friendly assistance offered to us by James Kreydt of Springer-Verlag, who arranged for us to make use of the numerous photographs reproduced here from the Springer archives. Finally, we must reserve two very special thank yous for our wives, Hilde and Carlie, those "ladies in waiting" who have patiently accompanied us through the ups and downs of this endeavor over the past four years.

David E. Rowe
John McCleary

The Context of Reception

Henri Poincaré (1854–1912)
Courtesy of Springer-Verlag Archives

A Theory of Reception for the History of Mathematics

John McCleary

> ...one can understand a text only when one has understood the question to which it is an answer.
>
> R. G. Collingwood

1. INTRODUCTION

Like mathematicians, historians solve problems. From the simply stated questions of priority and precise chronology to the thornier issues of influence and stature, the historian makes judgments by applying methods or assuming points of view. This paper is a contribution to historical method in the history of mathematics. Its goal is to outline a theory of reception that can provide the historian with a different point of view, and hence a new tool for the solution of historical problems.

Useful tools for the history of any subject are those that find their shape in the nature of the object of study. The subject of mathematics is unlike the sciences and unlike the arts in fundamental ways and so its nature presents new problems for the historian. The criterion for science of testing theory against observable natural phenomena does not apply to mathematicians, and the artist is most often exempt from the rigor demanded in mathematics. However, the subject retains a few important structural features of science and art and so some methods from the history of these areas may be adaptable. In another paper [9], an approach to the history of science is applied to certain questions in the history of mathematics. Here I turn to the arts for ideas, and focus on the *mathematical work* and its impact from the viewpoint of literature.

Historians of literature suffer some of the same maladies that historians of mathematics suffer—to reconstruct the viewpoint of a writer, one must consider the backdrop of a classical past that remains in view in contrast with a powerful present that informs and judges this past. This view is different for the scientist, who works with current paradigms and with less pressure from the past (e.g., physicists need not deal with phlogiston any more). The features common to literature and mathematics emerge more clearly through the theory of reception, here adapted to the history of mathematics from the work of Hans Robert Jauß [4]. To put it briefly, the history of the reception of a work is the analysis of the first

THE HISTORY OF
MODERN MATHEMATICS

3

readers' consideration of it. This requires knowing the author's intentions for his or her audience, and the audience's ability, inclination, and intentions toward the work.

The use of reception to study the history of mathematics is not new, and several papers in this collection treat their topics in this manner. Our goal is to present it as a clearly articulated methodology out of the context of a specific historical problem where reception would be only one of many approaches. In order to set the stage for our discussion, in §2 we present a brief critical review of some of the other points of view that are employed in the history of mathematics. Following that, §3 describes the adaptation of the theory of reception to mathematics. In §4 some features peculiar to mathematics are pointed out, and in §5 we focus on the reception of topology at the turn of the century as an instance of the method. The last section is devoted to a summary and concluding comments.

2. Points of View

Let us begin with a critical review of some of the methods or points of view with which historians construct their works. One common point of view is often associated with the term *historicism*. For our purposes, this notion can be summarized in the quote from Gervinus, a ninteenth century theorist of history—historicism is

> the fundamental law of writing history, according to
> which the historian should disappear before his object,
> which should itself step forward in full objectivity.[1]

This view is especially important to the historian of mathematics for whom the nature of the object of study is always in the context of the present state of its development, and so prejudiced by that development. Retaining a distance from the present and then reconstructing the past is important, but its further elaboration invites difficulties, here summarized in a quote of Ranke;

> the value *of a period of history* depends not at all on
> what followed from it, but rather on its own existence,
> on its own self.[2]

This approach is often unable to provide a satisfactory explanation of past events. No history of the parallel postulate, for example, is complete without reaching deep into the past. On the other hand, considering the apparent lukewarm acceptance of the pioneering work of Boylai and Lobachevsky, the historian must extend the discussion into the period following their work where other developments legitimated it.

The fundamental failure of a strict historicist view for mathematics is that it ignores one of the most compelling features of a work, that is, that problems are not only solved in it, but also posed by it. To borrow the language of literary theorists, the producer and the reader have been separated—the reader of a mathematical work is another mathematician in an active state of development of the subject and hence a paper is not read as the last word, even if it may be written as such.

Another method of historical analysis is termed *formalist* by Jauß and other the historians of literature (not to be confused with formalism in the philosophy of mathematics). From the formalist viewpoint a work is central and independent of its milieu. Notions like the evolution of a genre or a tradition become the focus of history and other roles are found for the background against which the works appear. This is similar to what Herbert Mehrtens refers to as *presentism*—"peeling out what is familiar mathematics and declaring the strange rest to be in need of explanation."([11]) In contrast to historicism, the present state of the subject determines the familiar and so a work finds its importance as it contributes to this current view of mathematics.

The fundamental force moving history from the formalist viewpoint is the process of *canonization*. History divides into the studies of the evolution of the canon and the evolution of taste. This solves the problem of historicity in that a work does not live in isolation. It now finds its place in the evolving canon at the heart of the subject, and the historian's job is to describe this process. This is philosophically satisfying (ahistorically) and for the purpose of describing mathematics it holds great appeal. In a paper with Audrey McKinney [9] we describe the notion of *reorientation* as one of the processes guiding changes in the canon of mathematics. Bourbaki's *Eléments des Mathématiques* is an ongoing representation of the canon which can best be described as historical reconstruction from this viewpoint.

The emphasis on the work in the context of the evolution of a system naturally leaves "the strange rest" unexplained. Left out are the social and economic details of a period which are not without decisive impact. Jauß describes the problem as the distinction between seeing a work *in history* as opposed to a *succession of systems*.[3] Furthermore formalism is at a loss to explain the sources for reorientations, that is, of works that test and eventually replace the canon; the viewpoint is unable to approach the question of why a mathematician introduces a new idea that is clearly outside the accepted norms of well-defined doctrine.

An opposing viewpoint to formalism that stresses the milieu and producers over the primacy of the work appears in sociological approaches to history. In sociological theories the producer and consumer (the writer and reader) are joined in a "real-life process" which reflects the push and pull of social and economic factors, that is, according to the concerns of the sociologist. Like the formalist approach to history, these viewpoints enrich the historicist view immensely, however, also like formalism, the application of a purely sociological view often misses essential parts of the picture. Mehrtens identifies the social history of mathematics as that

> of individuals, groups, institutions, and on a more so-
> phisticated level of social roles, of processes of differ-
> entiation, autonomization and professionalization.[4]

The social history of mathematics informs the history of the subject, but a practitioner of this view must be very careful not to leave out the subject from this vantage point. The role of tradition, or in the case of mathematics, its still lively past, can be lost to the sociological analysis. If one takes a goal of historiography to be the description of the progress of this body of knowledge, the sociological view is also less able to explain revolutionary steps, taken in opposition to the status quo.

3. RECEPTION

The foregoing discussion indicates that a more versatile, though firmly grounded, viewpoint that combines the good features of the others would be desirable. Borrowing from literary theory we consider the theory of reception as a possible candidate for such a viewpoint. The theory of reception begins with the relationship between a work and its intended audience. That a work is formed by its audience's expectations, and goes on to affect those expectations is the mechanism of reception of central importance for us. In his paper [4], "Literary history as a challenge to literary theory", Hans Robert Jauß outlines a number of points that define the methodology of reception. We adapt these points to the history of mathematics in what follows.

Jauß begins by describing the focus of this viewpoint.[5] The goal of avoiding the pitfalls of historical objectivism is to be achieved by grounding the representation of history in the ideas of reception and influence. For mathematics this implies that a work not be viewed as possessing an intrinsic and immediately recognized value or truth that remains clear for each subsequent generation of readers. A work can continue to have influence only if those who come after it respond to it, even if only to appropriate it in a larger scheme, imitate it or refute it.

In his second point, the task of analyzying the readers' response is made precise by identifying an "horizon of expectations" that arises out of the historical moment of a work's appearance.[6] An analogous notion in the history of science is that of the *paradigm* as introduced by Thomas Kuhn [7]. It is here that the broader issues of social order and canon alike can coexist in a dynamic system that provides a broad brush with which to consider a particular work or idea. Identifying the important features of the horizon of expectations for a given work becomes the focus for the historian. In mathematics, issues like the contemporary pressure of a given school, the support infrastructure, or the accepted philosophical framework in the educational system or in the mathematical community are key points in the reconstruction process. In Joan Richards' study of reception of non-Euclidean geometry in England [15], such factors are identified, centering around the place of geometry in the philosophical tradition at Cambridge, the breeding ground of most influential English mathematicians during this period. Here reception theory solves the problem posed by Richards of explaining Bertrand Russell's peculiar description of the development of geometry in 19th century England. Gregory Nowak's discussion of Riemann's *Habilitationsvortrag* and Gregory Moore's history of the Continuum Hypothesis in this volume follow a similar direction in their analysis.

Another key feature of the horizon of expectations is its ability to reflect the concerns of those working on mathematics. In particular, the current problems considered as important by a given individual, group or school are a necessary part of the horizon of expectations. The main work of mathematicians is the solution and extension of these problems, and so a work arrives for the horizon of expectations of its readers ready to change it by a new solution, counterexample or extension of a problem in a new direction. This is Jauß' third point, that how

> a work, at the historical moment of its appearance,
> satisfies, surpasses, disappoints, or refutes the expec-
> tations of its first audience provides a criterion for de-
> termination of its ... value.[7]

It is the stroke of a masterwork that it decisively changes the horizon of expectations of its audience or later audiences. Examples abound in mathematics: Hilbert's finite basis theorem, Dirichlet's introduction of analytic techniques in number theory, and Riemann's papers on the zeta function and on complex function theory. In contrast to these works which changed mathematics upon arrival, the negligible response accorded Lobachevsky's and Boylai's published accounts of non-Euclidean geometry can be explained by the contemporary concerns of geometers. The eventual recasting of the results in analytic (that is, differential-geometric) terms by Beltrami and Christoffel via Riemann's ideas made them available to the community seeking analytic questions and answers.

Jauß also calls attention to

> works that at the moment of their appearance are not
> yet directed at any specific audience, but that break

> through the familiar horizon of . . . expectations so com-
> pletely that an audience can only gradually develop
> for them.[8]

The familiar example of Galois' work comes to mind here.

Finally, Jauß could just as well be speaking of the work of mathematicians when he identifies the method of historical reception as the
"reconstruction of the horizon of expectations, which enables one . . . to
pose questions that the text gave an answer to."[9] He further considers the
problem in literature of the enduring nature of "classics". The analogous
problem in mathematics can be seen in terms of the educational profile of
a given age. For example, should we teach Euclid or Hilbert's *Grundlagen
der Geometrie* as geometry to successive generations? The reconstruction
of reception for each generation of students and the placement of each
classic work in the future horizon of expectations for them provides an
answer, and furthermore, allows us to reconstruct past horizons of expectations for similar bodies of students. The old notion of 'Ontogeny recapitulates Phylogeny' as a principle for pedagogy in mathematics takes
on new dimensions when we recast the work of educators as the crafting
of horizons of expectations for their students. Historical reconstruction is
only a form of strict phylogeny, while reception theory acts as a sort of
genetic engineering. The contrast of different views of elementary algebra in early American schools discussed in Helena Pycior's paper in this
volume gives an example of this process.

4. THE CONTEXT OF MATHEMATICS

Though we are arguing that it is fruitful to view mathematics as
literature and reap the benefits, it is prudent to mark the differences in the
application of reception theory to mathematics and literature. The first
obvious point that distinguishes mathematics from almost every other
enterprise is its ability to retain its literature. Though individual papers
and researchers may be forgotten, their work, perhaps recast, is part of
the present paradigm if only as part of the archive. Proper mathematics
may contribute to a then fashionable current in the subject and slip from
the mainstream, but it retains its correctness and potential applicability.
This feature of the subject makes it even better suited to the theory of
reception. Establishing the original questions to which a paper provides
the answer, and then describing its further influence, now perhaps far
removed from the original problem is a paradigm that obtains a handle
on a fair lot of exciting developments in the subject, and hence in its
history. For example, Riemann's investigations of the heat equation led
to the notion of curvature that lies at the heart of modern geometry.

Another feature of the application of reception theory to mathematics is the role played by the reader. In literature the readership ranges from the uninformed to the critic and the fellow writer. The receiver of a work of mathematics, on the other hand, is generally a fellow mathematician, a fellow producer, with criteria that are very different from the casual reader of literature. This active role played by the reader speeds the reception process on the appearance of a work, because one criterion for value set by a community of producers is the usefulness of a work or idea in their problems. This added pressure on a work of mathematics changes the analysis of its influence, and coupled with the long memory of the subject, makes reception studies an appropriate tool for historical analysis.

Finally, the possibility of applying mathematics adds new features to the horizon of expectations of any given audience of readers. The non-artistic, or non-aesthetic calling of mathematics to its usefulness distinguishes it from literature, and certainly figures prominently in the expectations of some audiences, offering a characteristic of a community whose identification can play an important role in determining the "questions answered by a work". See the contributions of Ivor Grattan-Guinness and Renate Tobies in volume 2 of this collection for examples of this phenomenon. This notion is especially helpful when studying how the role of mathematics changed from applied to pure. A result now received as abstract can have had its origins in questions of an applied nature, the likes of which have been forgotten in further developments.

5. THE RECEPTION OF TOPOLOGY

Having sketched a method let us now consider an application. Instead of focusing on a particular mathematical work we discuss the reception of the equivalent of a literary genre, a field of mathematics, namely topology at the turn of the century. The questions posed by *analysis situs* date back to Leibniz and Euler. Before the nineteenth century, research in topology focused on the question of clarifying and extending Euler's theorem on spherical polyhedra—a geometric problem. The program of developing analysis held the attention of nineteenth century mathematicians and determined their horizon of expectations. Indeed it was developments in integration theory that led figures like Cantor and Riemann to questions in topology and their work pointed in directions that would define analysis situs or topology for mathematicians in 1900. (See Pont [14] and Johnson [6] for details of this development.)

Understanding the reception of these ideas helps to clarify how mathematicians thought about topology at a time when it was just developing. That topology blossomed from its origins into a vital area is evidence of a favorable reception, but what made this possible? This question is not trivial as we see in the history of the subject up to 1900 and Poincaré's work. The developments within geometry (and without immediate application to questions of analysis) were scattered—e.g., in the combinatorial direction there was Lhuilier's work on Euler's theorem of 1813, Möbius's manuscripts of 1863-65, Listing's work of 1847-62, and Dyck's group-theoretic work 1883-90.

The answer to this question lies in the "heads of the readers" of this time. Classical analysis dominated mathematics. The reception of Cantor's work was mixed, although his introduction of point manifolds and their topological properties gained ground within the community of analysts developing his ideas. These currents led to the formation of the German community of topologists that included Hausdorff, Menger, and Schoenflies. The French community around E. Borel, Lebesque and Frechet developed a distinct line of research on problems related to function spaces. The reception in France was mixed, led by the work of Poincaré and Hadamard. In a series of general lectures given at Columbia in 1911, Hadamard refers to these developments as follows:

> *Analysis situs is* connected ... with every employment of integral calculus. *It* constitute(s) a revenge of geometry on analysis. Since Descartes, we have been accustomed to replace each geometric relation by a corresponding relation between numbers, and this has created a sort of predominance of analysis. Many mathematicians fancy that they escape that predominance and consider themselves as pure geometers in opposition to analysis; but most of them do so in a sense I cannot approve: they simply restrict themselves to treating exclusively by geometry questions which other geometers would treat, in general quite easily, by analytical means; they are of course, very frequently forced to choose their questions not according to their true scientific interest, but on account of the possibility of such a treatment without intervention of analysis. I am even obliged to add that some of them have dealt with problems totally lacking any interest whatever, this total lack of interest being the sole reason why such problems have been left aside by analysts.[10]

In short, the analysis situs that matters is that which bears directly on questions of analysis.

The principal source of well-received topological ideas at the turn of the century lies in the line of work begun by Riemann in his Göttingen *Dissertation* [16] and carried on and generalized in the work of Betti and Poincaré [13]. The questions of distinguishing surfaces, and determining the homological structure of a manifold bore directly on integration theory over the complex numbers for Riemann, and more generally for Poincaré. In algebraic geometry the notion of genus for an algebraic curve based on its associated Riemann surface proved to be of decisive importance (see J. J. Gray's article in this volume for details). The rich geometric ideas of Riemann and their analytic applications were not received well at first, however, having been based on Dirichlet's Principle which was considered questionable by Weierstrass and the Berlin school of analysis [12]. This reception changed with later progress on Dirichlet's Principle and its final resolution by Hilbert put Riemann's topological ideas on firm ground. By 1911, Weyl [18] was lecturing on algebraic functions and their integrals from the viewpoint of *analysis situs*. This was mainstream mathematics in the nineteenth-century tradition and researchers received the idea of the importance of the topological structures underlying analytic problems as worthy of development.

One of the sources for the determination of the reception of mathematical works and ideas is the survey article. With the proliferation of mathematical knowledge surveys became more important and a major project of this sort was the *Encyclopädie der Mathematischen Wissenschaften* that appeared at the beginning of this century. In it one finds *two* major articles dedicated to topology. The first is by M. Dehn and P. Heegaard [1] which treats combinatorial topology especially the topics pushed forward by Poincaré. The second by H. Tietze and L. Vietoris [17] appeared 25 years later after the subject had grown considerably. In particular, topology already contained three subfields—general topology, *n*-dimensional topology, and combinatorial topology. The bulk of this development was related to L.E.J. Brouwer's work, which made a profound impact in each of these areas. Indeed, the development of his work is signal to the reception of the topic. Although he began as a follower of Cantor and Schoenflies, Brouwer soon started pursuing questions closer to the mainstream, encouraged in this direction by Hadamard.[11] His success in answering questions important to the principal concerns of the larger mathematical community of the time (such as invariance of dimension) was decisive for the reception of topology during the early decades of this century. Moreover, his success in all areas of topology, and his appreciation and expansion of the subject helped legitimize the full spectrum of research in the field. This sets the tone for its development as a major subject area in this century.

Thus Brouwer's work can be considered in two ways—it treated questions that were current and recognized as important, in particular, those conforming to the horizon of expectations dominated by analysis; and further, through its success, it changed the horizon of expectations for later workers in topology in its breadth of application. Brouwer's place as the foremost topologist of his day reflects his importance, not only as a remarkable mathematician, but also as a superb example of the process of reception of ideas (see [6] for more details).

6. SUMMARY

Another example that can be approached fruitfully in the context of reception theory is the question of the introduction of rigor in analysis. The work of Judith Grabiner, especially her paper [2], focuses on a problem in reception. The shifting horizon of expectations of a mathematical work to include more rigor is the product of social forces (education) and the result of scientific efforts. Subsequent generations incorporated this rigorous focus to varying degrees in their horizons of expectations, and the degree of its dominance determines much of the approach to mathematical problems in the different communities of analysts.

This sort of analysis is applicable to other notions than rigor. The reorienting ideas of set theory, analytic number theory, and abstract algebra ([10]) all offer key moments in the history of mathematics where the reception of these ideas can be studied and the historical narrative enriched by the broader picture revealed. Reorienting ideas of a social nature, like the university, autonomization and professionalization are also susceptible to such analysis.

Among the uses of history is the support of the pursuit of the philosophy of an enterprise, like literature or mathematics. Jauß' title "Literary history as a challenge to literary theory", becomes, in the context of mathematics, "The history of mathematics as a challenge to the philosophy of mathematics". Reception theory as a tool for crafting the history of mathematics offers an escape from the objectivist traps, and more importantly, a paradigm that incorporates some of the key features of the practice of mathematics, for example, the focus on problems, as integral parts of the analysis. To translate Jauß again—to bridge the gap between mathematics and its history, historical practice should not simply describe the process of general history in the reflection of its works one more time, but rather it should discover in the course of the evolution of mathematics that properly socially formative function that belongs to the subject. If it is worthwhile for the scholar to jump over his or her ahistorical shadow for the sake of this task, then it might well provide us with the answers to many important questions.

NOTES*

1. Georg Gottfried Gervinus, from *Grundzüge der Historik* (1837) found in *Schriften zur Literatur*, Berlin, 1962.

2. Ranke, in *Geschichte und Politik—Ausgewählte Aufsätze und Meisterschriften*, ed. H. Hofmann, (Struttgart, 1940), p. 141. The quote dates from 1854.

3. Jauß [4], p. 18.

4. Mehrtens [11], p. 263.

5. Jauß [4], p. 20.

6. Jauß [4], p. 22.

7. Jauß [4], p. 25.

8. Jauß [4], p. 26.

9. Jauß [4], p. 28.

10. Hadamard [3], p. 33.

11. Johnson [5].

BIBLIOGRAPHY

[1] Dehn, M. and Heegaard, P., *Analysis situs*, section III AB 3 of "Encyclopädie der Mathematischen Wissenschaften". This article appeared June 25, 1907.

[2] Grabiner, Judith, *Is mathematical truth time-dependent?*, Amer. Math. Monthly, 81(1974), pp. 354–365.

[3] Hadamard, J., "Four lectures on Mathematics" (delivered at Columbia University in 1911), Columbia University Press, New York, 1915.

[4] Jauß Hans Robert, 'Toward an Aesthetic of Reception', University of Minnesota Press, Minneapolis, 1982.

[5] Johnson, Dale M., *L.E.J. Brouwer's coming of age as a topologist*, in "Studies in The History of Mathematics", Esther R. Phillips, ed., MAA Publications, 1987, pp. 61–97.

[6] _____, *The Problem of the Invariance of Dimension in the Growth of modern topology I., II.*, Archive for History of Exact Sciences 20(1979), 97–188, 25(1981), 85–267.

[7] Kuhn, Thomas, "The Structure of Scientific Revolutions", 1970, University of Chicago Press.

* My thanks to Patrick Will for first bringing the idea and the critical literature on reception aesthetics to my attention. My thanks also to David Rowe for encouragement in developing these ideas.

[8] Lakatos, Imre, "Proofs and Refutations. The Logic of Mathematical Discovery", Cambridge University Press, 1976.

[9] McCleary, J. and McKinney Audrey, *What Mathematics isn't*, Mathematical Intelligencer, 8(1986), 51-54.

[10] Mehrtens, Herbert, "Die Enstehung der Verbandstheorie", Arbor scientiarium, Reihe A, Band 6, Hildesheim, Gerstenberg, 1979.

[11] _____, *Social history of mathematics*, from "Social History of Nineteenth Century Mathematics", H. Mehrtens, H. Bos, I. Schneider, eds., pp. 257–280.

[12] Monna, A.F., "Dirichlet's Principle: A Mathematical Comedy of Errors and its Influence on the Development of Analysis", Oosthoek, Scheltema & Holkema, Utrecht, The Netherlands, 1975.

[13] Poincaré, H., *Analysis situs*, Journal de l'Ecole polytechnique, (1895), 1–211.

[14] Pont Jean-Claude, "La Topologie Algébrique des origines à Poincaré", Presses Universitaires de France, Paris, 1974.

[15] Richards, Joan, *The Reception of a Mathematical Theory: Non-Euclidean Geometry in England, 1868-1883* in "Natural Order: Historical Studies of Scientific Culture", B. Barnes and S. Shapin, eds. London (Sage), 1979, pp. 143–163.

[16] Riemann B., *Grundlagen für eine allgemeine Theorie der functionen einer veränderlichen complexen Grösse*, Göttingen, 1851.

[17] Tietze H. and Vietoris L., *Beziehungen zwischen den veschiedenen Zweigen der Topologie*, section III AB 13. of "Encyclopädie der Mathematischen Wissenschaften". This article appeared December 10, 1930.

[18] Weyl, Hermann, "The Concept of a Riemann Surface", Addison-Wesley Publ. Co., Reading, MA, 1955, 3rd edition.

The monument to Carl Frederick Gauss (1777-1855)
and Wilhelm Weber (1804-1891) in Göttingen.
Dedicated 17 June 1899.

Bernhard Riemann (1826–1866)
Courtesy of Springer-Verlag Archives

Riemann's *Habilitationsvortrag* And the Synthetic *A Priori* Status of Geometry

Gregory Nowak

THE *Habilitationsvortrag*

There are many instances in which philosophical preconceptions have affected the course of mathematical research. But what of the converse? Has new mathematical research ever led to a revision of philosophical preconceptions? Under what conditions can such communication between fields of inquiry occur? What happens when mathematics presumes to dictate to philosophy?

Bernhard Riemann submitted his *Habilitationsschrift* to the University of Göttingen in December of 1853. In order for him to receive his *Habilitation* and become a *Privatdozent* at Göttingen, entitled to lecture for student fees, he needed only to deliver a *Habilitationsvortrag* to the *Ordinarien* of the philosophical faculty of the University, demonstrating his ability to teach. As part of this process, Riemann submitted three possible topics for this lecture to the faculty. Going against tradition, Gauss passed over the first two topics, which Riemann had prepared in advance, and chose the third, "On the hypotheses which lie at the foundations of geometry." The *Habilitationsvortrag* was delivered on June 10, 1854, and evidently went well; even the normally taciturn Gauss spoke with excitement of the depth of Riemann's ideas.[1] The text of Riemann's lecture, however, was only published posthumously some fourteen years later.

What did Riemann say that so excited Gauss? Other authors have highlighted the mathematical tradition in differential geometry which started with the publication of the *Habilitationsvortrag*, or the possible connections between its physical speculations and later developments in physics. The mathematical content of Riemann's *Habilitationsvortrag* can be summarized as a generalization to higher dimensions of Gauss's work on the intrinsic geometry of surfaces, a generalization which Riemann presents in the second section of the *Habilitationsvortrag*.[2] Riemann's first step was the extension of the definition of the line element ds to arbitrarily many dimensions. He also presented a "counting argument" concerning the question of determining when one metric is equivalent to another after a coordinate transformation. Since the expression for ds on an n-manifold involves $n*(n+1)/2$ functions, but a change in coordinates involved only

n functions, Riemann concluded that there were $n*(n-1)/2$ functions which depended on the metric. He also saw that this meant $n*(n-1)/2$ functions were sufficient to determine the metric on any n-manifold, and he expressed these functions as the (Gaussian) curvatures of $n*(n-1)/2$ independent surfaces, or two-dimensional subspaces, at any point. Riemann also discussed the special features of flat manifolds and manifolds of constant curvature; his discussion of how space might exhibit these properties can be seen as a kind of mathematical physics. In addition to his generalization of Gauss's work, Riemann made a mathematical as well as a philosophical contribution by defining the concept of an n-dimensional manifold. Similarly, the discussion of the difference between unboundedness and infinitude have mathematical significance.

The study of differential geometry on n-dimensional manifolds was further advanced by Riemann's *Pariserarbeit* of 1861, in which he introduced quantities now known as the Christoffel symbols and the Riemann four-index symbol in order to discuss the question of isometric transformations between metrics, and the special case of determining when a given metric is equivalent to one with constant coefficients.

Publication of the *Habilitationsvortrag* in 1868 inspired a number of mathematicians to expand upon the mathematical ideas presented by Riemann. Eugenio Beltrami derived the general formula for the line element in spaces of constant curvature, and proved several other claims Riemann made in his paper. E. B. Christoffel considered the problem of isometric transformation of differential forms, generalizing some of Riemann's ideas in order to study differential invariants. In 1887 Gregorio Ricci-Curbastro discussed an operation later called covariant differentiation, based on techniques introduced by Christoffel and Rudolph Lipschitz, which allowed one to obtain higher-order differential invariants. In the 1890s, Ricci introduced what he called the absolute differential calculus, now known as tensor analysis, as his systematic generalization of the earlier work on differential invariants. His student Tullio Levi-Civita joined in the development of tensor analysis during the 1900s. The work of Ricci and Levi-Civita was soon to prove useful to Albert Einstein in the formulation of his theory of general relativity, a theme discussed by Dirk Struik in volume 2 of these proceedings.

While the mathematical tradition originating with the *Habilitationsvortrag* has proven to be enduring and fruitful, it does not constitute a complete explanation of the historical significance of the lecture as Riemann presented it. I shall attempt to demonstrate that treatment of the *Habilitationsvortrag* to date has been unnecessarily one-sided. Some authors have felt free to present pages of equations as Riemann's work with a

disclaimer that "Riemann did not give these equations explicitly," implying that since Riemann presented the results of these calculations in the
paper, we are justified in ascribing their derivations to him as well.[3] Close
readers of Riemann would benefit from a similar willingness to discern
the rich philosophical content lying just beneath the surface of the paper. As I do not propose to challenge the standard, mathematical view
of the *Habilitationsvortrag* in any respect, I shall from this point leave the
mathematical discussion aside in favor of another approach.

I wish to present an alternate view of Riemann's work, portraying it
as part of a philosophical tradition, a tradition which acknowledged Riemann's philosophical stance and responded to Riemann on philosophical
rather than mathematical grounds. Such a reading of the *Habilitationsvortrag* need not be based entirely upon demonstrating its reception by a
philosophical tradition. Riemann was aware of the philosophical implications of his mathematics and structured the *Habilitationsvortrag* as a
philosophical argument which used mathematics to demonstrate the un-
tenability of Kant's position that Euclidean geometry constituted a set of
synthetic *a priori* truths about physical space. The implications of Riemann's work for the Kantian conception of space have previously been
observed, although Riemann has not previously been recognized as explicitly making such an argument.

Riemann's philosophical interests were not confined to the topics
raised in the *Habilitationsvortrag*. He began his studies at Göttingen in
theology rather than mathematics, and took courses in philosophy and
pedagogy along the way.[4] Riemann's collected works include several
"Fragmente philosophischen Inhalts" which hint at the range of his philosophical speculations.[5] These fragments clearly demonstrate Riemann's
familiarity with Kantian philosophy.

Let us review those aspects of Kant's philosophy of space which
will be of importance in our reading of the *Habilitationsvortrag*. Kant
classified propositions along two lines, analytic–synthetic and a prior–
a posteriori.[6] A proposition is analytic for Kant if it is true by virtue
of the meanings of the words employed, and synthetic otherwise; for
example, "All bachelors are unmarried" is an analytic proposition. On
the other hand, "Fellows of Trinity College, Cambridge, are unmarried" is
synthetic. Kant labels a proposition *a priori* if it can be known prior to any
experience, and a posteriori if a certain kind of experience is necessary
to determine the truth of the proposition. Of our two examples, the first
is clearly *a priori*, since anyone able to form the proposition will assent
to its truth because of its meaning without requiring any confirmatory
experience. The second is a posteriori, because we require certain forms
of experience to determine whether this proposition might or might not

be true, and indeed its truth value will change depending on the decade in which our experience occurs.

A skeptic might observe that the pairs analytic/*a priori* and synthetic/*a posteriori* provide distinctions without differences, and this was exactly the view Kant wished to disprove. He sought to deny the radical skeptic view that all knowledge was either tautological (analytic and *a priori*) or contingent (synthetic and a posteriori). While he believed that analytic *a posteriori* propositions were impossible, he devoted considerable effort to demonstrating the existence of synthetic *a priori* propositions.[7]

Kant upheld (Euclidean) geometry as an example of a set of synthetic *a priori* truths about physical space.[8] The qualifier "about physical space" is necessary in order to distinguish the several meanings the term "space" had for Kant. There is space as pure intuition (*reine Anschauung*), with which we represent to ourselves the concepts of place and proximity. We can think of empty space, but not of the absence of space; this intuition is a "condition on the possibility of appearances" and thus *a priori*.[9] Kant identified this pure intuition with the physical space of ordinary experience, giving it the properties of both empirical reality and "transcendental ideality."[10] Some commentators have been concerned to interpret Kant as acknowledging the existence of a third kind of space, characterized by mathematical and logical possibility, relying only upon axioms for its definition.[11] It is true that Kant distinguished between the formal, logical possibility of a concept and the objective reality of a concept, and gave as an example "a figure which is enclosed by two straight lines"—a concept which was not self-contradictory and yet not objectively real. He defined the "possible" as that which was objectively real, denying that such concepts told us anything about space.[12] He went on to refer to them as "fictitious concepts" that could only be known *a posteriori* if at all—a classification introduced more to bring such propositions under his scheme than to allow the possibility that they might be true.[13]

We have discussed three kinds of space, the intuited, the physical, and the logically possible. Kant's approach was to identify the first two, and to deny that the third had anything to do with them, since it told us nothing about physical space. Statements about the first two were classed as synthetic *a priori*; those of the third would have been synthetic *a posteriori* if they were true. I shall endeavor to show that Riemann's approach was to identify intuited space with that which was logically possible (or "axiomatic"), and not with physical space. Physical space could be modeled by a combination of the other two. Statements about how features of physical space were modeled would be classed as synthetic *a posteriori*, while statements about the intuited, axiomatic spaces were analytic *a priori*.

Why should we consider the *Habilitationsvortrag* as a philosophy paper as well as a mathematical paper? For one thing, it makes much more sense of the paper's structure. The *Habilitationsvortrag* is divided into three sections, treating first the notion of multiply extended quantity, then the metric relations possible on such quantities, and finally, applications to physical space. These three sections neatly correspond to the three views of space mentioned above, beginning with an intuitive description of possible spaces, presenting the structure of metrics on these spaces, and then discussing the applicability of these spaces to physics. Only the second section is explicitly mathematical, but we shall see that all three main sections of Riemann's paper serve as an outline for his philosophical argument. In contrast, those wishing to portray the *Habilitationsvortrag* as a mathematical paper first admit that the paper is nearly devoid of explicitly mathematical content, and then claim that Riemann's analytical investigations were suppressed in the *Habilitationsvortrag* in order to make the paper suitable for a nonmathematical audience.[14]

Several questions, however, come to mind when we are asked to accept the idea that Riemann's intent was solely to develop a new approach to geometry: Why did Riemann publish no work on differential geometry in his lifetime?[15] Why did his paper on minimal surfaces, written mostly in 1860 and 1861, make no use of these techniques?[16] Riemann met Beltrami in the early 1860s, and they undoubtedly discussed geometry. If Riemann was interested in advocating a new approach to geometry, why do we find no mention of Riemann or any sign of the impact of his ideas upon Beltrami in Beltrami's work until 1868, after the posthumous publication of the *Habilitationsvortrag*?[17] Finally, why is only one of the three sections of the lecture mathematical in content?

Riemann's introduction made it clear that he saw himself involved in a philosophical as well as a mathematical enterprise:

> As is well known, geometry proposes the concept of space, as well as assuming the basic properties for constructions in space. It gives only nominal definitions of these things, while their essential specifications appear in the form of axioms. The relationship between these presuppositions [the concept of space, and the basic properties of space] is left in the dark; we do not see whether, or to what extent, any connection between them is necessary, or *a priori* whether any connection between them is even possible.

"This darkness," he went on, "has been dispelled neither by the mathematicians nor by the philosophers who have concerned themselves with it."[18] Thus Riemann immediately located himself in an area of common interest to mathematicians and philosophers—and if one were to have asked Riemann's audience which philosopher was most concerned with the connections between the concept of space and the axioms and methods of geometry, the answer would clearly have been Kant. In raising the question of whether the connection between the concept of space and the axioms of geometry is *a priori*, he was challenging Kant's answer to the question. Riemann's strategy for investigating this question was to bypass space entirely as an object of investigation in itself, subsuming it under the concept of "multiply-extended quantities."

In the introduction to his *Habilitationsvortrag*, Riemann distinguished the qualitative features of space, which he denoted by the term extension, from features relating to distance, or metric relations. He then proceeded to refute the idea that geometry (the determination of metric relations) is an inseparable part of the concept of space itself (extension) by conceiving of spatial objects—manifolds—bereft of any metric relations, concluding that since many metric relations were possible on the same manifold, they were not determined by the concept of extension alone. Therefore, physical space was merely a special case of triply-extended quantity. This meant, according to Riemann, that theorems concerning the geometry of space cannot follow merely from the concept of extension, but must be determined by experience. Finally, Riemann discussed the determination of metric relations by axiom systems, of which "the most important . . . is that laid down as a foundation of geometry by Euclid." He ended the introduction by insisting that these axioms were not logically necessary, and were empirically contingent in their description of space. Thus we see that Riemann's central concern was the nature of the connection between the concept of space and the axioms of the geometry of physical space. He did not wish his audience to lose sight of this theme during the delivery of the *Habilitationsvortrag*—at the beginning of the second part of the third section of the lecture, he reviews the structure of his argument, again concluding with the objective of determining the epistemological status of the geometry of physical space.[19]

Riemann appears to have organized the *Habilitationsvortrag* with this argument in mind. We shall see that viewing the paper as a work of philosophy provides a natural explanation for the presence of the non-mathematical sections. The first section, on multiply-extended quantity, contains Riemann's discussion of the intuitable [*anschauliche*] nature of space. Whereas for Kant—who identified intuited and physical space—geometry was primarily an activity (due to the constructive nature of geometrical proofs) performed in intuition, in the one space which exists,[20] Riemann discussed space as a general concept, capable of instantiation and of which physical space is merely a special case.

The second section begins by defining metric relations on manifolds, based upon the assumption that all curves are measurable and comparable. Riemann went on to discuss the functions determining metric relations, the difference between intrinsic and extrinsic curvature, and the special case of "flat" manifolds with zero curvature in all directions. The last two parts of the second section are devoted to a discussion of motion in flat manifolds; Riemann introduced them as a necessary precursor to his discussion of applications of his new geometry to space. He was arguing for the possibility that physical space is a three-fold extended manifold of minute but non-zero positive curvature by demonstrating that this would accord with our observations that bodies can move about freely in space.

In the third section of the lecture, Riemann discussed sets of conditions that, in addition to the concept of extension, allow one to determine metric relations. This is an intermediate step to the main objective of the third section of the paper: the demonstration that these conditions are only to be found empirically, through experience, and thus are *a posteriori*.

Before going further it will be useful to review Riemann's terminology. Riemann uses the term "Raum" to mean physical space, and "mehrfach ausgedehnte Grösse" (multiply-extended quantity) to mean the general concept of space, of which "Raum" is an instantiation. "Geometry" is a branch of knowledge dealing specifically with physical space; it is an instantiation of the concept "metric relations." These distinctions are clearly made in the *Habilitationsvortrag*:

> It will be shown that a multiply-extended quantity is susceptible of various metric relations, so that space constitutes only a special case of a triply-extended quantity. From this, however, it is a necessary consequence that the theorems of geometry cannot be deduced from general notions of quantity, but those properties which distinguish space from other conceivable triply-extended quantities can only be deduced from experience.[21]

One should not conclude from Riemann's language that he is prioritizing physical space or its geometry in any epistemological sense. For Riemann, the mathematical investigation of the concept of space was a logical process carried on in intuition in which physical space plays no part. Only after the general investigation was complete could one determine how some of its results might model physical space. This later, secondary investigation would then make use of some of the suggestions Riemann made in the third section of the *Habilitationsvortrag* about how observed qualitative features of physical space (such as the ability to move objects freely without distortion) might provide indications of its possible metric relations.

This aspect of Riemann's thought is captured by the title of his lecture if we make his meanings explicit and translate the term "geometry" as "the metric relations of physical space." The title then becomes "On the hypotheses which lie at the foundations of the metric relations of physical space." This is not to say that Riemann's usage of the term "geometry" is unique, but it is helpful to remember this interpretation when considering Riemann's understanding of these ideas. This version of the title is also more suggestive of the philosophical as well as the mathematical relevance of the "hypotheses at the foundation of geometry."

It is in the third section of the paper that Riemann outlined his position on the three varieties of space mentioned above. The concept of multiply-extended quantity is clearly available to our intuition, but so is the concept of different metric relations on the same space. Riemann demonstrated that this concept is imaginable through the use of various qualitative assumptions which determine the metric relations. Thus he identified intuitive and geometric space. He insisted that the qualitative assumptions that determine the geometry of physical space are not necessary consequences of the concept of multiply extended quantity, since there can be more than one set of metric relations on a threefold-extended quantity, and so they can only be known empirically. Thus these assumptions, which describe how physical space is modeled by geometry, are *a posteriori* propositions.

Riemann also uses his topological insight to criticize Kant's views on the infinity of space. As we have said, Kant believed that physical and intuited space were identical; since he thought that it was impossible to intuit the absence of space he was obliged to maintain the infinitude of space.[22] Riemann, however, uses his distinction between extension and metric relations to distinguish in turn the properties of unboundedness and infinitude. The extension property of unboundedness does not determine the metric property of infinitude, since a three-dimensional manifold of constant positive curvature is unbounded but finite:

> That space is an unbounded triply-extended manifold is an assumption which is employed for every apprehension of the external world, by which at every moment the domain of actual perception is supplemented, and by which the possible locations of a sought-for object are constructed, and in these applications it is continually confirmed. The unboundedness of space consequently has a greater empirical certainty

> than any experience of the external world. But its in-
> finitude does not in any way follow from this...[23]

Note Riemann's careful use of Kantian concepts; by describing the role of space in perception, he is inviting us to compare his view with Kant's. He differs from Kant in labelling the proposition that space is an unbounded triply-extended manifold as an "assumption;" for Kant it is an *a priori* intuition. Riemann grounds this assumption in experience; i.e., it is a posteriori. There would be no need for such a strong statement if Riemann were following a mathematical or physical agenda; such researches would be indifferent to the epistemological status of the infinitude of space. Only if we see Riemann as criticizing Kant's understanding of geometry does his insistence on the *a posteriori* nature of the infinitude of space make sense.

Riemann concluded his lecture with a discussion of curvature on a very small scale. He justified this concern for the sake of physics: investigations of causality pursue phenomena on a very small scale, at which point our large-scale assumptions about measurement may break down. If bodies are subject to small-scale distortions with motion, then space could be curved in the small, as long as the total curvature over measurable areas is near zero. Riemann further opened up this possibility by pointing out that the empirical bases for our assumptions about the geometry of physical space, rigid bodies and light rays, lose their validity in small-scale settings. We may interpret Riemann's observations in more modern language as suggesting that the idea of a rigid body may make no sense at the atomic scale, and a light ray may not have the quality of straightness on the scale of its wavelength or less.[24] This discussion was meant to indicate the ways in which the small-scale curvature of space would interact with the realm of physics, and thus become accessible to science and empirical verification.

Riemann returned to his plan to connect the concept of space and the axioms of geometry in the penultimate sentence of the *Habilitationsvortrag*, where he attempts to defend his method: "Investigations like the one just made, which begin from general concepts, can serve only to ensure that this work is not hindered by overly restrictive concepts, and that progress in comprehending the connection of things is not obstructed by traditional prejudices."[25] Riemann's more general concept of multiply-extended quantity helped him to comprehend the connections between the concept of space and the principles of geometry by giving up traditional Kantian prejudices about the necessity of Euclidean geometry.

At the close of the *Habilitationsvortrag*, Riemann asserted that he was stopping short of discussing the realm of physics. Section III of the paper developed no new mathematics, and Riemann explicitly said he was not doing physics. He was answering in the negative the question he set for himself in the first sentences of his lecture: whether there is any *a priori* connection between the concept of space, and the geometry of physical space. He was doing philosophy.

RIEMANN'S *Habilitationsvortrag* IN CONTEXT

A look at Riemann's audience, and those he cites as influences, is also relevant to a discussion of his intent in writing the *Habilitationsvortrag*. It is worth noting that Gauss chose the third of a list of three topics because he was interested in what Riemann would have to say on such a difficult topic; Riemann had prepared the first two topics but not the third. Although Riemann had recorded some thoughts about relations of extension earlier in his life, so far as we know he never organized this material in any form. We may conclude that except for the mathematical ideas used, Riemann wrote the paper in a few weeks during the spring of 1854, fully aware that it was to be delivered as a lecture rather than printed.[26] He referred to it beforehand as his *Probevorlesung* or "trial lecture," so we can assume that he had his expected audience—the Philosophical Faculty of Göttingen—in mind as he wrote. All commentators on the paper, from Dedekind on down, have suggested that Riemann simplified his discussion for the sake of his audience. If we do not interpret this consideration as merely a negative one—one that prompted him to reduce the mathematical content—but as a positive consideration—one that led him to speak to the philosophical interests of a large part of his audience—we can provide a better explanation for the structure and themes of the *Habilitationsvortrag*. Riemann was demonstrating that he could perform the functions of the ideal *Dozent*, conducting detailed, original research and relating it to the larger concerns of the culture.

Riemann's mention of his predecessors gives us some idea of how he could turn mathematical results to his philosophical purposes. At the beginning of the first section of the lecture, he names those who had influenced his approach to space as a multiply-extended quantity:

> In proceeding to attempt the solution of the first of
> these problems, the development of the concept of
> multiply-extended quantity, I feel particularly entitled
> to request an indulgent criticism, as I am little prac-
> ticed in these tasks of a philosophical nature where
> the difficulties lie more in the concepts than in the
> construction, and because I could not make use of any
> previous studies, except for some very brief hints on
> the subject which Privy Councillor Gauss has given
> in his second memoir on biquadratic residues in the

Göttingische Gelehrte Anzeigen and in his Jubileebook,
and some philosophical researches of Herbart.[27]

Gauss was in the audience, past the height of his powers but still universally regarded as the greatest mathematician in the world. If Riemann was interested only in highlighting his own mathematical skills, it would have been a politic tribute to Gauss and no dishonor to Riemann for him to introduce his lecture as a generalization to higher dimensions of Gauss's *Disquisitiones generales circa superficies curvas*.[28] Riemann does not do this, though the description is a good one-line summary of the mathematical content of the *Habilitationsvortrag*. He credits Gauss's paper only in the second section, as being helpful with the specific question of the metric relations possible on a multiply-extended quantity. Some authors have been too quick to perceive the mathematical influence which the *Disquisitiones* had upon Riemann, while ignoring his explicit references to other papers by Gauss which provided philosophical as well as mathematical inspiration on the subject of multiply-extended quantities.[29]

The German-language announcement of Gauss's second memoir on biquadratic residues that appeared in the *Göttingische Gelehrte Anzeigen* discussed, among other things, the definition of the complex plane and the progress this allowed Gauss to make in number theory. Gauss included an editorial on the lack of acceptance of complex numbers, and a claim that the "true metaphysics of the complex numbers have been placed in a new bright light" by his new spatial conception of the complex numbers. This was followed by a comment on the expansion of the number system, including, in turn, negative numbers, fractions, real numbers, and eventually complex numbers—though "this progress is always made at first with fearful and hesitant steps." Gauss may even have been the source of Riemann's metaphor of "darkness" for the state of geometry, since he described the study of complex numbers as similarly cast in darkness.[30] Gauss continued by suggesting that similar problems involving manifolds of more than two dimensions could find solutions in "general arithmetic" as well, a topic which was left for Riemann to develop.

Gauss's *Jubiläumschrift*, better known as the *Beiträge zur Theorie der algebraischen Gleichungen*, also discusses the geometry of the complex numbers, clearly separating the possibility of mathematics based on abstract spatial concepts from an approach limited by its derivation from our perceptions:

> I will present the argument in a form taken from the
> geometry of position, because in this manner one obtains the greatest clarity [*Anschaulichkeit*] and simplicity. Basically, however, the actual content of the entire

argumentation belongs to a higher domain of the general abstract theory of variables, independent of the spatial, whose subject is combinations of magnitudes that are linked by continuity, a domain which at the present time has been little developed, and in which one cannot even operate without language borrowed from spatial representations.[31]

Riemann was not concerned specifically with the complex numbers; he was drawing a deeper sort of inspiration from Gauss. He was following Gauss's lead in creating space-like objects, and citing Gauss as an authority for the validity of such expansions of the domain of mathematics beyond the limits usually assumed to be imposed by spatial intuition. He also followed Gauss in distinguishing between mathematical language which merely borrowed terminology from the vocabulary of sensible space and mathematics which was inherently dependent upon it. Gauss's influence on the first section of the *Habilitationsvortrag* is properly termed metamathematical or philosophical.

Riemann also acknowledged his debt to Herbart. Since Riemann had elsewhere spoken of the influence of Herbart in a somewhat contradictory manner—claiming to disagree with Herbart's "Synechology", or the study of continua, the question of Herbart's precise influence upon the *Habilitationsvortrag* poses somewhat of a problem.[32] Erhard Scholz has devoted an article to the question of Herbart's influence on Riemann, concluding that other branches of Herbart's philosophy had a stronger, more direct influence upon Riemann's philosophical speculations than did Herbart's synechology. Scholz records Riemann's notes on Herbart's concept of "continuous serial forms" produced by a process of "graded fusion", and concludes that Riemann drew from this not specific procedures of spatial construction but an inspiration to "transfer spatial concepts into a non-geometric context."[33]

Scholz claims that Herbart had little direct influence on Riemann's mathematics. He notes that Herbart saw continua everywhere, while Riemann thought continua were encountered only rarely outside of mathematics; Herbart never speculated about dimensions higher than three, while Riemann did; and "the essential points of Riemann's concept of manifold (multi-dimensionality, opposition between locally simple and globally complex behavior, separation of qualitative aspects of extended magnitudes from quantitative ones, and the separation and interdependence of structures on it...) had no connection with Herbart's geometric thoughts, and Riemann had to elaborate these from the mathematical material."[34]

These points are sufficient to demonstrate that Herbart had little direct influence upon Riemann's mathematics in general and the mathematics of the *Habilitationsvortrag* in particular. Unfortunately, Scholz does not consider the tension between Riemann's disavowal of synechology and the above-quoted sentence in which Riemann names Herbart as one of his influences in forming the concept of multiply-extended quantity. It is possible for Herbart's philosophy of space to have influenced Riemann without his having accepted it completely. Thus we should not conclude that Herbart had no direct influence upon Riemann's philosophy in the *Habilitationsvortrag*, especially given that Riemann explicitly credited Herbart in the first, "philosophical", section. Gauss's work on differential geometry also did not discuss dimensions higher than three, yet his views certainly exerted an influence on Riemann. Riemann may not have agreed with Herbart on the number of different manifolds we experience in everyday life, but he does give color as an example of one such, an example he borrowed from Herbart.[35]

Herbart's discussion of space inspired Riemann to create a more fruitful combination of higher-dimensional geometry and Gauss's differential geometry than he might otherwise have been able to. At least three points stand out. First, Herbart's constructive approach to space, already cited, mirrored the content of Riemann's reference to Gauss in that both discussed construction of spaces rather than construction in space. Second, Riemann followed Herbart in rejecting Kant's view of space as an *a priori* category of thought, instead seeing space as a concept which possessed properties and was capable of change and variation. Riemann copied some passages from Herbart on this subject, and the *Fragmente philosophischen Inhalts* included in his published works contain a passage in which Riemann cites Herbart as demonstrating the falsity of Kant's view.[36]

Finally, Riemann took from Herbart the view that the construction of spatial objects were possible in intuition and independent of our perceptions in physical space. Riemann extended this idea to allow for the possibility that these spaces would not obey the axioms of Euclidean geometry. We know from Riemann's notes on Herbart that he read Herbart's *Psychologie als Wissenschaft*; a few pages before the passage he copied on Kant's categories, he would have read the following:

> Strictly speaking, the objects of pure geometry are not part of the sensible world, which is partly filled by bodies, and partly empty between them. Geometric circles, squares, and polygons are, however, *nowhere* in it, have no place in it, and were not even extracted from it through delimitation; rather, the geometer creates from each of these, right from the start, and would

> produce out of each, a completely self-contained space
> as its background [*Umgebung*], if he needs to, so that
> this space would also have no particular position in
> relation to or in sensible space, but rather one would
> have to remove one of these from thought in order to
> think of the other.[37]

Here we see Herbart's idea of a plurality of spatial concepts, expressed in terms of geometry. This passage also displays his interest in the construction of spatial objects (here conceived of as the imagined plane containing imagined plane figures), distinct from experience and each other, an interest which is mirrored in Riemann's discussion of multiply-extended quantities. Particularly significant is Herbart's suggestion that in order to perceive these spaces in thought, one must remove sensible space from thought, and vice-versa. It is possible that this idea suggested to Riemann the distinction between embedded and unembedded manifolds. It is also worth noting that Riemann borrowed from Herbart such terms as "quanta" of a surface.[38]

Erhard Scholz has also uncovered some notes Riemann made on the manifold concept and the foundations of geometry which demonstrate Riemann's awareness of the connection between new approaches to geometry and Kant's view of space. If we accept Scholz's dating, which suggests that these notes were written in 1852-1853, shortly after Riemann participated in the philosophical and pedagogical seminars at Göttingen in which he most probably encountered Herbart's work, we can assume that the appearance of Herbartian ideas about space in these notes is not simply the result of coincidence. In these notes, Riemann disavowed the necessity of Kant's connection of the Euclidean axioms to spatial perception:

> From this definition of a straight line I would be able
> to derive all the propositions which occur in geome-
> try that concern straight lines. It is clear that one can
> proceed in this way without the least appeal to spatial
> intuition. Using this treatment of geometry or the the-
> ory of manifolds of three dimensions, all axioms which
> in the usual treatment of geometry are borrowed from
> spatial intuition would be dropped, as for example,
> that through two points only one straight line is pos-
> sible, the first axiom of Euclid, etc... [39]

Riemann, like Herbart, rejects any dependence upon spatial intuition in determining the axioms of geometry. It is also apparent that Riemann had no particular commitment to the truth of the Euclidean axioms in the sense that they might be descriptive of physical space. But his is not simply a negative program, aiming at casting doubt upon the Euclidean axioms; these notes also reveal that Riemann had the positive goal of developing a geometry with principles independent of spatial intuition:

> The concept of a manifold of several dimensions exists independently of our intuitions in space. Space, the plane, and the line are only the clearest [anschaulichste] examples of a manifold of two, three, or one dimensions. Without having the least spatial intuition, we would nevertheless be able to develop all of geometry.[40]

Riemann's use of the superlative "anschaulichste" is significant; it is a sign of his identification of intuited space with axiomatic space. By suggesting that there are degrees of intuitability, he is extending this property to manifolds other than space, the plane, and the line. These three are simply the "most intuitable" of the many manifolds which are available to the mathematician's intuition. Riemann's dismissal of the applicability of spatial intuition to creating a new geometry must be interpreted in the light of his identification of "Raum" with physical space, and "Mannigfaltigkeit" for the general concept of space. He was *not* saying that these manifolds are not intuitable. On the contrary, he was asserting their intuitability or conceivability without, or even despite, the judgments we derive from our intuitions of physical space. For example, Riemann believed that our intuition of physical space tells us that it is unbounded, yet he certainly claimed that one was able to conceive of bounded manifolds.

Riemann's audience clearly influenced his treatment of his topic, but not merely by leading him to simplify his discussion as some have supposed. His references indicate a concern with the construction of space-like mathematical objects unrelated to physical space and independent of spatial intuition, a concern which was reflected in his unpublished notes. These notes also reveal his interest in the epistemological status of the foundations of geometry, and in the new answers mathematics could provide to a problem first posed by Kant.

The Reception of the *Habilitationsvortrag*

The publication of the *Habilitationsvortrag* in 1868 met with both acclaim and criticism. Although the mathematical results eventually found perceptive readers, in most cases the work was praised or criticized for its philosophical content. The nature of the reactions depended in most cases upon the readers' previously formed opinions of Kant's views, and were influenced by their difficulties with its mathematical content.

One of the first readers with whom Riemann's paper found a sympathetic reception and advocacy was Hermann von Helmholtz. It was through Helmholtz that many (especially British) readers first became aware of Riemann's work. In his 1868 paper *Ueber die Thatsachen, die der Geometrie zu Grunde liegen*, Helmholtz mentioned Riemann's work briefly and discussed the necessity of avoiding mistakes that arise from depending upon spatial intuition.[41] His next paper dealt specifically with the philosophical issues raised by the new geometry.[42] In this paper, Helmholtz addressed the question of the possibility of synthetic *a priori* propositions, summarized Riemann's paper, repeated the warning about intuition, and tried to challenge neo-Kantians by "conceiving" of non-Euclidean spaces without using physical intuition. The orientation of this paper was entirely philosophical, concerned to demonstrate the empirical nature of the geometry of physical space. This philosophical analysis arose directly from reading Riemann's paper—it does not appear in Helmholtz's earlier paper, where Helmholtz clearly indicated that he had just learned of Riemann's work, and did not incorporate it into the body of his discussion. Only after Helmholtz had studied the *Habilitationsvortrag*, as is shown by his frequent references to Riemann in the second paper, did he take up the philosophical position that the geometry of physical space could only be known empirically.

Another of Riemann's readers was William K. Clifford, who translated Riemann's paper into English in the early 1870s. He also provided a brief abstract of the paper for the Cambridge Philosophical Society, stressing Riemann's conclusion that the geometry of physical space can only be known through experience.[43] In his lectures on the philosophy of the pure sciences to the Royal Institution in 1873, Clifford introduced the subject of the synthetic *a priori*, and in the third lecture, "The Postulates of the Science of Space", Clifford reduced to the form of a popular lecture many of the ideas contained in Riemann's *Habilitationsvortrag*.[44] For example, Riemann's restriction to line elements expressible as a square root was described as "elementary flatness". Clifford also repeated his view that the geometry of physical space must be determined empirically.

In the 1871 volume of *Nature*, W. Stanley Jevons replied to the second of Helmholtz's articles, which he described as "based on the latest speculations of German geometers", later naming Riemann.[45] Jevons took the Kantian position of identifying the truth of the Euclidean axioms with their validity in physical space, and claimed that even beings living in curved spaces would discover Euclidean geometry as the limit of behavior in the infinitely small. Given his assertion that the truth of a proposition of geometry depended upon its validity in physical space, Jevons was led to reject the notion that the "truths of geometry" cannot be obtained through experience. Instead, he was able to justify Euclidean geometry as the set of synthetic *a priori* truths valid on infinitely small scales no matter what the macroscopic situation, with propositions about a particular curved space presumably being synthetic *a posteriori* truths about that particular macroscopic situation. The fact that Euclidean geometry was equally accessible at infinitely small scales to geometers inhabiting universes of any given curvature convinced Jevons that its truths were known *a priori*.

At the close of his term as president of the London Mathematical Society in 1882, Samuel Roberts addressed the Society on the topic of Riemann's *Habilitationsvortrag*. Roberts was a mathematician concerned to reassure philosophers that mathematicians were not encroaching on their field:

> ... the progress of analysis has in no way affected the philosophical status of our notions of space and time, and that, consequently, metaphysicians need feel no alarm at what is going on behind the veil of mathematical symbolism.... Mathematics are neutral in Philosophy.[46]

Roberts regarded Riemann's empirical stance as serving to "determine his position in the philosophic ranks, but nothing more." He regarded other spaces as non-intuitable, mathematical objects without externality. Roberts went beyond Kant in ascribing to the various axiomatic spaces a kind of truth; he described Riemann's work as a kind of analysis using "quantitative conceptions", that have nothing to do with "geometry" or the mathematization of physical space. Since our everyday intuitions and physical measurements are clearly Euclidean, Roberts argued, the identification of intuited space with physical space could be preserved—thus for him, Riemann's work was mathematically valid but philosophically invalid.

A translation of Helmholtz's second paper in the journal *Mind* drew a response from J. P. N. Land. He directed his argument mainly at Helmholtz, yet the attack was clearly on the complex of ideas shared by Riemann and Helmholtz, and reveals the extent of Helmholtz's role in bringing Riemann's ideas to a wider public. Land supported the intuitive *a priori* nature of Euclidean geometry by denying that Helmholtz's experiential descriptions of non-Euclidean spaces made them conceivable. He argued that our perceptions are not in accord with Euclidean geometry— for example, receding objects appear to grow smaller—and yet we can affirm the truth of Euclidean geometry with sensory experience.[47] Land concluded that "geometrical axioms are synthetic propositions, because they are not to be deduced by pure logic from the definition of their subject-terms, but are found by intuition of the space offered to us as a form of our objective world."[48] While most opponents of the Riemann–Helmholtz view were concerned to show that the geometry of physical space is synthetic *a priori* and not *a posteriori*, Land wished to demonstrate that intuited geometric propositions are synthetic and not analytic *a priori*. Land denied the identification of axiomatic space with the intuitable space, correctly perceiving that this identification was part of Riemann's program to distinguish the status of propositions about intuitable space from those about physical space.

Hermann Lotze assumed Herbart's chair at Göttingen upon the latter's death in 1841. As Lotze outlived Riemann and remained in Göttingen for most of his life, it is highly probable that some of Riemann's instruction in philosophy and pedagogy came from Lotze, although I have been unable to uncover any direct evidence of this. Lotze was a committed Kantian who also admired Herbart, but he disagreed specifically with the latter's views on space.[49]

Lotze's approach to space was highly influenced by Kant; he believed in the priority of three-dimensional space and the subjectivity of spatial intuition, and used a perceptual justification of the parallel postulate. His comments many years later on the *Habilitationsvortrag* also reveal a Kantian perspective: Riemann's discussion of curvature leads him to conclude that Riemann has fallen prey to a "confusion of space with structures in space". It is interesting to note that Lotze's opposition hints at a resistance of philosophers to instruction by mathematicians: "I trust that in this point philosophy will not allow itself to be imposed upon by mathematics; space of absolutely uniform fabric will always seem to philosophy the one standard by the assumption of which all these other figures become intelligible to it."[50]

Fundamentally, Lotze's system was not falsifiable. He regarded the sufficiency of Euclidean geometry as a description of space as beyond discussion, claiming that if astronomical measurements seemed to demonstrate the existence of a triangle whose angles summed to less than 180 degrees (thus implying that the geometry of space was non-Euclidean), we should conclude that some new physical phenomenon, perhaps a new kind of refraction, was affecting the results, but not that space possessed a geometry which countered our spatial intuition.[51]

In 1877, Benno Erdmann published a book on Riemann's and Helmholtz's geometrical views.[52] Erdmann's aim was entirely philosophical; after surveying the concepts of space developed by Riemann and Helmholtz he presents his own version, in which Riemannian notions serve as a definition of space. Erdmann's account is interesting in that he maintained that Riemann's empiricism was compatible with but not derived from his geometry.[53] This claim was part of his effort to find common ground between Riemannian geometry and the Kantian position according to which our subjective representations of space consist of synthetic a priori knowledge. For the present discussion a consideration of the technical merits of Erdmann's argument is unnecessary; we should merely note that he easily discerned the philosophical position of Riemann's and Helmholtz's work, and found it worthy of extended analysis.[54]

Like all major topics of dispute, the debate over the *a priori* nature of geometry attracted its polemicists. Johann Bernhard Stallo was a German emigre to the United States who supplemented his legal career with much independent study of philosophy and science. In 1881, he published *The Concepts and Theories of Modern Physics*, which included an entire chapter on Riemann's *Habilitationsvortrag*. Stallo rejected Riemann's view on the grounds that it was not intuitive, arguing that it cannot be valid because it does not reflect our experience. Stallo's discussion was Kantian in that he accepted the doctrine of space as a form of intuition, used experience as a test of the validity of geometry, and adopted the Kantian identification of intuited and physical space.[55]

Stallo does have some conceptual difficulties with the material. For example, he concludes that "an inherent curvature of space presupposes differences between its several parts—heterogeneities in its internal constitution"; thus the curvature of space is impossible for Stallo because he saw it as violating the property of homogeneity, which was known to be true of space.[56] He rejects Riemann's empirical discussion because he believes that space cannot be an object of sensation, ignoring Riemann's suggestion that sensing objects in space could provide us information about space. His judgment of the *Habilitationsvortrag* may be summed up by his conclusion that it was not deserving of the attention that it had received, that its statements were "crude and confused", that Riemann had no knowledge of logical processes and terms, concluding that he was an "utter stranger to the discussions respecting the nature of space which have been so vigorously carried on by the best thinkers of our time ever since the days of Kant." Stallo also took at face value Riemann's assertion of lack of ability in philosophy, and included an extended footnote discussing the lack of philosophical ability among mathematicians.[57]

Bertrand Russell wrote *An Essay on the Foundations of Geometry* in 1897. He did ascribe a philosophical motive to Riemann, but rejected Riemann's view partly on Kantian grounds, asserting that we perceive physical space as Euclidean, and yet we know the truths of geometry *a priori*, and partly because he saw Riemann's presuppositions as misguided. He felt that Riemann was wrong to begin by assuming that space is a magnitude and then attempting to determine what attributes of space necessarily follow from this assumption. Russell suggested that the correct procedure was to determine what attributes of space must be presupposed in order to allow a quantitative science of space. He claimed that this led to a fallacy in the proof of the empirical nature of the axioms of physical space.[58] Thus Russell's dispute with Riemann did address Riemann's philosophical orientation—Russell disagreed with the assumptions Riemann chose to make in determining the hypotheses which he used to study the application of metric relations to physical space. Russell held substantially the same position as Roberts—he accepted the formal truth of Riemann's mathematics, but preserved the identification of intuited and physical space.

Ernst Cassirer, writing in 1921, was concerned to revise the history of mathematical physics in Kantian terms; for example, he wished to link the subjectivity of space found in the theory of relativity to the subjectivity of space as Kant's form of intuition. To this end, he portrayed Riemann as fundamentally interested, not in the existence of synthetic *a priori* concepts, but in the geometry of physical space. Thus he interpreted Riemann's demand for "full freedom for the construction of geometrical concepts and hypotheses" as "expressed . . . in the language of Herbartian realism". Space is "an ideal ground in the construction and progress of knowledge of reality". Cassirer depicted geometry as a body of synthetic *a priori* knowledge concerned with this ideal ground, while giving up the idea of propositions about physical space being *a priori*. He even attempted to describe Riemann's geometry as constructive (i.e., synthetic) *a priori*, prior to its application to space: "we are led to a system of pure *a priori* manifolds, whose laws thought lays down constructively, and in this construction we possess also the relation of the real structures of the empirical manifold."[59]

Conclusion

The impact Riemann's *Habilitationsvortrag* eventually had upon mathematics is unquestionable. But this should not lead one to assume too quickly that Riemann was only interested in presenting mathematics. Some authors have presented a "suppression" thesis about the *Habilitationsvortrag*: that Riemann wished to present detailed mathematics, but could not because there were non-mathematicians in his audience, and so he adopted a vague philosophical standpoint which actively detracted from the clarity of the presentation.[60] Several points, for example Riemann's failure to publish a paper surveying his results in differential geometry, and the rhetorical structure of the *Habilitationsvortrag*, do not so much support the "suppression" thesis as the possibility that Riemann had something other than geometry in mind as the purpose of the *Habilitationsvortrag*.

The suppression thesis also does a disservice to Riemann's reputation as a mathematical expositor: we are asked to believe that Riemann wished to present mathematics, but was forced to render his presentation vague and enigmatic in order to make this mathematics intelligible to nonmathematicians. Even if simplification was required for the sake of the audience, it does not account for either the extended investigation of the concept of space in the first section, or the discussion of applications to space in the third. If, on the contrary, we recall that the *Habilitationsvortrag* was the text of a lecture, delivered by a man well read in philosophy to an audience more learned in philosophy than in mathematics, and was part of an occasion whose purpose it was to demonstrate lecturing ability in a culture whose ideal researcher could relate his specific investigations to larger intellectual issues, we could well conclude that the paper was primarily philosophical rather than mathematical in intent. Such a view explains many features of the history of the *Habilitationsvortrag*—the fact that it was never published as a mathematical paper, its unusual organization, which placed most of the mathematical content in one section, and its reception, which took issue with it primarily on philosophical rather than mathematical grounds. In the *Habilitationsvortrag*, Riemann is functioning as more than a mathematician; we have a better understanding of the paper if we see him speaking primarily as a philosopher, possessed of a rather powerful mathematical methodology. Riemann was tactfully suggesting that new mathematical researches might have something to say to philosophy. His attempt to criticize the Kantian view of space on mathematical grounds presented mathematics as a significant and contributing component of intellectual culture.

NOTES

1. Dedekind, Richard, "Bernhard Riemann's Lebenslauf", in Riemann, Bernhard, *Gesammelte Mathematische Werke und Wissenschaftlicher Nachlass*. B.G. Teubner, Leipzig 1876, pp. 515–517. Second Edition 1892, pp. 547–549. The author would like to thank Michael Mahoney, Nancy Nersessian, Kathryn Olesko, and Joan Richards for their helpful comments, and Princeton University for a grant from the President's Discretionary Fund which made his attendance at the Vassar Conference possible.

2. Edna Kramer has produced a very readable introduction to differential geometry in the style of Gauss and Riemann. Kramer, Edna E., *The Nature and Growth of Modern Mathematics*, Princeton University Press, Princeton, NJ, 1982, pp. 449–473. A more technical account

is given by Morris Kline, who acknowledged Riemann's attempt to philosophize about the proper axioms for geometry, but missed the significance of Riemann's generalization of the concept of space from a unique physical space to a category of spatial forms. Kline, Morris, *Mathematical Thought from Ancient to Modern Times*, Oxford University Press, New York, 1972, pp. 889–902, 1122–1135. Kline also expressed surprise that Riemann treated the n-dimensional case "even though the three-dimensional case was clearly the important one." [Kline, op. cit., p. 890.] A comprehensive survey of differential geometry from Gauss to Riemann is given in Reich, Karin, "Die Geschichte der Differentialgeometrie von Gauss bis Riemann (1828-1868)" in *Archive for History of Exact Sciences*, Vol 11 (1973), No. 4, pp. 273–282. Surveys of the mathematical contents of the *Habilitationsvortrag* are also available in Portnoy, Esther, "Riemann's Contribution to Differential Geometry", *Historia Mathematica*, Vol. 9 (1982), pp. 1–18, and Spivak, Michael, *Differential Geometry*, Vol. II, 2nd ed., Publish or Perish, Inc., Berkeley 1979, pp. 132–204.

3. Kline, op. cit., p. 891.

4. Dedekind, "Riemann's Lebenslauf", *Werke* 1876, pp. 511–513, *Werke* 1892, pp. 543–545.

5. Riemann, *Werke* 1876, pp. 475–506; *Werke* 1892, pp. 507-538.

6. Kant, Immanuel, *Critique of Pure Reason*, tr. Norman Kemp Smith, The Modern Library, Random House, New York 1958. Kant introduces the terms a priori and *a posteriori* on pp. B2–B4, and associates with the *a priori* the qualities of necessity and universality. He identifies *a posteriori* knowledge with empirical knowledge. The analytic and synthetic classifications are introduced on pages A6-A10/B10-B14.

7. Kant characterizes synthetic *a priori* judgments as leading "to a genuinely new addition to all previous knowledge", while analytic *a priori* knowledge is given the role of clarifying the concepts being used. Kant, *Critique*, p. A10/B14. Kant explains the impossibility of analytic *a posteriori* judgments on p. A7/B11: "Judgments of experience [truths determined *a posteriori*] are one and all synthetic." It is an error to define synthetic judgments as those knowable only through experience; such a definition eliminates the possibility of synthetic *a priori* knowledge. The program of determining the existence of synthetic *a priori* propositions is presented as "The General Problem of Pure Reason" on pp. B19–20.

8. Kant's view of space and geometry is discussed in the *Critique* on p. A22–28/B36–44.

9. Kant, *Critique*, p. A24/B39.

10. Kant, *Critique*, p. A28/B44. Kant discusses the empirical reality of physical space and the objects in it on pp. B274–279.

11. Wiredu, J. E., "Kant's Synthetic *a priori* in Geometry and the Rise of Non-Euclidean Geometries," *Kantstudien*, Vol. 61 (1970), No. 1, pp. 5–6.

12. "Thus there is no contradiction in the concept of a figure which is enclosed within two straight lines, since the concepts of two straight lines and of their coming together contain no negation of a figure. The impossibility arises not from the concept in itself, but in connection with its construction in space, that is, from the conditions of space and of its determination. And since these contain *a priori* in themselves the form of experience in general, they have objective validity; that is, they apply to possible things." Kant, *Critique*, pp. A220–221/B268.

13. "Such fictitious concepts . . . can acquire the character of possibility not in *a priori* fashion, as conditions upon which all experience depends, but only *a posteriori* as being concepts which are given through experience itself. And, consequently, their possibility must either be known *a posteriori* and empirically, or it cannot be known at all." Kant, *Critique*, pp. A222/B269–270.

14. "Riemann's definition [of manifold] is vague and awkward by comparison [to modern ones], but it has an important advantage in being constructive rather than analytic. . . . We may regard his choice of an intuitive definition as an indication of the desire to be intelligible to his general scholarly audience," Portnoy, "Riemann's Contribution to Differential Geometry," p. 3. "Upon a first reading, Riemann's lecture may appear to have almost no mathematical content. But this is only because the analytic investigations, which occur in Part II, have been drastically condensed. . ." Spivak, *Differential Geometry*, Vol. II, p. 154. Also see Torretti, R., *Philosophy of Geometry from Riemann to Poincaré*, D. Reidel, Dordrecht, 1978, p. 83.

15. The *Habilitationsvortrag* first appeared in 1868, after Riemann's death; the only other work of Riemann's using these techniques was his response of 1861 to the prize problem proposed in 1858 by the Paris Academy of Sciences, *Commentatio mathematica, qua respondere tentatur quaestioni ab Illustrissima Academia Parisiensi propositae*, which first appeared in Riemann, *Werke* 1876, pp. 370–383, *Werke* 1892 pp. 391–404. Riemann's *Pariserarbeit* has recently been translated into German by Olaf Neumann, in *Gaußsche Flächentheorie, Riemannsche Räume und Minkowski-Welt*, Teubner Archiv zur Mathematik, Bd. 1, 1984.

16. Riemann, "Ueber die Fläche vom kleinsten Inhalt bei gegebener Begrenzung", *Werke* 1876, pp. 283–315, *Werke* 1892, 301–333. This point has been observed by Portnoy, op. cit, p. 10.

17. Dedekind, "Riemann's Lebenslauf", Riemann, *Werke* 1876, pp. 523–524; *Werke* 1892, pp. 555–556.

18. Riemann, Bernhard, "Ueber die Hypothesen, welche der Geometrie zu Grunde liegen." *Abhandlungen der Königlichen Gesellschaft der Wissenschaften zu Göttingen,* Vol. 13 (1866/1867) Göttingen, 1868, Abhandlungen der Mathematischen Classe, p. 133. *Werke* 1876, p. 254, *Werke* 1892, p. 272. Translations are available in Spivak, Michael, *Differential Geometry,* Vol. II, Second Edition, pp. 132–153, and W.K. Clifford, "On the Hypotheses which lie at the bases of Geometry", *Nature,* Vol. VIII, pp. 14–17, 36, 37. Also in his *Mathematical Papers,* MacMillan and Co., London, 1882, pp. 55–71. The author would also like to thank Michael Mahoney for supplying him with an unpublished translation of his own.

A surprising amount of confusion surrounds the dating of the first publication of Riemann's *Habilitationsvortrag,* and this confusion began only a few years after the paper appeared. Stallo in 1881 (see note 56) cites publication "by Dedekind in 1866, after Riemann's death". Felix Klein, writing in "Ueber die Nicht-Euklidischen Geometrie", *Mathematische Annalen* IV, 1871, p. 576, gives 1867 as the year of its appearance. These errors have been passed down or independently reproduced in many citations. They most probably result either from a confusion of Riemann's death date with the date of publication of the Abhandlungen, or a misreading of the various dates given in the Abhandlungen volume. Volume 12 mentions Riemann's death on July 20, 1866, but makes no other reference. Volume 13 includes the *Habilitationsvortrag* and two other papers by Riemann. It is supposed to contain papers submitted in the second half of 1866 and in 1867; the preface is signed by F. Wöhler and dated December 1867; the publication date on the title page is 1868. Thus there is some textual support for choosing any one of the dates 1866, 1867, and 1868; but if we take the date of publication to mean the year in which a work could be obtained by a general reader, 1868 would be the most accurate choice for Vol. 13 of the Abhandlungen and thus for the *Habilitationsvortrag.*

19. "In the course of the previous considerations, the relations of extension or regionality were first distinguished from the metric relations, and it was found that different metric relations were conceivable along with the same relations of extension; then systems of simple metric specifications were sought by means of which the metric

relations of space are completely determined, and from which all theorems about it are a necessary consequence. It remains now to discuss the question how, and to what degree, and to what extent these assumptions are borne out by experience" Riemann, "Ueber die Hypothesen...", *Abhandlungen*, p. 147, *Werke* 1876, pp. 265–266, *Werke* 1892, pp. 283–284, tr. in Spivak, Michael, *Differential Geometry* Vol. II, 2nd ed., pp. 149–150.

20. Kant is quite specific on this point: "Space is not a discursive or, as we say, general concept of relations of things in general, but a pure intuition. For, in the first place, we can represent to ourselves only one space; and if we speak of diverse spaces, we mean thereby only parts of one and the same unique space. Secondly, these parts cannot precede the one all-embracing space, as being, as it were, constituents out of which it can be composed; on the contrary, they can be thought only as in it. Space is essentially one, the manifold in it, and therefore the general concept of spaces, depends solely on [the introduction of] limitations. Hence it follows that an *a priori*, and not an empirical, intuition underlies all concepts of space" Kant, *Critique*, p. A25/B39.

21. Riemann, "On the Hypotheses...," tr. in Spivak, Michael, *Differential Geometry*, Vol. II, 2nd ed., p. 135, of "Ueber die Hypothesen...," *Abhandlungen*, pp. 133–134, *Werke* 1876, pp. 254–255, *Werke* 1892, pp. 272–273. For the sake of clarity I have used the modern terminology, referring to "physical space" explicitly when Riemann used "Raum"; to translate too literally would obscure rather than illuminate his differences with Kant.

22. "...the regulative principle of reason is grounded on the proposition that in the empirical regress we can have no experience of an absolute limit, that is, no experience of any condition as being one that empirically is absolutely unconditioned. The reason is this: such an experience would have to contain a limitation of appearances by nothing, or by the void, and in the continued regress we should have to be able to encounter this limitation in a perception which is impossible. ...Thus the first and negative answer to the cosmological problem regarding the magnitude of the world is that the world has no first beginning in time and no outermost limit in space." Kant, *Critique*, pp. A517,519/B545,547.

23. Riemann, "Ueber die Hypothesen...," *Abhandlungen*, pp. 147–148, *Werke* 1876, p. 267, *Werke* 1892, p. 285., tr. in Spivak, *Differential Geometry*, p. 150.

24. Riemann, "Ueber die Hypothesen...," *Abhandlungen*, p. 149, *Werke* 1876, p. 267, *Werke* 1892, p. 284, tr. in Spivak, *Differential Geometry*,

p. 152. Howard Stein has previously interpreted Riemann as speaking here about how the atomic structure of matter and the diffraction of light rays challenge our intuitive notions of straightness in "Some Philosophical Prehistory of General Relativity," in Minnesota Studies in the Philosophy of Science, vol. VIII, pp. 3–49.

25. Riemann, "Ueber die Hypothesen...,", Abhandlungen, p. 150, Werke 1876, p. 268,Werke 1892, p. 285, tr. in Spivak, Differential Geometry, pp. 152–153.

26. Dedekind, R., "Riemann's Lebenslauf", in Riemann, Werke 1876, pp. 515–516, Werke 1892, pp. 547–548.

27. Riemann, "Ueber die Hypothesen...," Abhandlungen, p. 134, Werke 1876, p. 255, Werke 1892, p. 273, tr. in Spivak, Differential Geometry, p. 136.

28. Gauss, C. F., "Disquisitiones generales circa superficies curvas" (1827) Werke, Vol. IV, Königlichen Gesellschaft der Wissenschaften zu Göttingen, Göttingen, 1873, pp. 217–258.

29. See, for example, Roberto Torretti, Philosophy of Geometry from Riemann to Poincaré, D. Reidel, Dordrecht, 1978, p. 107: "Riemann names Gauss and Herbart as his only authorities. His relations to Gauss ought to be plain by now." This follows a discussion of Gauss which refers to the Disquisitiones, but makes no reference to the preliminary note to Theoria residuorum biquadraticorum or the Beiträge zur Theorie der algebraischen Gleichungen.

30. "Von der andern Seite wird hierdurch die wahre Metaphysic der imaginären Grössen in ein neues helles Licht gestellt ... von dem Begriff der ganzen Zahlen hat [Arithmetik] ihr Gebiet stufenweise erweitert ... Dies Vorschreiten ist aber immer anfangs mit furchtsam zögerndem Schritt geschehen..." Gauss, C. F., Theoria residuorum biquadraticorum, commentatio secunda, Anzeige., Werke, Vol. II, Göttingen, 1876, p. 175, pp. 177–178.

31. "Ich werde die Beweisführung in einer der Geometrie der Lage entnommenen Einkleidung darstellen, weil jene dadurch die grösste Anschaulichkeit und Einfachkeit gewinnt. Im Grunde gehört aber der eigentliche Inhalt der ganzen Argumentation einem höhern von Räumlichem unabhängigen Gebiete der allgemeinen abstracten Grössenlehre an, dessen Gegenstand die nach der Stetigkeit zusammenhängenden Grössencombinationen sind, einem Gebiete, welches zur Zeit noch wenig angebaut ist, und in welchem man sich auch nicht bewegen kann ohne einem von räumlichen Bildern entlehnte Sprache." Gauss, C.F., "Beiträge zur Theorie der algebraischen Gleichungen",

Werke, Königlichen Gesellschaft der Wissenschaften zu Göttingen, 1876, p. 79.

32. In the "Fragmente philosophischen Inhalts" one reads: "Der Verfasser is Herbartianer in Psychologie und Erkenntnisstheorie (Methodologie und Eidolologie) Herbart's Naturphilosophie und den darauf bezüglichen metaphysischen Disciplinen (Ontologie und Synechologie) kann er meistens nicht sich anschliessen." Riemann, *Werke* 1876, p. 476, *Werke* 1892, p. 508.

33. Scholz, Erhard, "Herbart's Influence on Bernhard Riemann", *Historia Mathematica* Vol. 9, pp. 413–440. Herbart's theory of serial forms is laid out in *Psychologie als Wissenschaft neu Gegründet auf Erfahrung, Metaphysik und Mathematik*, (1825), Sections 109, 116, in Herbart, J. F., *Sämmtliche Werke*, Voss: Leipzig, 1850, pp. 114–150.

34. Scholz, "Herbart's Influence on Bernhard Riemann," pp. 423–424. Torretti, *Philosophy of Geometry from Riemann to Poincaré*, pp. 107–108, also disavows any significant influence of Herbart upon Riemann, laying great stress on the fact that Herbart's *Reihenformen* were dense pointsets, unlike Riemann's *ausgedehnte Grössen*, which were continua. He does not, however, discuss their similarity in having an anti-Kantian approach to space.

35. Riemann, "Ueber die Hypothesen...", *Abhandlungen*, p. 135, *Werke* 1876, p. 256, *Werke* 1892, p. 274.

36. See Scholz, "Herbart's Influence on Bernhard Riemann," pp. 421–422. Herbart's attack on Kant appears in *Psychologie als Wissenschaft*, Section 102, pp. 504–514. Riemann's citation of Herbart appears in Riemann, *Werke* 1876, p. 490, *Werke* 1892, p. 522.

37. "Genau genommen, liegen auch die Gegenstände der reinen Geometrie nicht in sinnlichen Weltraum; dieser letztere ist theils von Körpern erfüllt, theils liegt es leer zwischen ihnen; die geometrischen Kreise, Quadrate, Polygone aber sind *nirgends* in ihm, haben in ihm keinen Platz, wurden auch nicht durch Begrenzung aus ihm herausgehoben, sondern der Geometer macht jeden von ihnen ganz von vorn an, und würde aus jedem derselben einen ganz vollständigen Raum, als dessen Umgebung, produciren, wenn ihm daran gelegen wäre, so dass auch dieser Raum gar keine bestimmte Lage gegen oder in dem sinnlichen Weltraum hätte, sondern man einen davon sich aus dem Sinne schlagen müsste, um den andern zu denken." Herbart, *Psychologie als Wissenschaft*, Section 100, pp. 489–490.

38. Herbart, J.F., "Nachschrift zur zweiten Auflage Pestalozzi's Idee eines ABC der Anschauung", *Sammtliche Werke*, Vol. I p. 254.

39. "Aus dieser Definition der Geraden würde ich alle Sätze ableiten können, welche in der Geometrie über die Gerade stattfinden. Es ist klar, dass man auf diese Weise fortfahren könnte, ohne die mindeste räumliche Anschauungen zu Hülfe zu nehmen. Bei dieser Behandlungsweise der Geometrie oder der Lehre der Mannigfaltigkeiten dreier Dimensionen würden alle Axiome, welche bei der gewöhnlichen Behandlungsweise von der räumlichen Anschauung entlehnt werden, wie z.B. der Satz, dass durch zwei Punkte nur eine Gerade möglich ist, das erste Axiom des Euklid etc., wegfallen..." Scholz, Erhard, "Riemanns frühe Notizen zum Mannigfaltigkeitsbegriff und zu den Grundlagen der Geometrie", *Archive for History of Exact Sciences*, Vol. 27, p. 229.

40. "Der Begriff einer Mannigfaltigkeit von mehreren Dimensionen besteht unabhängig von unseren Anschauungen im Raum. Der Raum, die Ebene, die Linie sind nur das anschaulichste Beispiel einer Mannigfaltigkeit dreier, zweier, oder einer Dimensionen. Ohne die mindeste räumliche Anschauung zu haben, würden wir doch die ganze Geometrie entwickeln können." Scholz, E., "Riemann's frühe Notizen ...," p. 228.

41. Helmholtz, Hermann, "On the Facts underlying Geometry", pp. 39–71 in *Epistemological Writings*, Dordrecht, 1977. This paper originally appeared as "Ueber die Thatsachen, die der Geometrie zu Grunde liegen", in *Nachrichten der königlichen Gesellschaft der Wissenschaften zu Göttingen*, No. 9, June 3, 1868. Leo Koenigsberger records that Helmholtz presented an "Abriss" to the *naturwissenschaftlich-medicinischen Verein* in Heidelberg on May 22, 1868, entitled "Ueber die thatsächlichen Grundlagen der Geometrie"(*Hermann von Helmholtz*, Friedrich Vieweg und Sohn, Braunschweig, 1903, p. IX, pp. 125–126). Helmholtz's first reading of Riemann's *Habilitationsvortrag* can be dated rather precisely; noticing a memorial to Riemann by Schering in the *Nachrichten*, he wrote to Schering on April 21, 1868, asking for a copy of an essay on geometry by Riemann that Schering had referred to; on May 18 he writes again to Schering, thanking him for the copy of the *Habilitationsvortrag*, and describing the advances he was able to make. He also requested that his paper appear in the *Göttinger Abhandlungen*, perhaps as a tribute to Riemann; evidently it was submitted, and appeared in the *Nachrichten* instead. Helmholtz's contribution to the *Nachrichten* merely acknowledged Riemann's priority, changed the title as a tribute to Riemann since it was being published in a Göttingen journal, and stressed his originality in noting the consequences of the assumption that bodies were independent of position.

Helmholtz's letters to Schering appear to date Helmoltz's paper precisely, yet there has been some confusion in dating this paper as well. Some have suggested a date of May 22, 1866 (see Torretti, *Philosophy of Geometry from Riemann to Poincaré*, p. 155) but this is definitely ruled out, since Helmholtz refers to Riemann's *Habilitationsvortrag*, and we know that Dedekind obtained a copy of the paper for eventual publication only after Riemann's death on July 20, 1866. See Richards, Joan L., *Mathematical Visions*, Harcourt Brace Jovanovich, Academic Press, Cambridge, 1988, p. 74, n. 19. The author would like to thank Professor Richards for supplying him with a copy of her book in galley sheets.

42. Helmholtz, Hermann, "Ueber den Ursprung und die Bedeutung der geometrischen Axiome" (1868); tr. as "The Axioms of Geometry" in *The Academy*, Vol. I (1870), and as "On the Origin and Significance of the Axioms of Geometry" in *Epistemological Writings*, pp. 1–38.

43. Clifford, W. K., "On the Space-theory of matter", *Cambridge Philosophical Society Proceedings*, II, 1876, pp. 157–158, (read Feb. 21, 1870); also in *Mathematical Papers*, pp. 21–22.

44. Clifford, W.K., "The Postulates of the Science of Space", in *Lectures and Essays*, MacMillan and Co., London, 1879, pp. 295–323.

45. Jevons, W. Stanley, "Helmholtz on the Axioms of Geometry", *Nature*, Vol. IV, Oct. 19, 1871, pp. 481–482.

46. Roberts, Samuel, "Remarks on Mathematical Terminology, and the Philosophic Bearing of recent Mathematical Speculations concerning the Realities of Space." *Proceedings of the London Mathematical Society*, Vol. 14 (1882-83) pp. 5–15.

47. Land, J. P. N., "Kant's Space and Modern Mathematics", *Mind*, Vol. II, p. 42.

48. Land, "Kant's Space and Modern Mathematics," p. 45.

49. See Lotze, Hermann, *Philosophie seit Kant*, G. Hirzel, Leipzig 1882, Section 69, p. 94: "Die Speciellere Anwendung der Methode in Bezug auf Naturphilosophie ist nicht glücklich gewesen. Herbart schliesst sich nicht der Lehre Kant's von der Subjectivität der Raumanschauung an..." It is worth noting that Lotze was an *Ordinarius* at Göttingen, and thus most likely was present at the delivery of Riemann's *Habilitationsvortrag*. This fact might help to explain a slight difficulty in the present interpretation of the lecture—if it is supposed to be a criticism of Kant, why was Kant mentioned nowhere in the text? Riemann had a reputation for shyness, and the opinions of Lotze must have been known to him—it is easy to see how he could decide to

avoid offending Lotze (and others present) by describing his work as influenced by Herbart rather than critical of Kant. In this way Riemann tactfully avoided any suggestion that mathematics was being used to criticize philosophy, instead indicating that philosophy was inspiring mathematics.

50. Lotze, Hermann, *Metaphysic*. Clarendon Press, Oxford, 1887, 137, pp. 312–313. Translation of *Metaphysik*, G. Hirzel, Leipzig, 1879, 137, pp. 265–266.

51. Lotze, Hermann, *Metaphysik*, G. Hirzel, Leipzig, 1879, 131, pp. 248–249.

52. Erdmann, Benno, *Die Axiome der Geometrie: eine philosophische Untersuchung der Riemann-Helmholtz'schen Raumtheorie*, Leopold Voss, Leipzig, 1877.

53. Erdmann, *Die Axiome der Geometrie*, p. 119.

54. For a critique of Erdmann's book, see Torretti, *Philosophy of Geometry from Riemann to Poincaré*, pp. 264–272.

55. Stallo, J.B., *The Concepts and Theories of Modern Physics*, Belknap Press of Harvard University Press, Cambridge, Mass., 1960, p. 253. Percy W. Bridgman's introduction to this edition includes a biography of Stallo and a survey of the book.

56. Stallo, *The Concepts and Theories of Modern Physics*, p. 243.

57. Stallo, *The Concepts and Theories of Modern Physics*, pp. 259–260.

58. Russell, Bertrand, *An Essay on the Foundations of Geometry*, Cambridge University Press, Cambridge, 1897, 65, p. 69.

59. Cassirer, Ernst, *Substance and Function, and Einstein's Theory of Relativity*. Dover Publications, 1953, pp. 441–442.

60. "...we should be forewarned that Riemann's ideas as expressed in the lecture and in the manuscript of 1854 are vague. One reason is that Riemann adapted it to his audience, the entire faculty at Göttingen. Part of the vagueness stems from the philosophical considerations with which Riemann began his paper." Kline, op. cit., p. 890. See also note 14.

Foundations of Mathematics

Georg Cantor (1845–1918)
Courtesy of Springer-Verlag Archives

Cantor's Views on the Foundations of Mathematics

Walter Purkert

Georg Cantor (1845-1918) began his academic career as a Privatdozent at the university of Halle in 1869. The mathematical chair at Halle was held at that time by Eduard Heine (1821-1881), a deep and sharp thinker who is well known, for example, through the Heine-Borel theorem. Heine was then working on the uniqueness problem for the Fourier series development [24], and he stimulated Cantor to work on it as well. Cantor, for his part, quickly obtained a decisive result. He proved the following theorem in 1870 in "Beweis, dass eine für jeden reellen Werth von x durch eine trigonometrische Reihe gegebene Function $f(x)$ sich nur auf eine einzige Weise in dieser Form darstellen lässt" [9]: If

$$\frac{a_0}{2} + \sum_{n=1}^{\infty}(a_n \cos nx + b_n \sin nx) = 0 \tag{1}$$

for all $x \in (-\pi, \pi)$ then the coefficients a_i, b_i must be zero for all i. Cantor's proof was based on a fundamental idea introduced by Bernhard Riemann (1826-1866) in the theory of trigonometric series. With

$$\Omega(x) = \frac{a_0}{2} + \sum_{n=1}^{\infty}(a_n \cos nx + b_n \sin nx)$$

Riemann introduced the function

$$F(x) = \frac{a_0}{2}x^2 - \sum_{n=1}^{\infty}\frac{1}{n^2}(a_n \cos nx + b_n \sin nx) \tag{2}$$

and proved that
1) $F(x)$ is continuous in $(-\pi, \pi)$,
2) $\lim_{\alpha \to 0} \dfrac{F(x+\alpha) - 2F(x) + F(x-\alpha)}{\alpha^2} = \Omega(x)$. With these results in mind Cantor was able to reduce the uniqueness question to the following two problems:
(i) To show the uniform convergence of the series (2) under the condition (1).

THE HISTORY OF
MODERN MATHEMATICS

49

(ii) To show: If a function $F(x)$ is continuous in (a, b) and

$$\lim_{\alpha \to 0} \frac{F(x + \alpha) - 2F(x) + F(x - \alpha)}{\alpha^2} = 0$$

in (a, b), then $F(x) = c' + cx$ in (a, b).

Indeed, due to (1), $\Omega(x) = 0$, and according to Riemann's result together with (ii) we get

$$c' + cx + \frac{a_0}{2} x^2 = \sum_{n=1}^{\infty} \frac{1}{n^2} (a_n \cos nx + b_n \sin nx). \tag{3}$$

Since the right hand side is periodic we have $c = a_0 = 0$. Multiplying the remaining equation by $\cos n(x - t)$ and integrating over $(-\pi, \pi)$ (which is allowed due to (i)) one gets $a_n \cos nt + b_n \sin nt = 0$ for arbitrary t, which gives $a_n = b_n = 0$.

Concerning (i), Cantor went beyond a proof of uniform convergence: he proved that from $\lim_{n \to \infty} (a_n \cos nx + b_n \sin nx) = 0$ for all $x \in (-\pi, \pi)$ one can conclude that $\lim a_n = \lim b_n = 0$. This theorem is the content of his first publication on trigonometric series "Über einen die trigonometrischen Reihen betreffenden Lehrsatz" [8]; it was later generalized by Henri Lebesgue (1875-1941) [30[(cf. also [18]), and is today called the Cantor-Lebesgue theorem.

Concerning (ii), Cantor wrote a letter to his friend and former fellow student Hermann Amandus Schwarz (1843-1921) explaining to him that the question of uniqueness for the trigonometric development would be completely solved if (ii) could be shown. There arose along with this question a very interesting correspondence between Schwarz and Cantor, and eventually a rigorous proof was found by Schwarz [39]. Cantor used this proof with Schwarz' permission in his paper [9] (for details, see [37]).

Already at the time he was prepearing the paper [9] for publication, a time when set theory was not even in its infancy, Cantor followed debates about the foundations of mathematics and even took part in them. On March the 30th, 1870, he wrote to Schwarz concerning his uniqueness proof: "This proof was accepted by Herr Weierstrass as being completely rigorous; in particular, he checked your contribution and found it correct. He used the Bolzano theorem himself, and in his lectures he gave, for example, the same proof of the statement '$F'(x) = 0$ implies $F(x) = c'$ as you gave in your letter. Therefore he immediately accepted your proof for the second difference quotient. ... By the way, Herr Kronecker does not accept the Bolzano-Weierstrass theorem about the lower and upper limit z, but this won't keep me from publishing my proof, because I not only accept this theorem as being correct but regard it as being the basis of all the more important mathematical truths." [1, letter from 3/30, 1870].

The core of the differences between Karl Weierstrass (1815-1897) and his followers Schwarz and Cantor, on the one hand, and Leopold Kronecker (1823-1891), on the other, was their different viewpoints concerning mathematical existence. Kronecker's main aim was to ensure the existence of mathematical objects in a constructive manner, i.e. obtaining them eventually by a finite construction procedure. He was convinced that "all the results of the deepest mathematical research must be finally expressible in the simple forms of whole numbers." [29, vol. 3/1, p. 274] That meant they must be reducible to what he called "general arithmetics," by which he understood the theory of polynomial rings with whole numbers as coefficients. He especially wanted "to abandon irrational as well as continuous magnitudes" [29, vol. 3/1, p. 253]; (see also [33, 34]). It should be emphazised that the development of constructive methods meets a very important demand of modern mathematics, especially since electronic computers have been available, and that some of Kronecker's ideas have become of immediate interest in the last three decades (see the paper by H. M. Edwards in this volume). Kronecker's "only mistake" according to Hilbert [25, p. 487], was "to declare that transfinite conclusions were not allowed." Weierstrass and his pupils felt free to create new mathematical objects using the laws of logic for arbitrary sets of already defined mathematical objects. The only accepted corrective was that the created system must be consistent. A typical example of such a creation is Cantor's theory of real numbers. It was published in 1872 [10], but we know from his Nachlass that he presented it already in his 1870 calculus course at the university of Halle [35, p. 37]. (We shall return to Cantor's views on mathematical existence later in this paper in connection with the antinomies of set theory.)

The starting point for Cantor's set theory had nothing to do with speculations about infinity but rather was closely connected with the following concrete mathematical problem: Is it possible that the assertion of the uniqueness theorem $(a_i = b_i = 0, i = 1, 2, \ldots; a_0 = 0)$ remains true, if (1) is only valid in $(-\pi, \pi) \backslash P$, where P is a certain set of exceptional points in which either (1) does not converge or else does not equal zero? The answer is yes, and in order to describe the possible exceptional sets Cantor introduced the concept of the derivative of a point set. Let P be a point set, then the set of all condensation points of P is called the first derivative, P'. The nth derivative is defined by $P^{(n)} = (P^{(n-1)})'$. Cantor was able to show that the uniqueness theorem remains true if one allows exceptional sets P with $P^{(n)} = \emptyset$ (empty set) for a natural number n.

This result was published in "Über die Ausdehnung eines Satzes aus der Theorie der trigonometrischen Reihen" [10] in 1872.

The concept of the derivative of a point set led Cantor already in the years 1871-72 to the transfinite ordinals. He recognized that for any point set P, $P' \supseteq P'' \supseteq P''' \supseteq \cdots$ is true, and he formed $Q = \bigcap_n P^{(n)}$. Q is in some sense "derived" from P, but Q is in general not equal to $P^{(n)}$ for any natural number n. What is the derivation order of Q? Cantor wrote $Q = P^{(\infty)}$, where ∞ is the first transfinite ordinal number (later he replaced the symbol ∞ by ω). Now one can continue the process of derivation: $(P^{(\infty)})' = P^{(\infty+1)}$; $\ldots \bigcap_n P^{(\infty+n)} = P(\infty \cdot 2)$ and so forth.

The number $\infty + 7$, for example, is indeed a real mathematical concept, because one can construct a point set whose $(\infty + 7)$th derivative consists of a given set, say containing one given point x_0. We cannot find these ideas in Cantor's 1872 paper, but we know from his later testimony that he was in posession of these ideas in the early 1870s. In a footnote to part 2 of "Über unendliche lineare Punctmannichfaltigkeiten" [13, part 2, p. 358] published in 1880, he wrote concerning the generation of derivatives like $P^{(\infty)}$, $P^{(\infty+1)}, \ldots, P^{(\infty \cdot 2)} \ldots$ that he had developed this theory ten years earlier. Unfortunately, this footnote was not reprinted in his collected works and therefore went unnoticed for some time. Further evidence for the assertion that Cantor's first step in set theory was the development of the idea of transfinite ordinals in the early 1870's is given by a letter from Cantor to Gösta Mittag-Leffler (1846-1927) dated October 20, 1884 [2, Nr. 16, fol. 3]. (For a thorough discussion of the development of the idea of transfinite ordinals see J. Dauben [18]).

In a very interesting correspondence with Richard Dedekind (1831-1916) [17] during the fall of 1873, Cantor discussed the question whether or not the continuum of real numbers, say between 0 and 1, is countable. He found the answer to be "no" and published this result in "Über eine Eigenschaft des Inbegriffs aller reellen algebraischen Zahlen" [11] in 1874. This paper has sometimes been regarded as marking the birth hour of set theory [31]. We know now that this is misleading since at that time Cantor had already developed the basic concept of transfinite ordinals in the form of derivation orders of point sets.

If one tries to reconstruct the genesis of Cantor's ideas, one must seek a motivation for why he began to investigate the powers of sets in terms of one-to-one correspondences. A possible motive might have been that he felt there must be "more" of these newly discovered numbers than natural numbers. Already Bernard Bolzano knew ([6, p.28,29]) that the relation $A \subset B$ is not a useful gauge of size for infinite sets. He emphasized the fact (which can already be found for a special case in Galileo's "Discorsi" [21, p. 31]) that for infinite sets the relations $A \subset B$, $A \sim B$ are possible simultaneously; Dedekind used this in 1888 for a definition of infinite spaces [19]. Thus a natural first step consisted in checking whether or not there are infinite sets that cannot be put into a one-to-one correspondence with the set of natural numbers. Cantor soon found that the set of all algebraic numbers is countable, a result that appears in the first part of his paper [11]. The second part, as mentioned above, was his proof that $(0,1)$ is not countable (this proof, by the way, does not use the diagonal procedure but rather successive intervals). This latter result was often regarded as being a turning point with respect to what mathematicians understood by the term "infinite" inasmuch as here allegedly for the first time the actual infinite got its citizenship in mathematics. Cantor himself seemed to have felt no necessity to stress this point; neither in his paper [11] nor in his letters of that time can we find anything to indicate that he was pondering the implications this result had for the foundations of mathematics. The reason, in my opinion, is that working with infinite sets as mathematical objects was quite commonplace within the Weierstrassian school. When Weierstrass, for example, used the set of all power series $P(z - s)$ which one can get as analytic continuations of a given "element" $P(z - a)$ in order to define the concept of "analytic function," his definition clearly involved an infinite set. Even the following proof of the existence of transcendental numbers that Cantor deduced in [11] from the fact that $(0, 1) \not\sim \mathbf{N}$ does not go beyond Weierstrass' views about existence in mathematics: let T be the set of all transcendental numbers, A the set of all algebraic numbers between 0 and 1, then $(0,1) = A \cup T$, $A \cap T = \emptyset$; $A \sim \mathbf{N}$, $(0, 1) \not\sim \mathbf{N}$. Hence T cannot be empty. It is clear that such a nonconstructive proof could not have been accepted in this form by Kronecker.

A next logical step forward was Cantor's attempt to find larger powers by taking into account higher-dimensional continua, for example, the square $(0, 1) \times (0, 1)$ in $\mathbf{R}^2$. He eventually showed that all such continua are equivalent. Cantor was rather surprised by this result and wrote to Dedekind: "Je vois, mais je ne le crois pas." (I see it, but I don't believe it) [17, p. 34]. This result appeared in 1878 in *Crelle's Journal* [12] after considerable delay that was probably due to Kronecker (see some recently published letters from Cantor to Dedekind in [20]). It was thus here that Cantor faced Kronecker's real opposition for the first time.

The highpoint in Cantor's mathematical career was the extensive paper "Über unendliche lineare Punctmannichfaltigkeiten" [13] that appeared in *Mathematische Annalen* in 6 parts between 1879 and 1884. Ernst

Zermelo (1871-1953), the editor of Cantor's collected works, character-ized this sequence of articles as being the "quintessence of Cantor's life's work" [7, p. 246]. We shall concentrate here on part 5 (1883) which is in many respects the most interesting portion of [13] (concerning the other parts see [35, pp. 58-62; 69-72]). Cantor himself edited this fifth part for publication as a separate booklet entitled "Grundlagen einer allgemeinen Mannigfaltigkeitslehre" (1883).

Cantor's main aim in this paper was the introduction of transfinite ordinals, independent of the concept of derivation for point sets. Further he wanted to clarify the relation between ordinals and cardinals (the intro-duction of the aleph-sequence). He introduced the ordinals by means of two so-called generating principles ("Erzeugungsprinzipien"). The first one yields the transition from α to $\alpha + 1$. To describe the second generat-ing principle one needs the concept of a well-ordered set. Cantor defined this as follows: "By a well-ordered set we understand any well-defined set whose elements are related by a well-determined given succession ac-cording to which there is a first element in the set and for any element (if it is not the last one) there is a certain next following element. Further-more, for any finite or infinite set of elements there is a certain element which is the next following one for all these elements (except for the case that such an element which is the next following one to these elements does not exist)." [13, p. 168] The second generating principle yields the transition from an increasing well-ordered set of ordinals without greatest element to the corresponding limit ordinal. Cantor mentioned that the relation $\alpha < \beta$ establishes a well-ordering in any set of ordinals.

A key problem for Cantor was to establish a connection between transfinite ordinals and cardinals. He did so by applying his so-called restraint principle ("Hemmungsprinzip"). This principle consists in the demand "to create a new number by means of the two generating prin-ciples only if the totality of all the preceding numbers has the power of a well-determined number class that already completely exists" [13, p. 199]. The first number class is the class of the natural numbers. Can-tor later introduced for its power the symbol $\aleph_0$ (aleph zero). $\aleph_0$ is the smallest transfinite cardinal. The first number class N is characterized by the property that for any $\alpha \in N$ the set of the numbers preceding α is finite. The limit ordinal of $N = \{1, 2, 3, \ldots\}$ is the first number ω of the second number class. One gets this class by applying the two generating principles subject to the restraint principle that a number α only belongs to the second number class if the set of all the numbers preceding α has the power $\aleph_0$. The power of the second number class is denoted by $\aleph_1$. In the same way one can create the third number class, its power $\aleph_2$, and so forth. In the paper under discussion Cantor proved—and this is the fundamental connection between ordinals and cardinals—that $\aleph_1$ is the next power following $\aleph_0$ (meaning $\aleph_1 > \aleph_0$ and there is no power P with $\aleph_1 > P > \aleph_0$), $\aleph_2$ is the one following $\aleph_1$ and so forth. The well-ordering theorem was of essential importance for Cantor since only in the case of its validity do all cardinals belong to the aleph-sequence. Already in 1883 he was convinced that the well-ordering theorem is true and called it a "remarkable law of thought" [13, p. 169]. (Concerning another important theme of [13], part 5, the continuum hypothesis, see G. H. Moore's paper in this volume.)

Having sketched the development of Cantor's mathematical ideas up to 1883, we now go over to his philosophical views, especially concerning the foundations of mathematics. Cantor published some philosophical papers in the late 1880's [14, 15]. Indeed, it is a matter of record that he did not publish any mathematical papers between 1886 and 1890. The main reason, according to his own account, was his disappointment at Mittag-Leffler's refusal to accept his paper on ordinal types for publication in *Acta mathematica* [22, see especially pp. 101–104]. There is no direct connection between his change of interests, which became obvious in the second half of the 1880s and the first attack of his mental illness (manic-depression) in 1884, although this illness may have influenced his mathematical creativity.

We further note that Cantor had already developed deep philosophical interests by the early 1880s and especially in the above discussed part 5 of [13]. This paper did not meet commonly accepted standards for a mathematical research article, especially one to be published in the prestigious *Mathematische Annalen*. Cantor relates the main ideas, partly without proofs, and he links these with philosophical discussions on the actual infinite, on the nature of infinite numbers, on the continuum, and on the character of mathematical research. Moreover he tried to support his views by referring to sources taken from the history of philosophy. He himself felt that this was a most uncommon paper. On February 7, 1883 he wrote to Felix Klein (1849-1925) (the editor-in-chief of *Mathematische Annalen* at that time): "You would do me a great favour if you would accept the entire paper for the *Annalen.* ...I can assure you that the matter is thoroughly mathematical, notwithstanding the occurence of only a few formulae in it, and despite the fact that I was forced to speak about some philosophical issues that are related to this matter. These two aspects are, unfortunately, so closely connected in my paper that it would be very difficult for me to separate the mathematical elements from the other ones." [3, VIII, Nr. 432]

The foregoing analysis leads naturally to a number of questions: What led Cantor to untertake a thorough-going examination of the philosophical foundations of his new theories, or more generally all of mathematics? Why did he suddenly choose to publish such an unusual paper whose form and content went well beyond the conventional norms for mathematical research. The fact that Cantor was dealing with actual infinite sets and transfinite numbers cannot have been the only reason. There must have been some issue or circumstance that went beyond the issues raised by Weierstrass and Kronecker in reflecting on what constitutes an acceptable mathematical proof. There must have been something beyond this that rocked the very foundations of the whole building of mathematics. My thesis is that this underlying issue was brought out due to Cantor's discovery of the antinomies of set theory, or more precisely, his discovery of those facts that later mathematicians and logicians like Russell regarded as antinomies (for Cantor, there were no antinomies, as we will see). I shall show in the following, that already in 1882/83 (20 years before Russell's publication [38]) Cantor had the clear insight that the totality of all ordinals as well as the totality of all alephs are non-consistent sets. The evidence for this assertion is provided by Cantor's letters to David Hilbert (1862-1943) [4, Nr. 54] and by a new interpretation of some points in his publications that has become possible in the light of these letters.

Already in the first letter to Hilbert, dated Sept. 26, 1897, Cantor explained the antinomy of all alephs: " The totality of all alephs is namely one which cannot be regarded as a definite, well-defined set. If this were the case, then this totality would be followed by a definite aleph, which would belong to this totality and also not belong to it which would be a contradiction." [4, Nr. 54, fol. 2] Some lines further we find a clear hint that Cantor had been familiar with this fact for many years; he wrote: "Totalities that cannot be regarded as sets (an example is the totality of all alephs as is shown above) , I have already many years ago called absolute infinite totalities [*absolut unendliche Totalitäten*], which I sharply distinguish from transfinite sets." [4, Nr. 54, fol. 3]

In order to explore the question of just how many years ago Cantor had in mind in this quotation, we must focus our attention to his use of the term "absolute" or "absolute infinite" or"the absolute." We find the first use of this term in the notes to part 5 of [13], discussed above, where Cantor wrote: "It is without question for me that by pursuing this path [he had in mind the successive construction of ordinals and their classes] we will never reach an upper limit nor will we even approach the absolute. The absolute can only be accepted, but never be known, nor even approximately known. ... The absolute infinite series of ordinals appears to me therefore as an appropriate symbol of the absolute in a certain sense. In contrast to this, the infinity of the first class of ordinals appears to me as a completely vanishing void compared with the absolute." [13, p. 205, note to par. 4].

Cantor's concept of the absolute was influenced by thinkers like G.W. Leibniz and B. Spinoza who expressed by it the infinity of God (see e.g. Spinoza's Ethics, part 1 "De Deo" [40, pp. 23–24]). Cantor agrees with Spinoza that the absolute does not allow any determination, especially any mathematical determination. In a letter of 1886 to Albert Eulenburg (1840-1917) he wrote: "The transfinite with its plenty of figures and shapes refers with necessity to an absolute, to the true infinity, from which we cannot substract anything and to which we cannot add anything. Therefore the absolute must be regarded as being the absolute maximum. It exceeds the human intellect and, in particular, cannot be determined mathematically" [15, p. 405].

It is by no means a large step from the mathematical content of [13], part 5 to the statement that the set of all ordinals cannot be a consistent concept. Cantor knew, as we mentioned, that $\alpha < \beta$ establishes a well-ordering in any set of ordinals. He also knew that the "new numbers" (the ordinals) allow one to "count" the elements of any well-ordered set. As he put it: "Another great advantage of the new numbers consists for me in a new, till now undiscovered concept, the concept of the number of elements of a well-ordered set" [13, p. 168]. If the totality of all ordinals were to form a set, it would be well-ordered by the relation $<$ and, therefore, have an ordinal number Ω. But then this number would be the greatest ordinal in contradiction to Cantor's generating principles, which do not allow a maximum. It is almost unthinkable that Cantor, who reflected so deeply on these matters, would have taken no notice of such a simple consequence. Indeed, in the above quoted note to par. 4 of [13], part 5, he called the "absolute infinite series of ordinals [that is the totality of all ordinals] an appropriate symbol of the absolute." If we consider this remark along with his understanding of the absolute, we are forced to conclude that what Cantor expressed in his notes (*Anmerkungen*) to the 1883 paper was that any attempt to regard the totality of all ordinals as a mathematical object capable of being combined with the lower ordinals following the principles of ordinal arithmetic leads necessarily to a contradiction. This interpretation might at first seem speculative, but it can be further supported by another letter from Cantor to Hilbert from Nov. 15, 1899. In this letter Cantor emphasized the importance of his discovery of antinomic sets for Dedekind's *Was sind und was sollen die Zahlen?* [19]. He even referred to it as the foundation of his theory. He

then made the following telling remark: "You can find this, my fundament, in the *anno* 1883 published 'Grundlagen' [the separate edition of
[13], part 5], actually in the notes where it appears as already completely
clear, but also intentionally hidden." [4, Nr. 54, fol. 26] (Concerning the
connection with Dedekind's work and Cantor's letters to Dedekind from
the second half of 1899 [7, pp. 443 ff.], see [36] and [35, pp. 152–154]).

Having now shown that Cantor knew the antinomies of set theory by
1883, there now arise a number of questions: Why was he not disquieted
by his discovery? What enabled him to maintain his faith in his ideas
rather than abandoning this slippery field of set theory? Why did he
"intentionally hide" the very "fundament" of his theory?

The answer to the first two questions lies in his philosophical views.
Cantor was a representative of the Platonistic tradition and regarded objects such as, for example, natural numbers or transfinite ordinals as existing in a world of eternal and unchangeable ideas. His opinion in this
regard is especially clearly stated in a letter to Charles Hermite (1822-
1901) from Nov. 30, 1895. Hermite had asserted that numbers have the
same reality as those things in nature which are given to us by our senses.
Cantor answered: "Please, allow me to remark with regard to this point
that the reality and the ultimate inherent law of the whole numbers appear to me as being much stronger than the reality and the laws of the
material world. For this there is one reason only and a very simple one,
namely, that the whole numbers both as individuals and as an actual infinite totality exist in the divine intellect [*in intellectu Divino*] *as eternal ideas
in the highest grade of reality*" [2, Nr. 18, fol. 47/48].

Cantor never said that he created the transfinite numbers; he always
spoke about their perception. In a letter to Guiseppe Veronese (1857-1917)
from Nov. 17, 1890, in which he refused to accept Veronese's introduction
of infinitely small numbers, he proclaimed: "I never wrote that I suppose
it is possible that there are *transfinite* cardinal or ordinal numbers other
than those which I percieved" [2, Nr. 17, fol. 30]. For Cantor's Platonistic
ontology of mathematical objects, consistency was a necessary but not a
sufficient condition: "If I have known the internal consistency of a concept
that represents a being, then I am forced to believe by the idea of the
omnipotence of God that the being which is stated by the concept under
discussion must be realizable in some way. With regard to this, I call it
a *possible* being but this does not mean that it is realized somewhere and
some time and somehow in reality"([2, Nr. 16, fol. 52], letter to Eberhard
Illigens from May 21, 1886).

Cantor's Platonistic ontological foundation of mathematics had, in
particular, important consequences for his views about the infinite. Indeed, this is the deeper reason why he was not at all disquieted by the

antinomies, which, in fact, he did not regard as antinomies at all. The infinite was for Cantor an existent being, and the actual infinite existed for him in several gradations and forms. In Cantor's eyes the infinite could not be the logical counterpart of the finite (as it was for Dedekind or Frege), because absolute infinity, being identified with God, cannot be comprehended by the human intellect. Therefore, there cannot be general laws of thought which concern the infinite in general. Corresponding to this ontological understanding of the infinite, Cantor divided the infinite into three parts:

1) the transfinite, represented by transfinite sets, cardinals, and ordinals;
2) the infinite in the created world; and
3) the absolute or the so-called "Transfinitum in Deo" (transfinite in God).

Since, in his opinion, the infinite in the created world can be measured by transfinite ordinals or cardinals (see [35], pp. 109-110), we have only two completely disparate parts, the transfinite and the absolute or "Transfinitum in Deo." As an illustration of these concepts, consider the following passage from Cantor's paper "Mitteilungen zur Lehre vom Transfiniten" [15]:

> The actual infinite can be broken into three categories: first, the completely perfect infinite realized in Deo, which I call the absolute, second, the infinite existing in the created world, third, the infinite as a mathematical object, i.e. as a cardinal, ordinal or ordinal type which can be regarded in abstracto by the human intellect. The last two categories, which evidently represent a limited version of the actual infinite in that they are capable of enlargement, I call the Transfinitum which I place in the sharpest contrast to the Absolutum [15, p. 378].

Since Cantor regarded the totalities of all ordinals or of all alephs as being "appropriate symbols of the absolute in a certain sense" he most likely would have been very disquieted or even shocked had these totalities turned out to be consistent sets . But why did he "intentionally hide" his knowledge of the antinomies? There is no direct evidence to answer this question. In my opinion, he felt that the majority of mathematicians in the 1880s would hardly have agreed with such philosophical reasons, and that therefore the contradictions created by applying his own concepts to such "huge" sets like the set of all alephs would have been a serious blow against his theory, which was ignored and even under attack as it was.

The concept of the infinite has played an important role in philosophical and theological discussions for centuries. Cantor was very familiar with all these discussions. In particular, he was attracted to the argumentation of the church father Augustine (354–430), who regarded the natural numbers as an actual infinite entity. In *De civitate Dei* [5, p. 542] (*The City of God*, Book XII, chapt. 19), Augustine discussed the question of whether or not God can know all numbers. He argued that God, because of his absolute infinite intellect, can grasp any smaller infinity, or, in the language of Cantor, the absolute infinity realized in God can completely overview other infinite sets. Cantor commented on this argumentation as follows: "The *Transfinitum* cannot be more vigorously demanded, it cannot be more perfectly argued for and defended than it is done here by the Holy Augustinus" [15, p. 402].

Cantor himself believed that the philosophical importance of his theory of transfinite numbers was due to its service in filling in the huge gap between the finite and the absolute. According to G. Kowalewski [28, p. 201], Cantor regarded his sequence of alephs as being the steps to God's throne. Furthermore, he estimated his own contribution to philosophy very highly, and not without some justification considering how deeply he influenced subsequent thought about the infinite, even among theologicians (see [35, pp. 116–120]). Cantor even intended to abandon teaching mathematics at one point in order to lecture on philosophy instead. In a letter to Mittag–Leffler from Oct. 20, 1884, he wrote: "I shall presumably quit teaching mathematics within the next few semesters ..., instead I shall lecture on philosophy. That will not, according to my interests, be difficult for me, and I think I can be of more use to the students by working in this field..." [2, Nr. 16, fol. 3]. In connection with this passage, it should be remarked that until his retirement in 1913 Cantor taught a wide variety of mathematical subjects, but never set theory.

Another point closely connected with the foundations of mathematics that is illuminated by Cantor's letters to Hilbert in a very surprising way is the set concept itself. In his early works on set theory, Cantor never defined the concept of a set. We find the first definition of a set in the paper of 1883 discussed above, where in the endnotes one reads: "By a set I understand every multitude which can be conceived as an entity, that is every embodiment [*Inbegriff*] of defined elements which can be joined into an entirety by a rule" [13, p. 204] Twelve years later, in 1895, Cantor introduced his famous "Beiträge zur Begründung der transfiniten Mengenlehre" [16] with the following well–known definition of a set: "By a set we understand every collection [*Zusammenfassung*] M of defined, well–distinguished objects m of our intuition [*Anschauung*] or of our thinking (which are called the elements of M) brought together to form an entirety" [16, p. 282]. This definition, the basis of so–called naive set theory, was rightly regarded as being the source of the set-theoretic antinomies ([27, 31]) It is, therefore, most surprising indeed to learn from another letter to Hilbert, dated Oct. 2, 1897, that it was by means of this definition that Cantor *intended to avoid antinomic sets.* He begins this letter by repeating his assertion that the totality of all alephs cannot be regarded as a set, that is a collection of objects that form a new entity (for more details see [36, p. 324]). He then added: "I say of a set that it can be regarded as comprehensible [*fertig*], and call such sets, if they contain infinitely many elements, "transfinite". . .if it is possible (as is the case with finite sets) to conceive of all its elements as a totality without implying a contradiction. The set is thus to be thought of as a united thing in itself, or (in other words) it must be possible to conceive of a set as the actually existent totality of all its elements. . . . For that reason I translated the word "set" (if it is finite or transfinite) into French as "ensemble" and into Italian as "insieme." For that reason I also defined the term "set" at the very beginning of the first part of my paper "Beiträge zur Begründung der transfiniten Mengenlehre" as a collection (meaning either finite or transfinite). But a collection [*Zusammenfassung*] is only possible if it is possible to unite it [if a *Zusammensein* is possible]" [4, Nr. 54, fol. 5-6].

If we compare Cantor's motivation, as expressed here, with the generally accepted view of his naive set concept, we are forced to conclude that probably no idea of such fundamental importance for mathematics has ever been interpreted in a manner so contrary to its creator's own intentions. Why? In short, because no one among the following generation of mathematicians accepted Cantor's philosophical views concerning the foundations of mathematics.

The usual historical version of the events surrounding the antinomies of set theory that one encounters in the literature runs as follows. Cantor is said to have found the antinomy of all alephs in 1895 and mentioned it in a letter to Hilbert in 1896. It is further asserted that Burali-Forti found the antinomy of all ordinals in 1897. This latter assertion was recently repudiated by G. H. Moore and A. Garciadiego [32]. As for the above assertions regarding Cantor himself, these are, as we now know, also wrong. What are the sources for this version of the story? In 1904, in his paper "On the Transfinite Cardinal Numbers of Well-ordered Aggregates" [26], Phillip Jourdain mentioned for the first time the existence of a letter from Cantor to Hilbert concerning the antinomy of all alephs. Jourdain asserted that this letter was written in 1896.

Cantor's first letter to Hilbert of 1897 was concerned with the totality of all alephs, but he also wrote about his ideas for proving the well-ordering theorem. His strategy was to use the fact that the totality of all alephs does not form a set in proving the well–ordering theorem. The well–ordering theorem was known to be equivalent to the statement that all cardinal numbers occur in the aleph-sequence. Thus his basic idea was the following: Suppose a cardinal κ is not an aleph. Let A be a set representing κ. Since $\aleph_0$ is the smallest cardinal, $\aleph_1$ the next one, $\aleph_2$ the next following to $\aleph_1$ etc., there must be subsets in A which are equivalent to any given aleph. Hence Cantor concluded: The totality of all alephs must be a subset of A. But this totality is not a set. Originally Cantor wanted to complete his series "Beiträge zur Begründung der transfiniten Mengenlehre" [16] with this result [4, Nr. 54, fol. 2]. But it seems he himself finally regarded this proof as being insufficient. (Compare Zermelo's criticism in [7, pp. 450-451]). About six years later, Jourdain had the same idea for proving the well-ordering theorem, and he wrote to Cantor to obtain his opinion. Cantor answered on Nov. 4, 1903: "The doubtlessly true theorem that all transfinite cardinals are alephs is one I discovered more than 20 years ago intuitively. Later I formulated a proof similar to your proof. Already 7 years ago I mentioned this in a letter to Herrn Hilbert, 4 years ago to Herrn Dedekind. . ." [23, p. 116].

Jourdain evidently counted backwards and arrived at the year 1896, instead of 1897, as the date of the decisive letter to Hilbert. This mistake was due to the fact that Cantor himself had been off by one year in the above letter to Jourdain. The year 1895 as the date of the discovery of the aleph-antinomy, on the other hand, seems to have been nothing more than an invention, perhaps it was an estimation based on the assumed date of Cantor's letter to Hilbert. But whatever the source of this error might have been, it has been repeated over the course of the last eighty years. We now know that Cantor was aware of the inconsistency of the totality of all ordinals (or cardinals) at least 13 years before the date given by Jourdain and at least twenty years before Russell himself announced these results.

ACKNOWLEDGEMENTS

I am very grateful to Gregory Moore for some very useful suggestions, and to David Rowe for improving the language of the paper.

ARCHIVAL SOURCES

[1] Archive of the Academy of Sciences of the German Democratic Republic, Berlin. Nachlass Schwarz.
[2] Cod. Ms. Georg Cantor (Nachlass Cantor). Niedersächsische Staats- und Universitätsbibliothek Göttingen, Handschriftenabteilung.
[3] Cod. Ms. Felix Klein (Nachlass Klein). Niedersächsische Staats- und Universitätsbibliothek Göttingen, Handschriftenabteilung.

[4] Cod. Ms. David Hilbert (Nachlass Hilbert). Niedersächsische Staats- und Universitätsbibliothek Göttingen, Handschriftenabteilung.

REFERENCES

[5] Augustinus, A.: *De civitate Dei*. Ed. B. Dombart, vol. I, Leipzig 1921.
[6] Bolzano, B.: *Paradoxien des Unendlichen*. Leipzig 1851^1, Berlin 1889^2. Reprints: Darmstadt 1964, Hamburg 1975.
[7] Cantor, G.: *Gesammelte Abhandlungen mathematischen und philosophischen Inhalts*. Hrsg. von E. Zermelo nebst einem Lebenslauf Cantors von A Fraenkel. Berlin 1932. Reprint: Springer 1980.
[8] Cantor, G.: *Über einen die trigonometrischen Reihen betreffenden Lehrsatz*, Journal für reine und angewandte Mathematik, 72 (1870), 130–138. Reprinted [7], 71–79.
[9] Cantor, G.: *Beweis, dass eine für jeden reellen Werth von x durch eine trigonometrische Reihe gegebene Function f(x) sich nur auf eine einzige Weise in dieser Form darstellen lässt*, Journal für reine und angewandte Mathematik 72 (1870), 139–142. Reprinted [7], 80–83.
[10] Cantor, G.: *Über die Ausdehnung eines Satzes aus der Theorie der trigonometrischen Reihen*, Mathematische Annalen, 5 (1872), 123–132. Also reprinted in [7], 92–102.
[11] Cantor, G.: *Über eine Eigenschaft des Inbegriffs aller reellen algebraischen Zahlen*. Journal für reine und angewandte Mathematik, 77 (1874), 258–262. Reprinted [7], 115–118.
[12] Cantor, G.: *Ein Beitrag zur Mannigfaltigkeitslehre*, Journal für reine und angewandte mathematik, 84 (1878), 242–258. Reprinted [7], 119–133.
[13] Cantor, G.: *Über unendliche lineare Punctmannichfaltigkeiten*, Mathematische Annalen, 15 (1879), 1-7; 17 (1880), 355–358; 20 (1882), 113–121; 21 (1883), 51–58; 545–586; 23 (1884), 453–488. Reprinted without footnotes: [7], 139–246.
[14] Cantor, G.: *Über die verschiedenen Standpunkte in Bezug auf das aktuale Unendliche*. Zeitschrift für Philosophie und philosophische Kritik, 88 (1886), 224–233. Reprinted [7], 370–377.
[15] Cantor, G.: *Mitteilungen zur Lehre vom Transfiniten*, Zeitschrift für Philosophie und philosophische Kritik, 91 (1887), 81–125; 92 (1888), 240–265. Reprinted [7], 378–439.
[16] Cantor, G.: *Beiträge zur Begründung der transfiniten Mengenlehre*, Mathematische Annalen, 46 (1895), 481–512; 49 (1897), 207–246. Reprinted [7], 282–356.
[17] Cavailles, J.; Noether, E.: *Briefwechsel Cantor-Dedekind*. Paris 1937.
[18] Dauben, J.: *Georg Cantor. His Mathematics and Philosophy of the Infinite*. Cambridge, Mass./London 1979.
[19] Dedekind, R.: *Was sind und was sollen die Zahlen?* Braunschweig 1888.

64   Walter Purkert

[20] Dugac, P.: *Richard Dedekind et les fondaments des mathématiques*. Paris 1976.
[21] Galilei, G.: *Unterredungen und mathematische Demonstrationen. Erster und zweiter Tag*. Ostwalds Klassiker Nr. 11, Leipzig 1890. Reprint Darmstadt 1985.
[22] Grattan-Guinness, I.: *An unpublished paper by Georg Cantor: Principien einer Theorie der Ordnungstypen. Erste Mitteilung*, Acta Mathematica, 124 (1970), 65–107.
[23] Grattan-Guinness, I.: *The Correspondence between Georg Cantor and Philip Jourdain*, Jahresbericht der Deutschen MathematikerVereinigung, 73 (1971/72),111–130.
[24] Heine, E.: *Über trigonometrische Reihen*, Journal für reine und angewandte Mathematik, 71 (1870), 353–365.
[25] Hilbert,D.: *Die Grundlegung der elementaren Zahlentheorie*, Mathematische Annalen, 104 (1931), 485–494.
[26] Jourdain, Ph.: *On the Transfinite Cardinal Numbers of Well-ordered Aggregates*, Philosophical Magazine, Vol. VII, 6. ser., 1904, 61–75.
[27] Kamke, E.: *Mengenlehre*. Berlin 1962.
[28] Kowalewski, G.: *Bestand und Wandel*. München 1950.
[29] Kronecker, L.: *Über den Zahlbegriff*, Journal für reine und angewandte Mathematik, 101 (1887), 337–355. Reprinted: Werke, Bd. 3/1, Leipzig/Berlin 1899, 249–274.
[30] Lebesgue, H.: *Leçons sur les séries trigonométriques*. Paris 1906, 110–111.
[31] Meschkowski, H.: *Probleme des Unendlichen. Werk und Leben Georg Cantors*. Braunschweig 1967, 1983^2.
[32] Moore, G.; Garciadiego, A.: *Burali-Forti's Paradox: A Reappraisal of its Origins*. Historia Mathematica, 8 (1881), 319–350.
[33] Purkert, W.: *Leopold Kronecker*. Biographien bedeutender Mathematiker, Berlin 1973.
[34] Purkert, W.: *Elemente des Intuitionismus im Werk Leopold Kroneckers*, Mathematik in der Schule, 14 (1976), 81–86.
[35] Purkert, W.; Ilgauds, H.-J.: *Georg Cantor*. Basel 1987.
[36] Purkert, W.: *Georg Cantor und die Antinomien der Mengenlehre*, Bulletin de la Société mathématique de Belgique, 38 (1986), 313–327.
[37] Purkert, W.: *Cantors Untersuchungen über die Eindeutigkeit der Fourierentwicklung im Lichte seines Briefwechsels mit H. A. Schwarz*, NTM 24 (1987)2, 19–28.
[38] Russell, B.: *The principles of mathematics I*. Cambridge 1903.
[39] Schwarz, H. A.: *Beweis eines für die Theorie der trigonometrischen Reihen in Betracht kommenden Hülfssatzes*, Gesammelte Mathematische Abhandlungen, Bd.2. Berlin 1890, 341–343.

[40] Spinoza, B.: *Ethik*. Leipzig 1972.

Karl Weierstraß (1815–1897).
Courtesy of Springer-Verlag.

Leopold Kronecker (1823–1891)
Courtesy of Springer-Verlag Archives

Kronecker's Views on the Foundations of Mathematics

Harold M. Edwards

In the past I have lamented that Kronecker did not publish more about his views on the foundations of mathematics.[1] I no longer do so. It now seems to me that a rather clear picture of Kronecker's views emerges from his published works. Up until now, I wasn't looking hard enough, and, more importantly, I was looking for the wrong thing.

The Cantorian revolution has had such a profound and lasting effect on views of the the foundations of mathematics that we feel that the phrase "foundations of mathematics" *means* set theory, especially the treatment of infinite sets. It is not that Kronecker did not express his views on foundations, but that he did not respond—at least not in print—to the Cantorian views. (In the light of Walter Purkert's remarks, I should perhaps say "Weierstrassian" rather than "Cantorian." What I have in mind is not the metaphysics of infinite sets, but the use of arguments like the Bolzano- Weierstrass theorem and the Heine-Borel theorem.) And it is entirely reasonable that he would not. Kronecker's mathematical world-view was fully formed before the new ideas appeared. When they did appear, they bore almost no relation to his conception of the foundations of mathematics, and they did not answer any questions he had been asking.

Cantor's work on infinite point sets would have seemed to Kronecker to be far removed from the foundations of mathematics, and it is not surprising that he did not respond in print. He almost certainly did respond privately. Cantor believed Kronecker even expressed his opposition in his university lectures, at least in the summer of 1891.[2] But Kronecker's private response very likely was that Cantor's work did not deal with the foundations and was itself inadequately founded.

Dedekind's work was of course extremely close to Kronecker's, and it developed ideas much like the Bourbaki approach to the foundations—for example the conception of a ring as first and foremost an underlying *set* and an ideal as a certain kind of *subset* (although Dedekind used the word "system" instead of "set"). To Dedekind's ideas on the foundations Kronecker did reply, but his reply was brief: in essence, he said Dedekind's approach was not algorithmic and therefore not acceptable.[3]

THE HISTORY OF
MODERN MATHEMATICS

Fond as we are today of Dedekind, and steeped as we are in his ideas, Kronecker's reply seems harsh, dictatorial, and wrong, completely in accord with the despotic and reactionary image of Kronecker conveyed to us by E. T. Bell[4] and Constance Reid,[5] among others. Reconsidered in its historical context, however, Kronecker's reply to Dedekind has merit.

Dedekind's approach to ideal theory is, as I have argued elsewhere[6], *needlessly* nonconstructive, because he had a hidden agenda. He insisted it was unnecessary—and he implied it was undesirable—to provide an algorithmic description of an ideal, that is, a computation which would allow one to determine whether a given ring element was or was not in the ideal. The reason for his insistence was ideological. To admit the desirability of an algorithmic basis for algebra would open the door to a similar requirement for analysis, and Dedekind's approach to the foundations of analysis—the Dedekind cut—would not meet such a requirement. Without this hidden ideological justification, it is indeed hard to find any advantages that Dedekind's approach has over Kronecker's fully algorithmic development of the same theory. Dedekind's approach carried the day not because of its intrinsic merit, but because his ideology meshed with that of Hilbert, the most influential mathematician of the next generation, and, in no small degree, because Dedekind was such a good writer.

As we all know, Kronecker believed God made the natural numbers and all the rest was man's work. We only know of this opinion by hearsay evidence,[7] however, and his paper *Ueber den Zahlbegriff* indicates to me that he thought God made a bit more: *Buchstabenrechnung*, or calculation with letters.[8] In modern terms, Kronecker seems to envisage a cosmic computer which computes not just with natural numbers, but with polynomials with natural number coefficients (in any number of indeterminates). That is the God-given hardware. The man-made software then creates negative numbers, fractions, algebraic irrationals, and goes on from there. Kronecker believed that such a computer, in the hands of an able enough programmer, was adequate for all the purposes of higher mathematics. (Walter Purkert and Helena Pycior have spoken about mathematicians who sought the foundations of mathematics in the "mind of God." Perhaps instead of saying Kronecker envisaged a "cosmic computer" I should say he imagined "God's pocket calculator.")

Crucial to all the programming was the solution of the following problem. Two polynomials A and A' are equivalent modulo polynomials $M_1, M_2, \ldots, M_n$ if there exist $C_1, C_2, \ldots, C_n$ and $D_1, D_2, \ldots, D_n$ such that $A + \sum C_i M_i = A' + \sum D_i M_i$. Since equivalence mod $M_1, M_2, \ldots M_n$ is obviously consistent with addition and multiplication (if equivalent things are added or multiplied, the results are equivalent) the equivalence classes can be added and multiplied. They form a *semi-ring*—a ring except that subtraction is not always possible. I will call this *the semi-ring of polynomials mod* $M_1, M_2, \ldots, M_n$.

Kronecker repeatedly poses problems in terms of computation in a semi-ring of polynomials modulo some given set $M_1, M_2, \ldots, M_n$.[9] My greatest problem in understanding his approach to the foundations now lies in understanding his approach to the problem of computing in such a semi-ring, that is, the problem of deciding whether two given A's are equivalent modulo a given set of M's. On the one hand, based on my reading of Kronecker over the years, I regard it as a certainty that he would have regarded it as essential to show that this problem—which he puts at the heart of his formulation of mathematics[10]—can be solved algorithmically, that is, by a computational procedure which can be shown to terminate after a number of steps for which an *a priori* (finite) upper bound can be given. On the other hand, I find no such algorithm in his works. My best guess as to the explanation of this paradox is that he had an algorithm which he had not yet reduced to a form ready to publish, or, perhaps, that he had an algorithm in many cases but had not yet found one in the general case. Or, as is entirely possible, it lies somewhere in his voluminous collected works waiting to be found. In any event, the problem of finding such an algorithm is one which has attracted a good deal of attention from algebraists in recent years, who have introduced the notion of "Gröbner bases" to deal with it.

Since I am unable to solve this problem of Kronecker's approach to computation with polynomials modulo $M_1, M_2, \ldots, M_n$, I will circumvent it simply by *assuming* that an algorithm is known for deciding whether two given A's are equivalent modulo a given set of M's. Let it be called the "main algorithm."

Consider polynomials in one indeterminate e with natural number coefficients, and let $M = 1 + e$. Then $e^2 + (e + 1) = 1 + e(e + 1)$, so e^2 is equivalent to 1 modulo M, which I will denote $e^2 \sim 1$. It follows immediately that every polynomial is equivalent to one of the form $a + be$, where a and b are natural numbers. Two of these are equivalent, say $a + be \sim c + de$, if and only if $a + b + d + be \sim c + b + d + de$, which is true if and only if $a + d \sim c + b$. It is simple to prove that two natural numbers are equivalent mod $1 + e$ only if they are identical—this is essentially the statement that, when division is possible, the quotient and remainder are uniquely determined. Therefore, $a + be \sim c + de$ if and only if $a + d = c + b$. Thus, the semi-ring of polynomials modulo $e + 1$ is simply the ring of integers in the form it is normally constructed from the natural numbers.

Moreover, in any semi-ring of this type, if we throw in another indeterminate e and another relation $1 + e$, we get a possibly larger semi-ring which is a *ring* of equivalence classes of polynomials with *integer* coefficients.

Our impulse is to go on to construct the field of rational numbers, but Kronecker does not, and, in fact, one cannot. One can throw in another indeterminate t and another relation $1 + 10et$ (that is, two indeterminates e, t, and two relations $1 + e$ and $1 + 10et$) to get the ring of all terminating decimal fractions. Similarly, one can construct rings of rational numbers containing any finite set of denominators, so any computation with rational numbers can be described and carried out, but the field of rational numbers as a completed infinite set cannot be described in the way we are used to doing it. But that would not have bothered Kronecker, for whom even the *natural* numbers were not a completed infinite set.

Any algebraic number field can be constructed by adjoining to the rationals a root of an irreducible polynomial with integer coefficients. Any computation in such a field can therefore be described and carried out using the main algorithm in the case where there are two indeterminates e and x, and two relations $1 + e$ and $f(x)$ (f an irreducible polynomial) plus another indeterminate and another relation to provide needed denominators, if any.

Similarly, any computation in any algebraic *function* field can be described and carried out as a computation using the main algorithm. In particular, the field of rational functions on an algebraic curve can be handled in this way. As is well known—it was first spelled out in the classic paper of Dedekind and Weber[11]—the algebraic part of the theory of abelian functions, including Abel's theorem and the Riemann-Roch theorem, can be developed entirely in terms of the algebra of such function fields and can therefore be described algorithmically in terms of Kronecker's cosmic computer. I believe that Kronecker had this in mind when he spoke, in terms so wounding to Weierstrass, of showing that the so-called analysis was unnecessary and wrong.[12] It was unnecessary because it was not needed where it was most used, in the theory of abelian functions, wrong because, being nonconstructive, it was not rigorously and correctly founded. (It was "so-called" because the word "analysis" had only recently begun to be applied to the calculus of infinite processes and function theory, and because, I would guess, Kronecker disapproved of this appropriation of the word.)

But, and this is the perennial question, what about everything else? Does not Kronecker's approach exclude not just *major portions* of modern mathematics, but in fact the *majority* of modern mathematics, including the really powerful ideas? This is inevitably a matter of opinion. The dominant opinion today is certainly against Kronecker. I do not intend to go into the pros and cons of this question, but just to present what I think was Kronecker's viewpoint.

Some writers[13] believe that Kronecker, as editor of *Crelle's Journal*, delayed the publication of a paper of Cantor in 1877. However, there is *documentary evidence*[14] that Kronecker delayed the publication of a paper of Heine in 1870 until he had had a chance to meet with Heine and try to talk him out of certain ideas. Heine's paper[15] was published, a few months later, so we can read it to see what Kronecker objected to. To modern eyes, it appears altogether innocuous.

What I think bothered Kronecker, believe it or not, was Heine's consideration of an *arbitrary* trigonometric series. Kronecker did not believe in arbitrary infinite series, only in infinite series explicitly and constructively given. Our habit is to begin a theorem with "Let such-and-such be a trigonometric series . . ." or "Given an arbitrary real number satisfying . . .", but Kronecker condemned such ways of thinking. When one is dealing with the infinite, one must be more specific, so specific, in fact, that it is virtually finite; in his words:

> The *general* concept of an infinite series itself, for example a power series, is in my judgement permissible only with the reservation that in each particular case the arithmetical rule by which the terms are given satisfies, as above, conditions which make it possible to deal with the series as though it were finite, and thus to make it unnecessary, strictly speaking, to go beyond the notion of a finite series.[16]

In his works he is always specific. Formulas abound in his papers. Dedekind abhorred formulas and tried to avoid them. Kronecker was the opposite. He once said that he felt that the essence of mathematical truth lay in formulas, saying that, for example, although Lagrange's attempts to give foundations for calculus were already forgotten, the Lagrange resolvent remained[17]. The wish to get rid of formulas was, it seems to me, what brought set theory into being.[18] Set theory is what remains after formulas are banished. How can an *arbitrary* function be described, other than as a set of ordered pairs? Since Kronecker wished to place formulas at the heart of his mathematics, this motive for set theory would not have existed for him.

Let me give a specific example of the difference. Both Kronecker and Dedekind dealt, of course, with primitive cube roots of unity, that is, with solutions of the irreducible equation $x^2 + x + 1 = 0$ (a root of $x^3 - 1 = (x^2 + x + 1)(x - 1)$ which is not a root of $x - 1$). Sometimes it is useful to describe such a number using a square root of -3, and sometimes it is useful to describe the square root of -3 using a cube root of unity. Dedekind felt that it "marred" his presentation of the theory of algebraic number fields to use either the square root of -3 or the cube root of unity, and that the correct way to proceed was to talk about the field they generate over the rationals, which is the same in the two cases.[19] Kronecker would probably have said that it was ridiculous to let the ultimate reality be the field—a completed infinite set that could never be explained to anyone but a mathematician—and would simply have cited a formula like $(2x + 1)^2 + 3 = 4x^2 + 4x + 1 + 3 = 4(x^2 + x + 1)$.

Kronecker's insistence on the specific and the algorithmic led him to one position that is too extreme even for most modern constructivists. Kronecker believed that a mathematical concept was not well defined until it had been shown how to decide, in each specific instance, whether the definition was fulfilled or not.[20] It would seem much more reasonable to say that what is required of a definition is that it make clear what it would mean to prove that the definition was or was not fulfilled. I would like to know how Kronecker would have dealt with the question of the irrationality of Euler's constant. We feel that it is perfectly meaningful to say that Euler's constant is irrational (or that it is rational) because we have a clear idea what we would accept as a proof that this was the case—we have examples in the proofs of the irrationality of e and π. But, since we cannot prove either that Euler's constant is irrational or that it is rational, Kronecker's position seems to have been that the very concept of rationality is not well defined.

There is a problem here which most mathmaticians today are not aware of. Modern constructivists state it by saying that it is a fallacy, from the constructive point of view, to say that the real numbers are the union of the rational numbers and the irrational numbers, or, in other words, to say that every real number is either rational or irrational. Perhaps Kronecker would have said that in regarding the concept of rationality as being well-defined we open ourselves to the danger of committing this fallacy, and it would be better to avoid doing so. I believe that some garbled form of Kronecker's answer to this question survives in the otherwise absurd story that he reacted to Lindemann's proof of the transcendence of π by saying that it proved nothing because π did not exist.[21]

Poincaré once wrote, comparing Kronecker unfavorably to Weierstrass, "Kronecker too made many discoveries. But if he succeeded it was by forgetting that he was a philosopher and by voluntarily letting go of his principles, which were condemned in advance to sterility."[22] This passage is often quoted out of context and taken to mean that Poincaré believed Kronecker succeeded as a mathematician only by abandoning the principles he advocated for dealing with infinity. But the principles

Poincaré is referring to in this passage are not principles for dealing with infinity, but quite different principles which I do not believe Poincaré is even correct in ascribing to Kronecker. The preceding paragraphs of Poincaré's essay explain that *both* Weierstrass and Kronecker based their mathematics entirely on the whole numbers, so that all their work shared in the certitude of arithmetic. Describing the work of Weierstrass, he wrote, "The continuum itself goes back to this origin [the integers] and all the equations which are the object of analysis and which deal with continuous magnitudes are nothing but symbols, replacing an infinite collection of inequalities relating whole numbers." The difference he sees between Weierstrass and Kronecker is described as follows:

> Kronecker is above all concerned that the philosophical meaning of mathematical truths be put in evidence; whole numbers being the foundation of everything, he wants them to remain apparent throughout; for him, addition and multiplication are the only legitimate operations; it is only as a concession to contemporary prejudices that he admits division.[23] Such is not the point of view of Weierstrass. When he has made a construction, he forgets the materials of which it was made and wants to see only a new entity which he will use as an element of an even larger construction.

Thus, the principle Poincaré accuses Kronecker of abandoning is the principle that one can never place reliance on intermediate constructions but must always go back, at every step, to first principles. I seriously doubt that Kronecker ever espoused this principle, but I agree with Poincaré that success in mathematics must surely require that it be abandoned, at least temporarily. What must be emphasized, though, is that Poincaré was *not referring at all* to what I have described above as Kronecker's views on the foundations of mathematics.

This passage from Poincaré is seized upon by opponents of Kronecker's principles as evidence not only that Kronecker was inconsistent but also that his principles are, in Poincaré's words, "condemned in advance to sterility." In fact, there is a prevailing prejudice that Kronecker's *actual* views on the foundations of mathematics—not the ones Poincaré is referring to—are indeed condemned to sterility, and that one need not read Kronecker's works in order to know that he must violate his principles in order to achieve anything. Admittedly, I have mainly studied Kronecker's algebraic works, where one would expect constructivism to be easier, but his approach seems extremely consistent, and the same ideas and notations are constantly repeated and modified and improved, with frequent references to other papers, producing the impression that the work has extraordinary integrity. I would be surprised if one could point to any part of the work that is a violation of Kronecker's principles as he intended them. The key to his success lies not in abandoning his principles but in dealing always with specific constructions.

I believe that many modern critics misunderstand what Kronecker's principles are and believe, without making any profound study of his work, that Kronecker's results *a priori* could not be achieved without violating what they imagine to be his principles. For example, one recent paper states that Kronecker "was far from having any definite idea of how [to reconcile the] theory of limits" with his principles.[24] Of course, if one believes that Kronecker excluded limits from mathematics, then there is no need to study his work to know that it violates his principles.

Certainly Kronecker did not exclude limits. Already in Greek times, mathematicians understood very well that infinity was, as Gauss said, a *"façon de parler,"*[25] and that limits can be described—in fact are best described—without any mention of infinites. Of course Kronecker's mathematics is full of definite integrals and infinite series. What is missing, and what is excluded by his principles as he stated them, is integrals of "arbitrary" functions or sums of "arbitrary" infinite series. Not only did he not want to talk about the sum of an arbitrary infinite series, he did not even want to talk about an arbitrary infinite series. As long as he stuck to specific cases and dealt always with limits of specific series and integrals, it is hard for me to see how he could transgress against his principles, which, in his mind and in fact, were no different from the principles of his predecessors, from Archimedes to Gauss.

That Kronecker did not exclude limits—and did not exclude π—is shown by his description, in the introduction to his lectures on number theory, of Leibniz's formula

$$\frac{\pi}{4} = 1 - \frac{1}{3} + \frac{1}{5} - \frac{1}{7} + \cdots$$

as "one of the most beautiful arithmetic properties of the odd numbers, namely, that of determining this geometrical irrational number."[26] We try to convince even students in first semester calculus that a series like $1 - (1/3) + (1/5) - (1/7) + \cdots$ converges; the argument—usually based on a picture of nested intervals—is essentially to show that the Cauchy criterion holds: Once the terms up to and including $\pm\dfrac{1}{2n+1}$ have been added, all further sums differ from this sum by less than $\dfrac{1}{2n+1}$ in absolute value. Thus, the infinite sum describes a process which determines a quantity with any arbitrarily prescribed degree of accuracy. The "beautiful property" Kronecker refers to is that this quantity coincides with the quantity determined by the geometrical limiting process implicit in the notion of the area of a circle of diameter 1. As for this geometrical limiting process, we may not know how near Archimedes was to "having any definite idea how" to articulate a "theory of limits," but we do know that he spelled out, in *The Measurement of the Circle*,[27] a computation which showed that π was between $3\frac{1}{7}$ and $3\frac{10}{71}$. Moreover, Archimedes' method can be carried further, if desired, to determine π to *any* prescribed degree of accuracy.

Cantor complained to Mittag-Leffler that Kronecker merely repeated the old injunctions against completed infinites, and that he (Cantor) found these injunctions no more convincing coming from Kronecker than from anyone else.[28] Cantor's complaint accords with the picture I have just tried to give of Kronecker as the keeper of the flame, the defender of the classic views of the foundations as espoused by his predecessors. He was defending these views, in the first instance, against, in his words, "the various formulations of concepts with the aid of which people (starting with Heine) have recently been seeking to grasp in a completely general way and to give foundations for the concept of the irrational."[29] I would like to underscore the words *"ganz allgemein,"* completely general. Here is where Kronecker felt the problem lay. He believed one should stick with the specific, not trying to make statements about the totality of irrationals, and not trying to deal with arbitrary trigonometric series. To do otherwise would be to abandon the firm foundations of mathematics established by our predecessors, and *there is no need to do so.*

These, then, are what appear to me to be the main features in Kronecker's philosophy of mathematics. They have next to nothing to do with set theory and what goes by the name "philosophy of mathematics" today, because they proceed from a very different notion of what mathematics is and what it should attempt to do. That Kronecker published nothing on set theory and nothing on the treatment of general infinite processes is therefore not surprising. He published only a few footnotes on the new ideas and methods that were stirring in the mathematical world of his time. The bulk of the five volumes of his collected works is devoted to demonstrating what he thought was important in mathematics, and what sorts of arguments he thought should be used in dealing with it, which, after all, is what the foundations of mathematics are all about.

NOTES

1. H. M. Edwards, *Dedekind's Invention of Ideals*, Bulletin of the London Mathematical Society, 15(1983), 8-17, cited hereafter as *Invention; Kronecker's Place in History*, in W. Aspray and P. Kitcher, eds., History and Philosophy of Modern Mathematics, Minneapolis: University

of Minnesota Press, 1988; *An Appreciation of Kronecker*, Mathematical Intelligencer, 9(1987), 28-35, cited hereafter as *Appreciation*.

2. Letter from Cantor to Thomé, 21 Sept. 1891, published in W. Purkert and H. J. Ilgauds *Georg Cantor*, Basel: Birkhäuser, 1987, p. 217.

3. See volume 3, p. 156 of K. Hensel, ed. *Leopold Kronecker's Werke*, 5 vols. Leipzig: Teubner, 1895-1930, New York: Chelsea reprint, 1968, cited hereafter as *Werke*.

4. E. T. Bell, *Men of Mathematics*, New York: Simon and Schuster, 1937, p. 570. (See, however, p. 475 for a kinder picture of Kronecker.)

5. C. Reid, *Hilbert*, New York: Springer, 1970, p. 26.

6. *Invention*.

7. H. Weber, *Leopold Kronecker*, Jahresbericht der Deutschen Mathematiker-Vereinigung, 2(1892), 19.

8. *Werke*, vol. 3_1, p. 260.

9. Kronecker referred to the M's as a "Modulsystem." See, for example, vol. 3_1, pp. 145–208 or pp. 209–240.

10. See *Werke*, 3_1, p. 211.

11. R. Dedekind and H. Weber, *Theorie der Algebraischen Functionen einer Veränderlichen*, Journal für die reine und angewandte Mathematik, 92(1882),181–290.

12. *Appreciation*, p. 32.

13. J. W. Dauben, *Georg Cantor*, Cambridge, Mass.: Harvard Univ. Press, 1979, p. 69, p. 135.

14. Heine to Schwarz, 26 May 1870, published in Dauben *Georg Cantor*, p. 308.

15. E. Heine, *Ueber trigonometische Reihen*, Journal für reine und angewandte Mathematik, 71(1870), 353–365.

16. Translation from *Invention*, p. 13 of Kronecker's original in *Werke*, vol. 3, p. 156.

17. Kronecker to Cantor, 21 August 1884, published in H. Meschkowski, *Probleme des Unendlichen*, Braunschweig, 1967, 238–239.

18. *Invention*.

19. Letter from Dedekind to Lipschitz, 10 June 1876, published in *Rudolf Lipschitz Briefwechsel*, W. Scharlau, ed., Vieweg, 1986, p. 60.

20. See the beginning of §18 of "Grundzüge einer arithmetischen Theorie der algebraischen Grössen," (*Werke*, vol. 2, p. 316). This attitude of Kronecker's is what convinces me that he must have considered it a problem of the greatest importance to find a "main algorithm",

that is, an algorithm for determining whether two *A*'s are equivalent modulo a given set of *M*'s.

21. Repeated, for example, in Bell, *Men of Mathematics*, p. 568.

22. H. Poincaré, *L'Oeuvre Mathématique de Weierstrass*, Acta Mathematica, 22(1899), 1–18, esp. p. 17.

23. Note that in the description of the integers I have given above even *subtraction* is derived from addition and *Buchstabenrechnung*.

24. H. Stein, *Logos, Logic, and Logistike*, in W. Aspray and P. Kitcher, eds., History and Philosophy of Modern Mathematics, Minneapolis: University of Minnesota Press, 1988, p. 250.

25. This oft–quoted statement of Gauss is given in the original and in translation in *Invention*, p. 12.

26. L. Kronecker, *Vorlesungen über Zahlentheorie*, Leipzig: Teubner, 1901, republished by Springer, 1978, p. 4.

27. See, for example, T. L. Heath , ed. *The Works of Archimedes*, Cambridge: Cambridge University Press, 1897.

28. Letter from Cantor to Mittag–Leffler, 9 Oct. 1884, published by A. Schoenfliess, Acta Mathematica, 50(1927),12.

29. See the passage cited in note 3.

David Hilbert (1862–1943)
Courtesy of Springer-Verlag Archives

Towards A History of Cantor's Continuum Problem

Gregory H. Moore

1. INTRODUCTION

More than a century has passed since Cantor put forward his Continuum Problem and proposed, as a solution, his Continuum Hypothesis. Since that time, the Continuum Hypothesis (CH) has served as a primary focus of investigations in set theory and the foundations of mathematics. Since the First World War, it has come to be more and more widely used in a variety of branches of mathematics—a surprising development in view of the fact that most mathematicians were, and are, uncertain whether CH is true or false.

Despite the work of Paul Cohen [1963] establishing that CH is independent of the usual first-order axioms for set theory (including the Axiom of Choice), the Continuum Problem remains one of the great unsolved problems of mathematics—like Fermat's Last Theorem and the Riemann Hypothesis. And, like them, the Continuum Problem has led to the creation of significant new parts of mathematics. It remains a source of ongoing mathematical research.

The present paper explores the history of Cantor's Continuum Problem from its inception to the axiomatization of set theory in 1908. During this period the principal locus of research on this problem was Germany, but there were significant contributions from Sweden, France, the United States, and England. By contrast, in the three decades after axiomatization, research was carried on mainly in Germany, Poland, and the Soviet Union. After the Second World War, the principal loci were Poland and the United States. Perhaps these changes in loci reflect, to some degree, evolving national styles in mathematics. But they reflect more strongly the approaches to the Continuum Problem that were possible at different periods.

What were these approaches? The simplest approach was to prove CH or to refute it. By far the largest number of attempts were to prove it, rather than to refute it, and these attempts were made by several mathematicians besides Cantor (including Paul Tannery, Felix Bernstein, and David Hilbert). A more elaborate approach would be to prove CH for a restricted class of subsets of $\mathbb{R}$, and this approach was pursued over many decades, beginning with Cantor's 1883 proof for the closed sets. Another approach was to generalize CH and then to study the generalization, but this occurred only when Felix Hausdorff formulated the Generalized Continuum Hypothesis in 1908.

Surprisingly, some approaches that were possible in the era before axiomatization were not pursued with any vigor then, and the chief of these was to find some non-trivial consequences of CH. Cantor investigated the Continuum Problem for its own sake. He did not explore, in any detail, the consequences of CH for analysis or for any other part of mathematics. Nevertheless, he regarded CH as the most important proposition in set theory. *Why* he did so is not altogether clear, but *that* he did so certainly increased the importance of CH in the eyes of many later mathematicians.

The rigorous study of consequences of CH demands that one consider CH, in a formal fashion, specifically as a hypothesis. This did not occur until Nikolai Luzin did so, while investigating a problem posed by René Baire, in [1914]. Soon afterward, Waclaw Sierpiński began to explore the consequences of CH in great detail, and continued to do so for more than three decades. During these years he contributed more than anyone else to understanding the ramifications of CH. It was only after the strength of CH and the variety of its consequences were known that someone proposed an *alternative* to CH. Again, the mathematician who did so was Luzin [1935].

Some approaches to the Continuum Problem could not be pursued rigorously (and were not pursued at all) before set theory was axiomatized by Ernst Zermelo [1908]. One of these was the problem of finding propositions equivalent to CH. This search for equivalents made sense only relative to a particular axiom system for set theory, since a different system might well have different consequences (possibly including CH itself). A second problem of a similar sort was to derive CH from a stronger axiom. This approach was only taken in [1938], by Kurt Gödel, in another context: mathematical logic.

It was mathematical logic that provided, with Gödel's work, still other possible approaches to the Continuum Problem. These were—in a spirit that owes much to Hilbert's work on the foundations of geometry— to prove the consistency of CH, or at least its relative consistency, and to prove the independence of CH. Such problems could *only* be pursued in a framework where the underlying logic was specified, since different underlying logics turned out to give different answers. About 1932, Zermelo showed that CH is decided in second-order logic. More precisely, he established that all models of Zermelo-Fraenkel set theory (as formulated in second-order logic) have CH true in them or all such models have CH false in them. But Zermelo's result remained unpublished until 1980, and after 1930 second-order logic was increasingly unfashionable.

Gödel pursued the problems of consistency and independence, in the wake of Hilbert's unsuccessful attempt to prove CH, and established in [1938] the relative consistency of the Generalized Continuum Hypothesis (and hence of CH in particular) within first-order logic. He could not hope to establish consistency *tout court* for CH, since by his Incompleteness Theorem [1931] the consistency of set theory cannot be proved except in a stronger system. Soon after Gödel established the relative consistency of CH, he turned to the problem of showing its independence (see [Moore 1988], 149–151). Despite intensive attempts to do so, he did not succeed, and turned increasingly to philosophy.

The independence of CH was finally proved by Cohen in [1963], again wihtin first-order logic. In particular, he showed that the cardinal of R can be arbitrarily large among infinite cardinals. The technique that Cohen discovered to do so, called forcing, was exploited in a vast number of papers over the next two decades. In regard to the Generalized Continuum Hypothesis, Easton [1964] quickly established that it can be consistently violated at *regular* cardinals. But, since he could not establish its behavior at singular cardinals, there arose the Singular Cardinals Problem: Determine the possible behavior of cardinal exponentiation at singular cardinals. This problem continues to stand at the center of set-theoretic interest.

Within this axiomatic context, and the context of first-order logic, interest shifted back to the problem of deciding CH by deducing it (or its negation) from a "natural" axiom. At Gödel's suggestion [1947], the main candidates have been axioms that postulate the existence of extremely large cardinals. One other candidate was Martin's Axiom [Martin and Solovay 1970], which decided many of the same questions as CH, though not always in the same way. Finally, D.A. Martin and J. Steel have quite recently completed the program of establishing CH for various subsets of R (begun by Cantor a century earlier) by showing that if there is a supercompact cardinal, then CH is true for all projective subsets of R.

It is the author's eventual aim to treat the history of the Continuum Problem, from Cantor to the present, in a book. For now, as part of that aim, he offers the above discussion and an appendix containing a chronology of major developments. Yet he hopes that the chronology will provide a larger, if limited, historical context for the Continuum Problem.[1] The remainder of this article is more limited in scope, being devoted to the period 1877-1908.

2. The Origins of the Continuum Hypothesis

Cantor first proposed the Continuum Problem at the end of his article [1878], which showed the set R of all real numbers to be uncountable but that of all algebraic numbers to be countable. Formulated with all the generality that his concepts and results allowed him at the time, this problem remained close to analysis by referring only to point-sets, that is, subsets of R^n. On the one hand, he could now reduce any question concerning the cardinality of a point-set to a question about the power of some subset of the unit interval. On the other hand, after December 1873 he knew of two distinct infinite powers, that of N (the set of all natural numbers) and that of R, but of no others. In [1878] he inquired whether there are any other cardinalities possible for infinite subsets of R. More specifically, he posed the question: How many equivalence classes exist among infinite subsets A and B of R if the equivalence relation is taken to be the existence of a one-one mapping between A and B? He opted for the smallest possible number of such classes—a solution that was the most natural one under the circumstances:

> Through a process of induction, which we do not describe further at this point, one is led to the theorem that the number of classes of linear point-sets yielded by this partition is finite and, indeed, that it is equal to *two* .... We defer to a later occasion the exact investigation of this question. [1878, 258]

Although Cantor returned to the subject on numerous occasions, he never specified in print the nature of this "process of induction". But, from a letter that he wrote to Giulio Vivanti on 6 November 1886, he clearly had in mind the fact that the cardinal of N is not increased by adding it to a smaller cardinal, or multiplying it by a smaller cardinal, or exponentiating it by a smaller cardinal, and that the cardinal of R is obtained every time that a smaller cardinal is exponentiated by the power of N (see §11 below).

Thus the Continuum Problem (still unnamed at the time) entered mathematics as the question how many cardinalities exist among infinite subsets of R. What we shall call the *Weak Continuum Hypothesis*, for reasons that will shortly become clear, answered: only those of N and R. In other words, the Weak Continuum Hypothesis asserted that:

(1) Every infinite subset of R is denumerable or has the power of R.

At the time Cantor knew of no other infinite cardinalities. He had not yet discovered the alephs, nor the diagonalization procedure which would eventually enable him to generate a sequence of ever higher infinite powers by taking the set of all subsets (*i.e.*, the power set) of a given infinite set. In 1878 neither Cantor nor anyone else understood how difficult it would be to determine whether the Weak Continuum Hypothesis was true, much less to solve the Continuum Problem if this hypothesis turned out to be false. Nor was it clear at the time, though it would eventually become so, that the Continuum Problem raises fundamental questions about the foundations of mathematics.

Why did Cantor formulate his Continuum Problem and the Weak Continuum Hypothesis in 1877?[2] He could have formulated them as early as December 1873 when he became aware that no one-one correspondence exists between N and R. Perhaps what he lacked was the general concept of cardinal number, a concept that he first used in 1877. All the same, Cantor could easily have stated the Continuum Problem with the concepts already at hand. One version might have asked how many infinite sets of real numbers can exist such that there is no one-one correspondence between any two of the sets. Perhaps the lack of any direct application in analysis inhibited Cantor from considering the Continuum Problem at an earlier date. The problem would have lacked much interest for him before he had shown, as he did in 1877, that all questions concerning cardinality in n-dimensional Euclidean space can be reduced to such questions for R. In any case the surviving Cantor–Dedekind correspondence, so informative on other topics, sheds no light on the Continuum Problem prior to 1882.

3. DERIVED SETS: AN INTERLUDE

During January 1879 Cantor completed the first of a series of six articles (each entitled "On Infinite Linear Point-Sets") which formed the bulk of his research until 1885. These articles introduced a number of Cantor's topological concepts and furthered the study of derived sets. During the same period he began to investigate the notion of abstract set and to discuss philosophical questions raised by the actual infinite. Among the concepts introduced for abstract sets was that of well-ordering, which enabled him late in 1882 to give a second and stronger form to the Continuum Hypothesis.

In 1878, when Cantor first stated the Weak Continuum Hypothesis in print, he presented it not as a conjecture but as a *theorem*. At the time, indeed, he believed himself to have a proof for it. Nevertheless, as he attempted to spell out this proof in detail, some uncertainty overtook him. For, in the first article of the series, he returned to the question of partitioning the infinite subsets of R into equivalence classes by means of one-one mappings: "Whether these two classes [of N and R] are the *only ones* into which the linear point-sets are partitioned shall not be investigated here; instead, we wish to prove that they are *really distinct classes*

... " [1879, 4]. He devoted the remainder of the article to presenting a variant of his 1878 proof that R is uncountable.

The next three articles in the series [1880; 1882; 1883] did not mention the Continuum Problem. Nevertheless, they introduced concepts and results which later aided Cantor in clarifying this problem. The first such concept, that of infinite ordinal number, appeared in [1880] as a means of investigating point-sets. At this time Cantor did not construe the infinite ordinals as a kind of number but rather as "symbols of infinity", which merely provided indices for derived sets.

In the next article [1882] Cantor discussed for the first time the notion of an abstract set, whose members might belong to any conceptual domain (*Begriffssphäre*). In keeping with the new breadth of his thinking, he extended the notion of power to abstract sets. He specifically considered sets whose elements were curves, surfaces, or solids, and conjectured that the power of such a set might be larger than that of R [1882, 115]. A decade was to pass, however, before he offered a proof for this conjecture [1891]. Finally, his article of 1882 introduced the notion of a denumerable set, a concept already implicit in his paper [1874], and asserted that the power of N is the smallest of the infinite powers. In other words, every infinite set includes a denumerable subset, and every infinite subset of a denumerable set is denumerable. In the same vein he demonstrated the Countable Union Theorem: The union of a countable[3] family of countable sets is countable [1882, 117]. Such results underlined the importance of the notion of denumerable set and rendered the Weak Continuum Hypothesis more plausible.

In the fourth article [1883], completed in September 1882, Cantor demonstrated a number of theorems about the structure of point-sets. His results were based on the notion of isolated set and on some related decomposition formulas which used derived sets.[4] His first theorem indicated that he was still occupied with the partition of point-sets into equivalence classes by means of equipollence: "Every isolated point-set is countable and hence belongs to the first class" [1883, 52].

4. MITTAG-LEFFLER AND INFINITE ORDINALS

In April 1882 Cantor acquired a new correspondent, the Swedish mathematician Gösta Mittag-Leffler, who soon replaced Dedekind as Cantor's principal mathematical confidant. Mittag-Leffler, whose research was in complex analysis and who had close ties with both Hermite in Paris and Weierstrass in Berlin, was about to initiate a new mathematical journal, *Acta Mathematica*. This journal was intended to act as an intermediary between French and German mathematicians (in the aftermath of the Franco-Prussian War) and to stimulate mathematical research in Scandinavia.

On 25 October 1882 Cantor sent Mittag-Leffler a lengthy letter, heralding a major breakthrough in set theory. For the first time he considered as a single collection all of his ordinal numbers having countably many predecessors. He asserted that this collection, which he then called (α) and which was later called ω_1, had a higher power than that of N and, in fact, the next higher power. He regarded (α) as a natural extension of N and had proved that:

(2) Every infinite subset of (α) is denumerable or has the power of (α).

This theorem was extremely important to him, as he indicated in his letter:

> I believe that I can rigorously prove that the set of all real numbers ... can be put in one-one correspondence with the collection (α). If this theorem is combined with [(2)] above, one obtains without further ado the *long* sought proof for the theorem mentioned in the conclusion to the memoir [1878] ... : "Every *infinite* linear point-set ... is either denumerable ... or can be put in one-one correspondence with the collection of all real numbers ....

In that letter Cantor then introduced the next well-ordered power after (α), the power now called $\aleph_2$.

Thus this letter contained the first statement of what was eventually called the Continuum Hypothesis, often abbreviated CH:

(3) R has the power of the set of countable ordinals.

At the time, CH was important to Cantor only because it allowed him to deduce the Weak Continuum Hypothesis. Before very long, however, his emphasis would shift to CH itself.

In a letter of 3 November, Cantor changed his terminology, now calling N the first number-class and (α) the second number-class, as well as referring to the third and fourth number-classes. While in his previous letter he had believed that no number-class higher than the third was relevant to point-sets, now he thought that the fourth might prove useful as well.

Mittag-Leffler was impressed by Cantor's new results, and wrote to him on 9 November about the paper he would submit to *Acta Mathematica*:

> Will you now also prove, in the memoir which you promised me, that "every infinite linear point-set is either denumerable or can be put in one-one correspondence with the collection

of all real numbers ..."? I have long wondered what astonishing methods you intended to use to prove such a theorem, but now I begin to have an idea how you will do it. The theorem is obviously one of the most important in all of analysis, and one of the most beautiful that has ever been found. I heartily congratulate you on having made such a tremendous advance.

On 12 November, Cantor informed Mittag-Leffler that the set of all analytic functions of n variables has the power of the second number-class, while the set of all real functions has the power of the third number-class.

5. WEIERSTRASS

The role that Weierstrass played in Cantor's set-theoretic research has not received much scholarly attention. By 1870 Cantor had sought Weierstrass's professional advice and approval, and continued to do so. In turn, Weierstrass used Cantor's result on the countability of the algebraic numbers in his own research in analysis.[5] Cantor, as he informed Mittag-Leffler on 10 November 1882, had written to Weierstrass about his number-classes, or infinite well-ordered cardinals,

which I have been driven to introduce almost *against my will*, by the force of reason .... I find myself scarcely in a position any more to make advances in the theory of sets or of functions without using these strange things.

Cantor did not receive a reply from Weierstrass, and so, as he wrote Mittag-Leffler on 6 February 1883, he went to Berlin, where Weierstrass expressed no reservation about infinite ordinal numbers; as Cantor put it, "they seemed rather to interest him."

Garrett Birkhoff has stated in print [Birkhoff and Aspray 1984] that Weierstrass believed himself to have proved the Continuum Hypothesis. Birkhoff, however, gave no source for this claim. Correspondence with the author revealed that Birkhoff had heard this statement in the 1970s from an oral source whom he considered reliable but whom he can no longer determine.[6] Thus it remains a mere possibility, and not a very likely one, that Weierstrass concerned himself at all with the Continuum Hypothesis.

6. DEDEKIND AND THE EQUIVALENCE THEOREM

On 5 November 1882 Cantor wrote Dedekind (who had been his chief mathematical correspondent during the 1870s) a long letter on the same subject and informed him of many of the results found in the October letter to Mittag-Leffler. But, in regard to the Continuum Hypothesis, he made two additional comments. First, he referred to the Weak Continuum

Hypothesis as the "Two-Class Theorem" ("Zweiclassensatz"), although did not use the term "Continuum Hypothesis" then or ever.

Second, his letter revealed that the Equivalence Theorem (which had not yet been proved) was intimately connected in his mind with CH. When talking with Dedekind in September 1882, Cantor had been uncertain how to prove the Equivalence Theorem for point-sets:

(4) If $A \subseteq B \subseteq C$ and A has the same power as C, then A has the same power as B.

"Now I have found the source of this proposition", he wrote to Dedekind, "and can prove it with the necessary generality, thereby filling an essential gap in the theory of point-sets" [Noether and Cavaillè's 1937, 55]. The "necessary generality" was for subsets of $\mathbb{R}^n$, and the source was the Continuum Hypothesis. Indeed, when Cantor first published the Equivalence Theorem in 1883, he stated it only under the hypothesis that C in (4) has the power of the second number-class [1883a, 574, 582]. On the other hand, at the end of his letter to Dedekind he posed the Equivalence Theorem for arbitrary sets as an open problem.

7. CANTOR'S "GRUNDLAGEN"

Cantor developed the ideas found in his letters to Mittag-Leffler and Dedekind in a lengthy article, the "Grundlagen" [1883a], which he also published as a separate monograph. In the "Grundlagen" there first appeared in print the concept central to his later work in set-theory: well-ordering. By showing that the set of rational numbers (as well as the set of algebraic numbers) is denumerable, he established that these two sets can be well-ordered. It was a natural next step to introduce the Well-Ordering Principle, which asserts that every set can be well-ordered. He first stated this principle, which he had not mentioned in his letters to Mittag-Leffler and Dedekind, in the "Grundlagen" [1883a, 550]:

> The concept of *well-ordered set* turns out to be essential to the entire theory of point-sets. It is always possible to bring any *well-defined* set into the form of a *well-ordered* set. Since this law of thought appears to me to be fundamental, rich in consequences, and particularly marvelous for its general validity, I shall return to it in a later article.

After discussing the arithmetic of his infinite ordinals, Cantor added: "Furthermore, it will now be possible for me, with the aid of my new discoveries, to provide a firm foundation for the theorem [Weak Continuum Hypothesis] that I stated at the end of my article [1878] ..." [1883a, 551]. Later in the article he returned to this theme:

The matter of ascertaining the power of R^n reduces to the same question specialized to the interval (0,1). I hope that very soon I shall be able to answer the question with a rigorous proof that the desired power is none other than that of our second number-class. From this it will follow ... that the collection of all [real] functions of one or several variables representable by an infinite series has *only* the power of the second number-class .... [1883a, 574]

Keeping this promise in mind, Cantor twice used the Continuum Hypothesis in the article without stating that he had done so. One of these uses concerned his notion of a semicontinuum, *i.e.*, a connected point-set having the power of the second number-class. As an example of this concept, he cited the space M obtained from R^n by removing any countable set of points. Without CH, however, it only followed that M has the power of the continuum, not that M has the power of the second number-class.

Cantor's second implicit use of the Continuum Hypothesis occurred when he first stated in print what became known as the Cantor-Bendixson Theorem. There he asserted that, for any point-set P, the derived set $P^{(1)}$ is countable or has the power of the second number-class [1883a, 575]. Thus he presupposed that any uncountable closed set has the power of the second number-class, an assumption equivalent to the Continuum Hypothesis. On the other hand, he had already communicated the Cantor-Bendixson Theorem to Mittag-Leffler, on 10 November 1882, in a form that does not assume CH; likewise, when he published this theorem in Mittag-Leffler's journal, he did not use CH [Cantor 1883b, 405–406].

Finally, Cantor stated in the "Grundlagen" his only generalization of the Continuum Hypothesis. After remarking that the collection of all continuous real functions (and even of all Riemann-integrable functions) has the power of the second number-class, he asserted that:

(5) The set of all real functions has the power of the third number-class. [1883a, 590]

If (5) is restated in his later aleph notation and c is the power of the continuum, then (5) asserts that $c^c = \aleph_2$. Thus (5) implies the Continuum Hypothesis, but is stronger than it. But with the conceptual tools Cantor possessed in 1883, there was no natural way to extend CH beyond (5). Any further generalization required the notion of cardinal exponentiation, which he did not develop until the 1890s.

8. BENDIXSON AND PERFECT SETS

The first person whom Cantor's work inspired to prove theorems in set theory was the Swedish mathematician Ivar Bendixson, a student of Mittag-Leffler. Cantor was impressed by Bendixson's discovery of an error in the original statement of the Cantor-Bendixson Theorem. On 5 May 1883 Cantor asked Mittag-Leffler to tell Bendixson that

> I have not yet found a completely satisfactory proof for the proposition that the linear continuum has the power of the second number-class, although at various times I believed that I had climbed the mountain. It would be good if he were willing to bring his ingenuity to bear on this important question. Unfortunately, I am prevented by *many* circumstances from working regularly, and I would be fortunate to find, in you and your distinguished students, coworkers who probably will soon surpass me in "set theory".

Two days later Cantor wrote again to Mittag-Leffler about one form of CH:

> My principal interest concerns the proposition: If S is a perfect set, then $S \sim$ (II), that is, S has the second power. A great deal depends on this proposition. You know that I have been working on the proof; several times I believed that I had finished it, but no, there was always a gap somewhere, and I had to be patient.

Mittag-Leffler replied to these comments on 14 May:

> I understand quite well how extremely important it would be to prove that the linear continuum has the power of the second number-class (II). Bendixson too has worked a great deal on this question, but he scarcely hopes to achieve a definitive result.

Cantor, on 17 May, praised Bendixson's insight and added: "I am not yet in a position to prove the *one* important proposition ... that every perfect set has the power of (II). The proof I have of this theorem is still inadequate at one point."

Several months later, after a lapse in the correspondence, Cantor wrote again. On 2 September he returned to the theme of perfect sets, their power, and their connection to the Continuum Hypothesis:

> Again and again I have labored at finding a complete and rigorous proof that all *perfect* sets have the same power. It is very important to prove this. For, by means of the theorem

$$P^{(1)} = R + S,$$

it can then easily be shown that no point-set can have a higher power than that of the *second* class (II). I can rigorously prove, for a large class of perfect sets, the proposition that perfect sets have the same power. There remains, however, a class of them for which the proof is inadequate.

On 26 November he informed Mittag-Leffler that he had proved this theorem on perfect sets and, on 6 December, that Bendixson had independently done so as well.[7]

But the November letter claimed more, for Cantor once again believed that he had proved the Continuum Hypothesis:

I now have two proofs for the theorem [Weak Continuum Hypothesis] that I found seven or eight years ago .... But soon, I think, I will find at least a third proof for it. From this theorem it follows easily that point-sets within G_n [R^n] have *at most* the second power and hence that there exist no *point-sets* of the third power. In the first proof, as I already wrote you in the summer, I reduce that theorem to the following: *All perfect sets* have *the same power*.

It remains unclear how he hoped to reduce CH to proving that all perfect point-sets have the same power.

Bendixson wrote Cantor again on 18 January 1884, remarking that

The proof of the theorem that there do not exist more than two powers [in R^n] interests me above all because I do not believe it possible to penetrate completely into the nature of irrational numbers before this question has been resolved in a satisfactory manner.

The final letter from Bendixson, dated 1 April, expressed his joy that Cantor, according to the end of his article [1884], had found a proof of the Weak Continuum Hypothesis.

9. The Continuum Hypothesis and Closed Sets

What actually happened was quite different. Cantor had made his first major advance towards establishing the Continuum Hypothesis in [1884] by proving a special case of the Weak Continuum Hypothesis:

(6) Every closed subset of R is countable or has the power of the continuum.

His proof followed easily from two results. The first was the Cantor-Bendixson Theorem that every closed subset of R can be partitioned into a perfect set and a countable set. The second was his result that every perfect subset has the power of the continuum.

Once again, Cantor ended his article with a promise to demonstrate the Weak Continuum Hypothesis for all subsets of **R**. This time his strategy for obtaining a proof was to extend the Cantor-Bendixson Theorem: "That this marvelous theorem is valid also for linear point-sets which are *not closed* ... will be proved in later paragraphs" [1884, 488]. He then noted that, by means of (2), he would be able to use the Weak Continuum Hypothesis to prove CH itself.[8]

On 26 August 1884, believing that he had at last succeeded, Cantor sent his purported demonstration of CH to Mittag-Leffler: "I am now in possession of an extremely simple proof for the most important theorem of set theory, that the continuum has the power of the second number-class." The crux of Cantor's argument was the assertion that:

(7) There exists a closed subset of **R** with the power of the second number-class.

Such a closed set M could be decomposed into a perfect set P and a countable set. Since P has the power of the continuum, then M does as well, and so the power of the continuum is the same as that of the second number-class. Thus it only remained, Cantor assured Mittag-Leffler, to define a single closed set with the power of the second number-class. When all was in order, he would forward the details.

10. The Negation of the Continuum Hypothesis

Cantor failed in his repeated attempts to define such a closed set, but he soon attacked the problem from a different direction. As he refined his research on derived sets, he discovered another argument for the Continuum Hypothesis. On 20 October 1884 he communicated the underlying ideas to Mittag-Leffler. Central to the argument was a decomposition theorem of his:

(8) If $M \subseteq \mathbf{R}$, then M can be partitioned into a set A with no subset dense-in-itself, a homogeneous set B of first order, and a homogeneous set B of second order.

(Here a point-set was homogeneous of the $(n + 1)$st order if it was dense-in-itself and intersected every neighborhood in the empty set or in a set with the power of the nth number-class.) It followed from his previous results that A and B were countable and that C, if non-empty, had the power of the second-number class. Thus the Continuum Hypothesis followed at once from (8).

On 14 November Cantor excitedly wrote to Mittag-Leffler that this "proof" of the Continuum Hypothesis was erroneous, since there was a flaw in the demonstration of (8):

You know that I have often believed myself to possess a rigorous proof that the linear continuum has the power of the second number-class. Time and again gaps have arisen in my proofs, and each time I exerted myself anew in the same direction. When I believed I had reached the ardently desired goal, suddenly I was startled because in some hidden corner I noticed a false conclusion.

These days, as I again strove toward the goal, what did I find? I found a *rigorous* proof that the continuum does *not* have the power of the second number-class, and, what is more, that it has no power specifiable by an ordinal number. No matter how dreadful the error, which has been fostered for so long, to eliminate it definitively is that much greater a victory.

Thus a chastened Cantor came to believe that both the Continuum Hypothesis and the Well-Ordering Principle were false, since the real numbers could not be well-ordered. Consequently, there existed no closed subset of ℝwith the power of the second number-class. All of this meant that set theory had a structure vastly different from what Cantor had believed hitherto. Uneasily he concluded the letter: "For today I ask you to be content with this brief announcement. I am more astonished over it myself than I can tell you."

Yet the reversals had not run their course. The next day Cantor wrote Mittag-Leffler to reject his own previous reasons for abandoning the Continuum Hypothesis and the Well-Ordering Principle. On the contrary, Cantor now had every reason for believing the Continuum Hypothesis to be true, and he hoped to find a definitive proof in the near future.

A day later, on 16 November, he explained what had occurred. While writing the article [1884a] on point-sets, he had discovered his theorem characterizing the rational numbers as an order-type: Any two denumerable dense linear orders U and V without endpoints are isomorphic. Such sets U and V were clearly homogeneous of the first order. On 14 November he had found, or so he thought, a proof for an analogous isomorphism theorem where U and V were homogeneous sets of the second order. But this last "theorem" would imply that ℝ does not have the power of the second number-class. For, assuming ℝ has such a power, we let U be the open interval $(0,1)$ and V be the set of all irrationals in U; then the order-isomorphism between U and V would yield that V, together with its endpoints, would be a perfect set—an absurdity for a set of irrationals. Now, however, he was convinced that this "theorem" could not be demonstrated, and was in fact false. Thus was found the first non-trivial proposition that (given the Axiom of Choice) implies the negation of the Continuum Hypothesis:

(9) Any dense linear orders of power $\aleph_1$ are isomorphic.

With this anticlimax came to an end Cantor's most intensive period of research on the Continuum Hypothesis.

Nevertheless, Cantor continued to advocate a kind of Continuum Hypothesis in mathematical physics. In his letter of 16 November to Mittag-Leffler, he discussed atomism and, like Boscovich and Cauchy, opted for extensionless point-atoms. Cantor believed that there are two kinds of point-atoms, the first being the material atoms of which bodies consist and the second being the aether atoms. What he called his "First Hypothesis" was that there are denumerably many material atoms and that the set of aether atoms has the power of the second number-class.[9] Since he held that continuous motion occurs in the aether, his First Hypothesis included the Continuum Hypothesis as a special case. His "Second Hypothesis" asserted that the aether can be given the kind of decomposition (agreeing with CH) found in (8) above. Two months later he finished an article [1885] which concluded with similar views.

11. TANNERY'S ATTEMPTED PROOF

In 1884 Paul Tannery, stimulated by the French translation in *Acta Mathematica* of [Cantor 1878], found an argument for the Weak Continuum Hypothesis and published it in a reputable journal, the *Bulletin de la Société Mathématique* de France [1884]. Unfortunately, this argument was erroneous.

Tannery was convinced that any such argument must set up an algebraic symbolism representing the powers of infinite sets, and especially of the second number-class relative to the first. Thus he supplied a symbolism for representing finite and denumerable sequences whose terms could vary over the natural numbers or be restricted to a finite set A. He then concluded, for reasons that remain uncertain, that the power of the set represented by this symbolism, with A finite, was the power immediately following $\aleph_0$ [1884, 93].

On 6 November 1886 Cantor wrote to Giulio Vivanti concerning this attempted proof:

> I read P. Tannery's note of more than two years ago . . . at the time. There he believed himself to furnish a proof for the proposition, which I had first stated nine years ago, that only two classes occur among the linear point-sets (Borch. J. Vol. 84, p. 257) or, what comes to the same thing, that the power of the linear continuum is precisely the *second*. But he is surely mistaken. The facts that he cites in support of this theorem

were all known to me at that time, as anyone can see, and
form a part of the induction which I ... said there had led
me to this theorem. I was then convinced that this induction
is *incomplete*, and I still have this same conviction today.

Here, as in [1878], Cantor used the term "induction" in the sense of evidence for a conclusion rather than in the sense of mathematical induction.
His letter introduced his first notation for transfinite cardinals—using o,
the Gothic o, for what he designated in the 1890s by $\aleph$:[10]

In my previously unpublished papers I have designated this
[first] power by the sign o_1, the power of the second number-
class or *second* transfinite power by o_2, the third by o_3, etc.

If we understand by $\mathfrak{a}$ the power of the linear continuum,
then the proposition to be proved is this:

$$\mathfrak{a} = o_2.$$

The facts on which Tannery believes he can found this theorem are now these:

$$|\nu| + o_1, o_1 + o_1 = o_1, o_1.|\nu| = o_1,$$
$$o_1^{|2|} = o_1, \ldots, o_1^{|\nu|} = o_1;$$
$$o_1^{o_1} = \mathfrak{a}; |2|^{o_1} = \mathfrak{a}; |3|^{o_1} = \mathfrak{a}; |\nu|^{o_1} = \mathfrak{a}$$

(by $|\nu|$ is meant the cardinal number or *finite* power corre-
sponding to the finite ordinal number ν). These facts indeed
suggest that $\mathfrak{a}$ is the *next* power following o_1, that is, o_2, but
they are far from providing a rigorous proof of this. There-
fore such a proof remains a desideratum after Tannery's note
just as much as before it appeared.

12. MITTAG-LEFFLER AND THE EARLY RECEPTION OF SET THEORY

How mathematicians and philosophers responded, during this period, to the Continuum Problem depended largely on their reaction to
Cantorian set theory in general. The initial reception of set theory was
mixed.

As early as 1881, Mittag-Leffler had reason to believe that set theory
would be well received in France, by Hermite in particular. Thus on 9
May 1881 Hermite wrote to Mittag-Leffler: "Could you satisfy my lively
curiosity concerning the new kind of infinities that Monsieur Cantor has
considered in the theory of functions?"[11]

On 10 January 1883 Mittag-Leffler wrote to Cantor, after reading the "Grundlagen":

> And now permit me, as your true friend, to add some remarks that are *only for you*. I believe that the philosophical part of your work will create a sensation in Germany, but I do not believe that the same will happen to the mathematical part. Except for Weierstrass—and perhaps for Kronecker, who, nevertheless, basically has little interest in these questions and, for that matter, will scarcely share your views—there are no mathematicians in Germany with the fine sense for difficult mathematical investigations that is necessary for the proper conception of your works. Klein, for example, said to me just a few years ago—this is naturally just between us—that he could not see the purpose of all this. Of Weierstrass's students in Germany, Schottky is perhaps the only one who will have some understanding of your work. How our mutual friend Schwarz will rail against you I can well imagine.
>
> But in France things are quite different. There, a very active and lively movement exists at present in the mathematical world. Poincaré, Picard, Appell ..., Goursat, and Halphen (to name only the best of Hermite's students) are all extremely gifted men, endowed with much feeling for the finest mathematical investigations .... All of them, as well as Hermite himself, are touched most keenly by your discoveries precisely because they need such investigations now and because, in their beautiful works on the theory of functions, they are beset by difficulties that can be overcome only through your works. But all of them, except Appell, understand German very poorly.

Mittag-Leffler then proposed to have Cantor's earlier papers on set theory translated into French and printed in *Acta Mathematica*.

But matters did not turn out in the way that Mittag-Leffler had intended. Hermite, while the papers were being translated with his assistance, wrote to Mittag-Leffler on 5 March 1883 expressing doubts about the philosophical parts of Cantor's work. Two weeks later Hermite added: "Appell and Picard are both resistant to the methods used by Monsieur Cantor, but Poincaré, while judging them to be quite premature in the present state of analysis, believes, as you do, that they have some importance".[12] That same day Poincaré also wrote to Mittag-Leffler, remarking that "these numbers in the second, and especially in the third, number-class have the appearance of being form without substance, something repugnant to the French mind".[13]

A climax was reached with Hermite's letter of 13 April to Mittag-Leffler:

> The impression that Cantor's memoirs make on us [Appell, Hermite, Picard, and Poincaré] is disastrous. Reading them seems, to all of us, to be an utter torture .... While recognizing that he has opened up a new field of research, none of us is tempted to pursue it. For us it has been impossible to find, among the results that can be understood, a single one having a *real and present interest*. The correspondence between the points of a line and those of a surface leaves us completely indifferent, and we think that this remark, as long as no one has deduced anything from it, stems from such arbitrary methods that the author would have done better to withhold it and wait.[14]

Hermite showed a draft of this letter to Appell, who added to it that Cantor's articles [1879], [1880], [1882], and [1883] seemed to be merely a new system of notation without real substance. Hermite, with Appell's approval, concluded his condemnation of Cantor by praising "Jacobi, whose writings, marvelously rich in ideas, contain neither a new symbol nor a strange notation".[15]

No doubt Mittag-Leffler had these severe criticisms in mind when, on 9 March 1885, he asked Cantor to withdraw a paper on order-types, "Principien einer Theorie der Ordnungstypen", that had already been partly typeset for *Acta Mathematica*:

> I find the new fundamental ideas that you develop in it very beautiful .... But I do not wish to conceal from you that it appears to me to be in your best interest not to publish these investigations until you can obtain new and very positive results from your new way of proceeding. If you were to succeed, for example, in deciding by means of your theory of order-types the question whether or not the linear continuum has the same power as the second number-class, then your new theory would certainly have the greatest success among mathematicians.[16]

Surely this reference to the Continuum Hypothesis, which Cantor had labored so long and so hard to prove, cut him to the quick. But Mittag-Leffler had seen Cantor claim too many times to have a proof of CH, only to retract it later. And Mittag-Leffler no doubt heard in his own ears the howls that Hermite, and his French colleagues, would make against this new paper of Cantor's as all form and no substance. Cantor withdrew the paper, and turning his back on Mittag-Leffler and, for five years, the mathematical community as well.

13. THE EARLY RECEPTION OF CH IN GERMANY AND FRANCE

In Germany, during the 1880s, the Continuum Hypothesis was discussed primarily within the debate whether set theory was a part of mathematics. Thus Paul du Bois-Reymond [1882] analyzed Cantorian set theory, and WCH in particular, in a quasi-philosophical book devoted to what he called "Empiricist" and "Idealist" conceptions of the theory of functions (roughly, constructive and Platonistic conceptions). Du Bois-Reymond claimed to be neutral, but his sympathies clearly lay with the Empiricists and against Cantor's set theory. Nevertheless, du Bois-Reymond inclined toward accepting WCH:

> Cantor considers it to be highly probable that only these two powers exist [those of N and of R]. The considerations in the following sections contain many indications, not easy to make precise, that things happen this way in sets of linear magnitudes [1882, 199].

Among the "indications" that du Bois-Reymond had in mind, for example, was the fact that R is not reduced in power by removing from it a denumerable set.

During the 1890s various mathematicians, lulled by Cantor's repeated statements that he was about to publish a proof of CH, regarded CH (or WCH) as proven or at least as true. In particular, Ernst Schröder did so when he remarked that the set of points on a circle has the power of the second number-class [1898, 304-305].

Another instance of assuming WCH to have already been proven reached Cantor from a mathematician who knew that this was not so. On 27 December 1892 Heinrich Bruns wrote to Cantor after reading the German translation [1892] of Dini's book [1878] on real analysis. In their translation, Jacob Lüroth and Adolf Schepp had added a few passages, and one of these passages concerned Cantor's work. There [1892, 27] it was stated that the Weak Continuum Hypothesis is true, and references were given to [Cantor 1884a], [Bendixson 1884], and [Tannery 1884]. Bruns' letter mistakenly took Dini, rather than the translators, to task for this statement:

> What appears in Dini is also not completely correct; through conversations with you I already have a higher standpoint than the author on p. 27, where it states that all sets not of the first power have the power of the continuum. Here it

is ... falsely taken as proven that between the two powers mentioned there lie no others.

During this period set theory received mixed reviews from the philosophers. Benno Kerry gave a very sympathetic account [1885] of Cantor's work, regarding it as of great significance. He was more sensitive than most to what Cantor had actually proven. Thus he carefully noted that, although Cantor had sought for a long time a proof of the "conjecture" ("Vermuthung") that the continuum has the power of the second number-class, such a proof had not yet been produced [1885, 221]. He observed that it would suffice, for such a proof, to find a perfect set which had the power of the second number-class. Kerry also mentioned Cantor's assertion that the set of all real functions has the power of the third number-class [1885, 231].

The German neo-Scholastic philosopher Constatin Gutberlet largely agreed with Kerry's assessment of Cantor's work. Gutberlet was primarily concerned to establish the existence of the actual infinite in mathematics, but to do so by theological arguments—above all, the omniscience of God [1885, 215]. However, the first part of his article was devoted to an exposition of Cantor's work, and here he uncritically accepted CH [1885, 188].

In France the neo-Kantian philosopher Arthur Hannequin wrote a book [1895] criticizing various kinds of atomism in physics and mathematics. He regarded Cantor's treatment of the continuum as atomistic, and hence regarded it sceptically. For he was convinced that the notion of number did not suffice to treat the geometric continuum. Hannequin, who relied heavily on Kerry's [1885] account, discussed the Continuum Problem at some length, and saw no convincing reason that a numerical continuum might not have the power of the third or fourth number-class—contradicting CH. Despite his praise for Cantor's work, Hannequin saw it as doomed to failure by what he called "the first fundamental contradiction: the necessary divisibility of the indivisible element" [1895, 69]. As for CH, he concluded by noting "the mathematical imperfection of a system that is still waiting for the direct demonstration of its principal theorem ... that is, the theorem that would establish directly that the power of continua is none other than that of the second number-class" [1895, 64].

The philosopher Louis Couturat was much more favorably inclined toward Cantor's work and toward CH in particular. Couturat discussed that work sympathetically in his book *De l'infini mathématique*. He was one of those who, like Lüroth and Schröder, presupposed CH to be true during the last decade of the nineteenth century [Couturat 1896, 655].

14. CANTOR'S LAST WORK

Cantor's final published work on set theory appeared as a two-part article, the "Beiträge" [1895; 1897], in *Mathematische Annalen*. The connection between CH and this article is evident from a letter that Cantor wrote to Valerian von Derwies on 17 October 1895 concerning the Continuum Hypothesis:

> You are right to stress the importance of a rigorous proof of the theorem
> $$2^{\aleph_0} = \aleph_1,$$
> which I expressed in another form in 1877 .... I hope in this respect to satisfy you in a continuation of the work whose first installment [1895] you have at hand.

Thus Cantor hoped to establish CH in the "Beiträge". But when its second installment appeared [1897], it made no mention of the Continuum Hypothesis.

In the first installment [1895] he had stated the equation $o = 2^{\aleph_0}$, where o was now his notation for the cardinal of the real numbers and where $2^{\aleph_0}$ used his new cardinal exponentiation to represent the real numbers as binary sequences. But he did not state CH in print, then or ever, in the form $2^{\aleph_0} = \aleph_1$ found in his letter to von Derwies. Nevertheless, that form suggested a straightforward generalization if zero were replaced by an arbitrary ordinal A: $2^{\aleph_\alpha} = \aleph_{\alpha+1}$. Would Cantor have made such a generalization in the third installment (which, as he told Hilbert in his 1897 letters, he intended to write)? We do not know.[17] The third installment, if he ever drafted it, is lost.

But we do know, from his letter to Hilbert of 26 September 1897, that Cantor intended to include in the third installment his proof using "inconsistent multitudes" (such as the class of all alephs) that every set can be well-ordered. The letter reaffirms Cantor's belief that "the power c of the linear continuum is equal to a definite aleph ($c = \aleph_1$), as I hope to show." Thus he continued to believe in the truth of CH, and no doubt viewed his "proof" that the real numbers can be well-ordered as a step toward establishing CH.

Cantor's final surviving statement on the subject may be what he wrote to Hilbert on 20 September 1912. There he considered set theory to have several parts—the first being pure set theory while the second "consists of the *applications* of *set theory* founded in the *first* to *number theory, geometry, and analysis,* hence to what we call *pure mathematics.*" Cantor regarded as belonging to the second part what he now called "the Continuum Theorem, *i.e.,* the rigorous proof that the power or cardinal number of a continuum is equal to aleph-one." A decade later Hilbert would also use the term "Continuum Theorem" for the Continuum Hypothesis.

15. RUSSELL'S CHANGING PERSPECTIVE

Bertrand Russell first became familiar with Cantor's work in 1896, as a result of reviewing Hannequin's book. Stimulated by the book, Russell read Cantor's various articles of 1883 in *Acta Mathematica*; Russell's notes on these articles show his confusion at the time.[18] In June 1896 he drafted for *Mind* an article, "On Some Difficulties of Continuous Quantity", which he never published. This article criticized Cantor severely, rejecting both his transfinite ordinals and his treatment of the continuum:

> Cantor's transfinite numbers, then, are impossible and self-contradictory; ... the exhaustion of continua remains a pious belief. Cantor hopes to prove that the second class of numbers will exhaust continua, and there seems as yet no reason to suppose the contrary. But the proof is not as yet forthcoming.

Russell concluded that numbers, which remain discrete, cannot measure arbitrary parts of a continuum. Thus his scepticism toward CH mirrored his scepticism toward set theory in general.

By 1899, however, Russell reversed himself and adopted Cantor's ideas on the continuum, ideas discussed in the draft that he wrote (in 1899 and early 1900) of *The Principles of Mathematics* [1903].[19] Yet in the draft he remained sceptical of Cantor's infinite cardinal and ordinal numbers.

The first published sign that Russell had accepted the Continuum Hypothesis as true occurs in his article "Sur les axiomes de la géométrie" of November 1899. In a footnote to that article, which on 21 October he had asked Couturat to have inserted, he stated that "the class of all points on a linear continuum is of the second power (in Cantor's language)" [1899, 699].

On the other hand, his 1899-1900 draft of the Principles reveals quite clearly, by changes in the manuscript, that he came to abandon Cantor's Continuum Hypothesis sometime before November 1900 when he composed this same part of the *Principles* as published in 1903. In the draft he wrote, following Cantor, that "the series of real numbers ... is of the second power" (Book V, Chapter III). But he has crossed out "the second power" and written over it "a higher power than the first".[20] Thus in 1900, before November if not (as is likely) before July, he came to regard the Continuum Hypothesis as uncertain.

This situation reflects the fact that Cantor had given no proof for CH, but may also reflect Russell's doubts about Cantor's Well-Ordering Principle and its consequence, the Trichotomy of Cardinals. Certainly in the *Principles* as published, Russell questioned an important case of Trichotomy, wondering whether the power of the continuum is even comparable with $\aleph_1$ or $\aleph_2$ [1903, 323]. On the other hand, he appeared to accept WCH [1903, 310n].

In 1905, in the wake of König's purported disproof of the Well-Ordering Principle and the Continuum Hypothesis (see section 20 below), Russell did believe that CH is false. For, in a letter to Jourdain of 15 June 1905, Russell insisted that König "does prove $2^{\aleph_0} \neq \aleph_1$, so that Cantor's expectation was in any case mistaken" [Grattan-Guinness 1977, 53].

But the Continuum Hypothesis was a less important question to Russell than was the Well-Ordering Principle (I.e., the Axiom of Choice, or, in Russell's terminology, the Multiplicative Axiom). Russell and Whitehead devoted considerable discussion in *Principia Mathematica* [1913] to the Multiplicative Axiom but apparently none at all to the Continuum Hypothesis.

As late as 1919, Russell continued to be sceptical of the Trichotomy of Cardinals and to entertain the possibility that $\aleph_1$ and $2^{\aleph_0}$ are incomparable. *A fortiori*, he remained sceptical of the Continuum Hypothesis [1919, 124].

16. C.S. Peirce and the Beth Hierarchy

Charles Sanders Peirce was much influenced by Cantor's work. In a letter of 5 December 1908 to Philip Jourdain, Peirce recalled that he "cannot well have made the least acquaintance with Cantor's work until the winter of 1883-4 or later" [Peirce 1976, 882]. In his unpublished *Grand Logic* of 1893, which was only printed in 1933, Peirce did not find Cantor's definition of a continuum satisfactory but was intrigued by his notion of cardinal number. With his usual penchant for neologisms, Peirce renamed Cantor's term "cardinal number" as "multitude". At this time Peirce was familiar with Cantor's proof that a line and a plane have the same cardinal number [Peirce 1933, 91–92].

Peirce returned to the theme of transfinite cardinals in an unpublished article of about 1895, "On Quantity, with Special Reference to Collectional and Mathematical Infinity" (printed in [1976]). There he was particularly concerned to prove the Trichotomy of Cardinals, but in effect his argument assumed what it meant to prove [1976, 50]. In that same article, on the other hand, Peirce introduced a hierarchy, which he later named the beths, of transfinite cardinals. In the beth notation of his letter of 23 December 1900 to Cantor, this hierarchy is as follows:

$$\beth_0 = \aleph_0, \beth_1 = 2^{\aleph_0}, \text{ and } \beth_{n+1} = 2^{\aleph_n} \text{ for finite } n.$$

Although Peirce did not introduce this notation until 1900, he defined this sequence of cardinals in detail in the 1895 article, calling $\beth_1$ "the first abnumeral", $\beth_2$ "the second abnumeral", and so on [1976, 52].

In that article Peirce immediately went on to make two assertions: "It remains to be shown that there can be no multitude [cardinal] intermediate between these multitudes, and none greater than them all." The first assertion included the Weak Continuum Hypothesis as a special case and generalized it below $\beth_\omega$:

(10) For finite n, there is no cardinal M such that $\beth_n < M < \beth_{n+1}$.

A quarter century later, Lindenbaum and Tarski [1926] recognized that, even in the absence of the Axiom of Choice, (10) implies (11), which is the Generalized Continuum Hypothesis below $\aleph_\omega$:

(11) For every natural number n, $2^{\aleph_n} = \aleph_{n+1}$.[21]

Peirce's second assertion was, in effect, that there is no cardinal greater than $\beth_n$ for every finite n. In other words, $\beth_\omega$ does not exist.

As a first step toward proving (10), Peirce gave an algebraic argument for the case when $n = 0$, that is, for the Weak Continuum Hypothesis. Thus he observed that a finite cardinal multiplied by itself a finite number of times gives a finite cardinal, that the cardinal of a denumerable set multiplied by itself a finite number of times still is denumerable, and that the cardinal of a denumerable set multiplied by itself denumerably many times is equal to the first abnumeral (that is, to $2^{\aleph_0}$). From these facts he claimed that "the first [abnumeral] multitude is the multitude *next greater* than the denumerable multitude" [1976, 53]—that is, that the Weak Continuum Hypothesis holds. What he actually showed, however, was only that no cardinal between $\aleph_0$ and $2^{\aleph_0}$ can be obtained by taking products of finite cardinals or of $\aleph_0$. Thus, while attempting to prove the Weak Continuum Hypothesis, Peirce made an error similar to that made by Paul Tannery in 1884 (see section 11 above).

In order to show that "there are no multitudes intermediate between higher abnumerals" (ibid., 54), Peirce argued in an analogous fashion, adding the case of exponentiation to that of products. Along the way he attempted to show that, for any finite n, a set of power $\beth_n$ can be arranged in a linear order. Unfortunately, his attempt was not successful (ibid., 55–56). Moreover, his arguments did not establish (10) but only the much weaker result that no cardinal between $\beth_n$ and $\beth_{n+1}$ can be obtained from beths by products and exponentiations.

In an unpublished article of about 1897, Peirce again gave an argument for the Weak Continuum Hypothesis, as follows:

That there is no multitude [cardinal] greater than the de-numerable multitude and less than the first abnumeral mul-titude [$2^{\aleph_0}$], I argue as follows. The demonstration does not seem to be perfect; but I think it is only because I do not suc-ceed in stating it quite right. The addition, or aggregation [union], of denumerable collections is entirely without effect upon the multitude, even if the multitude of collections ag-gregated is denumerable. The multiplication of denumerable collections ... is also without effect, if the multitude of denu-merable factors is enumerable [finite]. But if the multitude of factors is denumerable, the result is the first abnumeral [$2^{\aleph_0}$]. Now since there is no multitude greater than every enumerable multitude and less than the denumerable mul-titude, it follows that there is no multitude greater than the product of every enumerable collection of denumerable fac-tors and less than the product of a denumerable collection of denumerable factors which is the first abnumeral multitude [$2^{\aleph_0}$]. [Peirce 1976, 84]

Peirce's argument here for the Weak Continuum Hypothesis is quite sim-ilar to that discussed above, except that, on the one hand, he now consid-ered sums as well as products and, on the other, he asserted a proposition about cardinal products which can be restated as follows:

(12) There is no cardinal M such that, for every finite n,

$$\prod_{1}^{n} \aleph_0 < M < \prod_{1}^{\infty} \aleph_0.$$

Unfortunately, his argument for (12) is not valid. In fact, (12) is simply another form of the Weak Continuum Hypothesis.

In his published article [1897] (a review of Schröder's *Algebra der Logik*) Peirce returned to the assertion he had made circa 1895 that there is no cardinal number greater than $\beth_n$ for every finite n. Peirce argued that if there were such a cardinal, then it would have the same cardinality as its power set—a contradiction.

Peirce spelled out his reasoning on this matter when he wrote to Cantor on 23 December 1900 [Peirce 1976, 778], and there he argued as follows: Suppose there is a set A of cardinality $\beth_\omega$; then any subset of A of cardinality $\beth_n$ has its power set of cardinality $\beth_{n+1}$ and hence is itself a subset of M; therefore the power set of M has the same cardinality as A—a contradiction. It should be pointed out, however, that Peirce's conclusion does not follow. In particular, he did not consider those subsets of A

having the same power as A (and that is almost all of them); for such subsets, his argument does not apply.

Nevertheless, in an unpublished article of about 1909 [1933, 572], Peirce asserted that his argument was correct and that (10) was true, so that the only infinite cardinals are the beths of finite index. Consequently, he concluded, there are many more ordinals than cardinals—contradicting Cantor's [1883a] claim that there are just as many alephs as ordinals.

17. BOREL

In 1898 Emile Borel discussed set theory at length in his textbook on the theory of functions. He emphasized the notion of cardinal number, particularly those of N and R. Drawing on the work of du Bois-Reymond, Borel introduced the notion of a "scale" of real functions. If f and g were functions, then $f < g$ if $\lim_{x \to \infty} f(x)/g(x) = 0$ and $f = g$ if that limit is finite but not zero. Borel took as an axiom (which he regarded as analogous to the Archimedean Axiom) that for any collection A of real functions linearly ordered by $<$ there is a set S, having power $\aleph_1$, of functions such that for any $f \in A$ there is some $g \in S$ such that $f < g$. Any such set S was called a *scale* [1898, 116]. He was convinced that his axiom was true, at least in the case where the functions in A are "effectively defined" [1898, 116-117].

The existence of a scale is a consequence of CH. But Borel made no reference to CH and gave only a heuristic argument for the existence of a scale. Much later, in an unpublished paper of 1970, Gödel introduced a group of axioms that assert the existence of various scales and argued (mistakenly) that these axioms put a bound on the cardinality of R.[22]

18. SCHOENFLIES AND HILBERT

In his 1900 report on set theory to the Deutsche Mathematiker-Verein-igung, Arthur Schoenflies was more sceptical of the Continuum Hypothesis than Russell had been in 1899. "It remains an open question, however," Schoenflies wrote, "whether there can be still other point-sets than the countable ones and those of the power c" [1900, 18-19]. His scepticism was connected to his distrust of Cantor's Well-Ordering Principle. "If the proposition were true," he wrote of Cantor's principle,

> then this would be of the most far-reaching consequence, since the power of every set must be included in the sequence A [of the alephs]. In my opinion, ... the determination of the place which c has in the sequence A has always been one of Cantor's chief goals and ... we owe a large part of his general set-theoretic investigations to this circumstance. If c really occurs in A, then it must be possible ... to put the totality

of [real] numbers into the form of a well-ordered set, and in
case $c = \aleph_1$ in particular, as is Cantor's conviction, then this
set must be similar to the set of all numbers of the second
number-class. It cannot be denied that various ... factors
suggest assuming such a relation between the continuum
and the numbers of the second class. [1900, 49–50]

Three factors that he had in mind were the use of the notion of limit
in both c and $\aleph_1$, that for investigating $\mathbb{R}$ by derived sets the countable
ordinals are sufficient, and that all known uncountable point-sets have the
power c. "Nevertheless," he concluded, "the question of the power of the
continuum remains unsolved today, just as when Cantor first recognized
that the continuum is uncountable" [1900, 50].

The Continuum Problem was thrust into the limelight when David
Hilbert discussed it in his celebrated Paris lecture of 1900 at the International Congress of Mathematicians. Indeed, he posed the Continuum Problem as the first of 23 problems central to the development of
twentieth-century mathematics. He treated the Continuum Problem as
consisting of two parts: (1) the Weak Continuum Hypothesis and (2)
the existence of a well-ordering for the real numbers [1900, 263–264].
Hilbert maintained an active interest in the Continuum Problem during
the decades that followed, and he made a serious attempt to prove it in
the period 1925–1928.

19. BERNSTEIN, HAUSDORFF, AND LEVI

Felix Bernstein, in his doctoral dissertation written under Hilbert
[1901], observed that two problems presently stood at the center of set-
theoretic interest. The first of these was what he named Cantor's Contin-
uum Problem: How many distinct powers $\overline{\overline{A}}$ exist, where A is an infinite
subset of $\mathbb{R}$? This was the first time that the term "Cantor's Continuum Problem" ("das Cantorsche Continuumproblem") appeared in print.
Hilbert in his Paris lecture had entitled his first problem "Cantor's Problem of the Power of the Continuum", and presumably Bernstein obtained
the name "Continuum Problem" by abbreviating Hilbert's title.

One line of attack on this problem, Bernstein observed, was to resolve it for more and more inclusive families of subsets of $\mathbb{R}$, thereby
extending Cantor's result that the answer is two for the family of closed
sets. Nevertheless, Bernstein insisted, Cantor's conjecture of two for *all*
subsets of $\mathbb{R}$ remained uncertain. As it happened, later research on CH
vigorously pursued this line of attack.

A second way to approach the problem, Bernstein added, was via the
alephs, and here the relationship of $\aleph_1$ to the power $2^{\aleph_0}$ was fundamental.
Then he offered a proof that $\aleph_1 \leq 2^{\aleph_0}$ by using a parallelism between

the order-types of denumerable sets and the order-types of well-ordered denumerable sets [1901, 7-11, 40]. In order to establish this result, he communicated a previously unpublished proof of Cantor's that the power of all the order-types of denumerable sets is at least as great as the power of the continuum.

In December 1900 Bernstein sent Hilbert a letter containing this and other results on set theory. One point in the letter directly concerned the Continuum Hypothesis:

> Perhaps there has come to your attention a work by Beppo Levi [1900], in which he claims to have solved Cantor's [Continuum] Problem and intends to communicate a proof of this later. That work gave me a reason to put forward the following theorem:
>
> 4. The totality of all linear closed sets has the power of the continuum.
>
> Therefore one can, for example, characterize each perfect set by a definite real number. It follows easily—from Theorem 4 and from Cantor's theorem that the totality of *all* subsets of the continuum has a power, 2^c, which is greater than c—that B. Levi's main theorem is false.
>
> The interest of Theorem 4 lies in the fact that it shows how small the domain of subsets [of $\mathbb{R}$] is for which, up to the present, the Two-Class Theorem [WCH] has been proved.
>
> I have communicated all these theorems to Professor Cantor, and he has confirmed them.

The paper [Levi 1900] that Bernstein cited here was devoted to the general notion of real function and, in particular, to finding properties possessed by all real functions. His main theorem stated that every subset A of $\mathbb{R}$ can be represented as the union of $F_i - B_i$, where i is a a natural number, the F_i are closed sets, and the B_i are of first category. He restated his theorem in the form that every subset A of $\mathbb{R}$ can be decomposed into a closed set augmented by a set of first category and diminished by another set of first category; stated in later terminology, his assertion was that every subset A of $\mathbb{R}$ has the Baire property.

Levi then added that it was "an immediate consequence" of this theorem that WCH holds. He distinguished sharply between WCH and CH—unlike mathematicians before him and most of those after him. But he gave no argument as to why WCH holds if every subset of $\mathbb{R}$ has the Baire property. He promised a forthcoming article, to be entitled "Recherches sur le continu et sur sa puissance", would contain the proofs omitted from his 1900 paper (and this proof in particular). But this promised article never appeared.

One of those influenced by Bernstein's dissertation was Felix Hausdorff. One theorem in particular, due to Bernstein and Cantor, struck Hausdorff as important: The class A of all denumerable order-types has the power of the continuum, while the class B of all denumerable ordinals has the power $\aleph_1$. This result suggested to Hausdorff [1901] that one should look for a class C of denumerable order-types intermediate between A and its subclass B as a way of possibly solving the Continuum Problem. He proposed such a class C and explored some of its properties, but it did not provide the solution that he was seeking.

20. KÖNIG'S ATTEMPTED REFUTATION

At the International Congress of Mathematicians held at Heidelberg in August 1904, the Hungarian mathematician Julius König offered a "proof" that the Continuum Hypothesis is false since the power of the continuum is not an aleph. Thus, he claimed, the Well-Ordering Principle is false. His argument depended on the assumption that the power of the continuum is an aleph, and so did not affect the Weak Continuum Hypothesis.

What Cantor, who was present at the Congress, undoubtedly found devastating about König's argument was that it was based on Cantorian tools. It was not simply another argument put forward by the opponents of set theory. The crux of König's argument was the proposition that the sum of a countable increasing sequence of cardinals has a power less than the same sum raised to the power $\aleph_0$. This proposition was correct, and the error in König's argument lay elsewhere, namely in Bernstein's proposition that

(13) $\aleph_\alpha^{\aleph_0} = \aleph_\alpha \cdot 2^{\aleph_0}$ for every ordinal α.

The error was apparently found by Zermelo the day after König's lecture [Kowalewski 1950, 202], and consisted of Bernstein's proposition when α was a limit ordinal.

It was unclear to a number of mathematicians at the time just what König's argument did say about the power of the continuum. For example, Schoenflies, in the second installment of his report on set theory, believed König's argument to establish that the power of the continuum cannot be $\aleph_\alpha$ for any limit ordinal α [1908, 25]. But Schoenflies was mistaken. The crux of his mistake was in presupposing that any limit ordinal α is equal to $\beta + \omega$ for some ordinal β. It might happen that the power of the continuum is, for example, $\aleph_{\omega_1}$.

The matter was clarified by Hausdorff's [1907a] notion of cofinality, so that, as Lindenbaum and Tarski pointed out [1926, 310], what König's argument actually established was that the power of the continuum cannot be $\aleph_\alpha$ for any limit ordinal α of countable cofinality.

Meanwhile, when König's Heidelberg paper was published [1905], he regarded his argument as showing that $\mathbb{R}$ cannot have the power $\aleph_{\beta+\omega}$ for any ordinal β. He soon published another paper [1905a] which argued against the existence of a well-ordering for $\mathbb{R}$ on grounds of finite definability similar to Richard's paradox.

21. BERNSTEIN'S ATTEMPTED PROOF

König's lecture was not the only talk at the Heidelberg Congress devoted to the Continuum Problem. Bernstein had announced a lecture to be given there, but he was unable to do so. Instead he published the text of his lecture separately [1905], claiming to sketch a proof that the continuum can be well-ordered and has the power $\aleph_1$. He considered the continuum as the collection of "calculable" number-theoretic functions, and gave a sequence of $\aleph_1$ sets of such functions. In the same vein, he argued that the notion of arbitrary function was not "sufficiently justified". It was not clear, however, whether the union of this sequence of sets has power $\aleph_0$ or $\aleph_1$, and he relied on the uncountability of the continuum to argue that the union has power $\aleph_1$.

This fallacious proof of CH evoked little response, although it bore a certain resemblance to Hilbert's later attempts [1926; 1928] to prove CH. W.H. and G.C. Young described Bernstein's result, with undue generosity, as "a sketch of the method by which he hopes to obtain a proof that $\aleph_1 = c$" [1906, 145].

22. HAUSDORFF AND ORDER-TYPES

Hausdorff was one of those very much affected by König's argument at Heidelberg. In the aftermath of the Heidelberg Congress, Hausdorff met Cantor, Hilbert, and Schoenflies at Wengen, where they discussed König's purported proof and Bernstein's ill-fated proposition (13) [Schoenflies 1922, 100–101]. On 29 September 1904, shortly after Hausdorff returned home to Leipzig, he wrote to Hilbert: "After the Continuum Problem had plagued me at Wengen almost like a monomania, naturally I looked first here at Bernstein's dissertation." The error in the proof of (13) lay where it was suspected—in Bernstein's claim that if $\alpha < \beta$, then every subset A of power $\aleph_\alpha$ taken from a set B of power $\aleph_\beta$ lies in an initial segment of B. This was false, Hausdorff observed, when A was of power $\aleph_0$ and B was of power $\aleph_\omega$. (The notion waiting to be discovered here was that of cofinality, which Hausdorff would make central to his researches on order-types two years later.) Hausdorff was inclined to regard König's proof as false and König's proposition as the height of improbability. On the other hand, you will scarcely have received the impression that Cantor has found, in the last weeks, what he has sought in vain for 30 years. And so your Problem No. 1 appears, after the Heidelberg Congress, to stand precisely where you left it at the Paris Congress.

But perhaps, while I write this, one of the parties to the dispute is already in possession of the truth. I am very anxious to see the printed proceedings of the [Heidelberg] Congress.[23]

Much of Hausdorff's extremely insightful work on order-types [1906; 1907; 1907a] was motivated by a desire to solve the Continuum Problem. In particular, he put forward a new way of posing the problem, in the hope that this would lead to a solution. Namely, CH is true if the following proposition holds [1906, 156]:

(14) There is a dense order-type of power $\aleph_1$ having no $\omega\omega^*$-gaps.

The following year Hausdorff again considered (14) as a form of the Continuum Problem, and, influenced by du Bois-Reymond's notion of "Pantachie", introduced the notion of η_α type.

The place of the pantachie-types in the H-types is narrowly tied to the Continuum Problem. If CH is true, then there is only one pantachie-type. If there is a pantachie-type without $\Omega\Omega^*$-gaps, then the power of the continuum is $2^{\aleph_1}$ and hence is greater than $\aleph_1$ [1907, 151].

23. By Way of Conclusion: The Situation in 1908

When Zermelo axiomatized set theory in 1908, his work had no immediate effect on the Continuum Problem. Hausdorff, who was the mathematician then most actively engaged with this problem, continued to operate in a pre-axiomatic framework, as did most other set theorists. Nevertheless, Hausdorff formulated in 1908 an assumption that would eventually be view axiomatically, the Generalized Continuum Hypothesis.

But 1908 saw one avenue of development closed. Bernstein, who in [1901] had stressed proving WCH for more and more subsets of $\mathbf{R}$, showed that the approach used to do so would not work in general. This approach was to show that such subsets, if uncountable, had a perfect subset. Bernstein, in the course of research on the uniqueness of trigonometric series, showed (by implicitly using the Axiom of Choice) that there is an uncountable set A of real numbers without a perfect subset [1908]. The proof used transfinite induction to construct two disjoint uncountable sequences of numbers, each intersecting every perfect set; thus either sequence was the desired set A.

Although Bernstein's result showed that the perfect set property (an uncountable set has a perfect subset) did not hold for all subsets of $\mathbf{R}$, it left open the possibility that it held for a very extensive class of subsets of $\mathbf{R}$. Within a decade, Hausdorff [1914; 1916] and Alexandrov [1916] independently established that this was so for the Borel sets.

APPENDIX

A CHRONOLOGY OF THE CONTINUUM PROBLEM

The following chronology, by no means complete, gives the main significant events in the history of the Continuum Problem from its origin to the present. Since events after 1908 are not discussed in the body of the paper, they are stated in more detail than earlier ones. All events refer to publications unless otherwise indicated.

1878–Cantor: proposes WCH as theorem, not as hypothesis.
1882–Cantor: (in letters) introduces concept of well-ordering and ordinal number-classes; proposes CH in terms of second number-class; generalizes CH to set of all functions.
1883–Cantor: states CH and its generalization to set of all real functions; states Well-Ordering Principle; proves Cantor-Bendixson Theorem; (in letters) proves that all perfect sets have the same power.
1884–Cantor: proves WCH for closed sets; (in letters) seeks closed subset of $\mathbf{R}$ with power $\aleph_1$, then uses decomposition "theorem" into dense sets of different powers, temporarily believes CH to be false.
   –P. Tannery: "proves" WCH.
1885–Kerry, Gutberlet: positive philosophical reception in Germany.
1895–Cantor: rephrases CH as $2^{\aleph_0} = \aleph_1$ as a result of formulating cardinal exponentiation.
1896–Hannequin, Couturat: philosophical discussion in France.
   –Russell: suspicious of infinite cardinals and ordinals.
1899–Russell: accepts CH.
1900–Hilbert: CH as first of his 23 problems (draws attention of the mathematical world to CH).
   –Levi: "proves" WCH (using Baire property).
1901–Bernstein: two approaches to Continuum Problem; $\aleph_1 \leq 2^{\aleph_0}$.
1903–Russell: rejects CH and Well-Ordering Principle.
   –W.H. Young: WCH for G_δ sets.
1904–Bernstein: "proves" CH.
   –König: "refutes" CH.

–Zermelo: Well-Ordering Theorem, by means of the Axiom of Choice.

1906–Luzin: (unpublished) attempts to prove CH.

1907–Hausdorff: introduces cofinality, shows $\aleph_{\alpha+1}$ regular; considers the possibility of weakly inaccessible cardinals.

–Brouwer: intuitionistic critique of CH.

1908–Hausdorff: formulates GCH.

–Bernstein: shows perfect set property to fail for some subset of **R**.

1914–Hausdorff: proves WCH for $G_{\delta\sigma\delta\sigma}$ and $F_{\sigma\delta\sigma}$ sets.

–Luzin: CH used as hypothesis to obtain Luzin set.

1915–Jourdain: doubts that CH can be proved or disproved.

1916–Alexandrov, Hausdorff: each proves WCH, and perfect set property, for the Borel sets.

1917–Suslin: for the analytic sets, shows WCH and perfect set property.

–Hilbert: (unpublished) lectures on set theory and CH.

1919–Sierpiński: establishes an equivalent to CH.

1920–Luzin: poses problem (in Fund. Math.) of determining possible powers of CA sets (complements of analytic sets).

1923–Skolem: CH probably undecidable in (first-order) Zermelo set theory.

–Hilbert: asserts that Beweistheorie will prove CH.

1925–Tarski: reformulates and names GCH.

1926–Hilbert: publishes attempted sketch of proof of CH.

–Lindenbaum and Tarski: If WCH and there is no cardinal between that of **R** and its power set, then CH.

–Sierpiński: every Σ_2^1 point-set (*i.e.*, projection of a CA set) is a union of $\aleph_1$ Borel sets.

1928–Hilbert: second published attempt to prove CH.

–Fraenkel: rejects Hilbert's argument for CH.

1929–Luzin: rejects Hilbert's argument for CH; discusses a constructive version of CH.

1930–Baer: use of GCH regarding algebraic extension fields.

1930 to present—Many equivalents to CH found; many consequences of CH found in diverse branches of mathematics (algebra, analysis, topology).

c.1932–Zermelo (unpublished): CH decided in second-order set theory.

1933–Kuratowski: every Σ_2^1 set of power greater than $\aleph_1$ has the perfect subset property; every uncountable Σ_2^1 set has power $\aleph_1$ or $2^{\aleph_0}$.

1934–Sierpiński: publishes book on equivalents and consequences of CH.

1935–Luzin: two alternatives to CH (any point-set of power $\aleph_1$ is CA; $2^{\aleph_0} = 2^{\aleph_1}$).

1938–Bernstein: new axiom implies CH (no response).

–Gödel: constructible sets and relative consistency of GCH; $V = L$ implies that GCH holds and that perfect set property fails for some CA set.

1942–Gödel: unpublished attempts to show independence of CH from set theory in first-order logic.

1947–Gödel: CH will be shown undecidable in set theory; need for new axioms to decide CH, especially large cardinal axioms; GCH determines cardinal exponentiation and infinite cardinal products.

1963–Cohen: consistency of ZF implies consistency of ZF + not CH; $2^{\aleph_0}$ can be arbitrarily large and can be any aleph not of countable cofinality.

1964–Easton: GCH can fail wildly at all regular cardinals (general belief of set theorists that the same is true for singular cardinals; Solovay tries to establish this, but fails).

–Mycielski: shows that Axiom of Determinacy implies WCH and not AC (soon: Axiom of Projective Determinacy implies WCH for projective sets).

1964-1965–Levy and Solovay: each shows that all known large cardinal axioms do not decide CH.

1967–Solovay: GCH must hold at all regular limit cardinals above a supercompact cardinal.

1969–Solovay: if there is a measurable cardinal, then every Σ_2^1 set has the perfect subset property.

1907–Gödel: (unpublished) propsoes axioms, including his square axioms, to settle the Continuum Problem by showing that $2^{\aleph_0} = \aleph_2$.

–Martin and Solovay: Martin's Axiom as alternative to CH.

1974–Silver: if GCH holds everywhere below $\aleph_\alpha$, then GCH holds at $\aleph_\alpha$, provided that α is singular of uncountable cardinality.

1975–Jensen: Covering Theorem (Singular Cardinals Hypothesis holds, and the universe of sets is much like L, unless there is a measurable cardinal).

1977–Martin: if there is a measurable cardinal, then every uncountable Σ_3^1 set is the union of $\aleph_2$ Borel sets, and so has power $\aleph_1$ or $\aleph_2$ or $2^{\aleph_0}$.

1986–Martin and Steel: if there is a supercompact cardinal, then WCH holds for all projective sets.

NOTES

1. Some of these later developments are discussed in [Moore 1989].
2. An upper bound on their formulation is set by the date (11 July 1877) at the end of Cantor's paper [1878]. A probable lower bound is 29 June 1877, when Cantor sent Dedekind a letter containing certain related ideas used in that paper.
3. A set is said to be countable if it is finite or denumerable.
4. For Cantor, a point-set is isolated if it contains none of its limit points.
5. Letters from Weierstrass to H.A. Schwarz (16 December 1874 and 28 May 1875), in [Dugac 1973], 140–141.
6. Letters from Birkhoff to Moore, December 1983 and January 1984. Pierre Dugac, who studied Weierstrass's published and unpublished work extensively, wrote to Moore in March 1984 that, to the best of his knowledge, Weierstrass had no interest in the Continuum Problem.
7. Both Bendixson [1884] and Cantor [1884] published proofs of this result.
8. Cantor reiterated these claims in an article [1884a], dated 15 November 1883 and published in April 1884, in *Acta Mathematica*.
9. The passage of the November letter discussing these matters can be found in [Purkert and Ilgauds 1987], 204–206.
10. In his December 1893 letter to Vivanti, Cantor began the infinite cardinals with $\aleph_1$ (cf. [Meschkowski 1965], 508), just as in 1886 he had begun them with o_1. But in his article [1895] he began with $\aleph_0$, as has been the standard practice ever since.
11. Hermite in [Dugac 1984], 122.
12. Hermite in [Dugac 1984], 204.
13. Poincaré in [Dugac 1984], 278.
14. Hermite in [Dugac 1984], 209.
15. Hermite in [Dugac 1984], 210.
16. Mittag-Leffler in [Grattan-Guinness 1970], 101.
17. These letters are printed in [Purkert and Ilgauds 1987], 222–224. When an earlier version of the present paper was given as a talk at Vassar College in June 1988, Ivor Grattan-Guinness proposed that Cantor probably formulated the Generalized Continuum Hypothesis in the 1890s. We responded that, at present, no evidence is known showing that Cantor actually did so.
18. These notes are found in the Russell Archives and will be printed as an appendix to Volume 2 of the *Collected Papers of Bertrand Russell*, a volume edited by Nicholas Griffin and Albert Lewis.
19. This draft of the *Principles* will be published in Volume 3 of the *Collected Papers of Bertrand Russell*, a volume edited by the author.

20. He has made analogous changes two more times on folios 25 and 26 of the manuscript of Book V.
21. Peirce gave a statement close to (11) in a letter of 21 November 1904 to E.H. Moore: "The $\aleph_1$ or $2^{\aleph_0}$ I call the simply-abnumerable multitude. The $\aleph_2$ or $2^{\aleph_1}$ I call the doubly-abnumerable. These run up through the finite orders of abnumerable; and I prove that these are all the multitudes that are possible" [Peirce 1976], 916.
22. Gödel's unpublished paper is discussed in [Moore 1989], 173–175.
23. This letter is kept in Hilbert's *Nachlass* (Hilbert 136) at the Niedersächsische Staats- und Universitätsbibliotek in Göttingen.

ACKNOWLEDGEMENTS

We gratefully acknowledge the permission of the Bertrand Russell Archives in Hamilton to quote from Russell's unpublished manuscripts, from the Institut Mittag-Leffler in Stockholm to quote from unpublished letters of Ivar Bendixson, Georg Cantor, and Gösta Mittag-Leffler, as well as that of the Niedersächsische Staats- und Universitätsbibliotek in Göttingen to quote from unpublished letters of Felix Bernstein, Heinrich Bruns, Georg Cantor, and Felix Hausdorff.

REFERENCES

ALEXANDROV, PAVEL
  1916   *Sur la puissance des ensembles mesurables B*, Comptes rendus hebdomadaires des séances de l'Académie des Sciences, Paris 162, 323–325.

BAER, REINHOLD
  1930   *Eine Anwendung der Kontinuumhypothese in der Algebra*, Journal für die reine und angewandte Mathematik 162, 132–133.

BENDIXSON, IVAR
  1884   *Sur la puissance des ensembles parfaits de points*, Svenskavetenskapsakademien (Stockholm), Bihang till handlingar 9, no. 6.

BERNSTEIN, FELIX
  1901   Untersuchungen aus der Mengenlehre (Doctoral dissertation: Göttingen); reprinted with revisions in Mathematische Annalen 61, 117–155.
  1905   *Die Theorie der reellen Zahlen*, Jahresbericht der Deutschen Mathematiker-Vereinigung 14, 447–449.
  1908   *Zur Theorie der trigonometrischen Reihe*, Königlich Sächsischen Gesellschaft der Wissenschaften zu Leipzig, Mathematisch-Physische Klasse, Sitzungsberichte 60, 325–338.
  1938   *The Continuumproblem*, Proceedings of the National Academy of Sciences (U.S.A.) 24, 101–104.

BIRKHOFF, GARRETT, AND WILLIAM ASPRAY
1984   Review of [Moore 1982], Isis 75, 401–402.

BOREL, EMILE
1898   Leçons sur la théorie des fonctions (Paris: Gauthier-Villars).

CANTOR, GEORG
1874   Über eine Eigenschaft des Inbegriffes aller reellen algebraischen Zahlen,
        Journal für die reine und angewandte Mathematik 77, 258–262;
        reprinted in [Cantor 1932], 115–118.

1878   Ein Beitrag zur Mannigfaltigkeitslehre, Journal für die reine und
        angewandte Mathematik 84, 242–258; reprinted in [Cantor 1932],
        119–133.

1879   Über unendliche, lineare Punktmannichfaltigkeiten. I, Mathemati-
        sche Annalen 15,1–7; reprinted in [Cantor 1932], 139–145.

1880   Über unendliche, lineare Punktmannichfaltigkeiten. II, Mathemati-
        sche Annalen 17, 355–358; reprinted in [Cantor 1932], 145–148.

1882   Über unendliche, lineare Punktmannichfaltigkeiten. III, Mathemati-
        sche Annalen 20, 113–121; reprinted in [Cantor 1932], 149–157.

1883   Über unendliche, lineare Punktmannichfaltigkeiten. IV, Mathemati-
        sche Annalen 21, 51–58; reprinted in [Cantor 1932], 157–164.

1883a   Über unendliche, lineare Punktmannichfaltigkeiten. V, Mathemati-
        sche Annalen 21, 545–591; reprinted as Grundlagen einer allge-
        meinen Mannichfaltigkeitslehre (Leipzig: Teubner, 1883);
        reprinted again in [Cantor 1932], 165–209.

1883b   Fondements d'une théorie générale des ensembles, Acta Mathematica
        2, 381–408; revised French translation of part of [1883a].

1884   Über unendliche, lineare Punktmannichfaltigkeiten. VI, Mathema-
        tische Annalen 23, 453–488; reprinted in [Cantor 1932], 210–246.

1884a   De la puissance des ensembles parfaits de points, Acta Mathematica
        4, 381–392; reprinted in [Cantor 1932], 252–260.

1885   Über verschiedene Theoreme aus der Theorie der Punktmengen in
        einem n-fach ausgedehnten stetigen Raume G_n. Zweite Mitteilung,
        Acta Mathematica 7, 105–124; reprinted in [Cantor 1932], 261–
        277.

1891   Über eine elementare Frage der Mannigfaltigkeitslehre, Jahresbericht
        der Deutschen Mathematiker-Vereinigung 1, 75–78; reprinted in
        [Cantor 1932], 278–281.

1895   Beiträge zur Begründung der transfiniten Mengenlehre. I, Mathema-
        tische Annalen 46, 481–512; reprinted in [Cantor 1932], 282–311;
        translated in [Cantor 1915], 85–136.

1897   Beiträge zur Begründung der transfiniten Mengenlehre. II, Mathe-
        matische Annalen 49, 207–246; reprinted in [Cantor 1932], 312–
        356; translated in [Cantor 1915], 137–201.

1915 Contributions to the Founding of the Theory of Transfinite Numbers (Chicago: Open Court), translated and edited by Philip Jourdain; reprinted (New York: Dover).

1932 Gesammelte Abhandlungen mathematischen und philosophischen Inhalts. Mit erläuternden Anmerkungen sowie mit Ergänzungen aus dem Briefwechsel Cantor-Dedekind, edited by Ernst Zermelo (Berlin: Springer).

COHEN, PAUL J.

1963 *The Independence of the Continuum Hypothesis. I,* Proceedings of the National Academy of Sciences, U.S.A 50, 1143–1148.

COUTURAT, LOUIS

1896 De l'infini mathématique (Paris: Alcan).

DEVLIN, KEITH I., AND RONALD B. JENSEN

1975 *Marginalia to a Theorem of Silver,* ISILC Logic Conference. Proceedings of the International Summer Institute and Logic Colloquium, Kiel 1974 (Springer Lecture Notes in Mathematics, Vol. 499), 115–142.

DINI, ULISSE

1878 Fondamenti per la teorica della funzioni di variabili reali (Pisa).

1892 Grundlagen für eine Theorie der Functionen einer veröndlichen reellen Grösse (Leipzig: Teubner).

DU BOIS-REYMOND, PAUL

1882 Die allgemeine Functionentheorie (Tübingen: Laupp).

DUGAC, PIERRE

1973 *Eléments d'analyse de Karl Weierstrass,* Archive for History of Exact Sciences 10, 41–176.

1984 *Lettres de Charles Hermite à Gösta Mittag-Leffler (1874-1883),* Cahiers du Séminaire d'Histoire des Mathématiques 5, 49–285.

EASTON, WILLIAM B.

1964 *Powers of Regular Cardinals* (doctoral dissertation: Princeton University).

FRAENKEL, ABRAHAM

1928 Einleitung in die Mengenlehre, third edition (Berlin: Springer).

GÖDEL, KURT

1931 *Über formal unentscheidbare Sätze der Principia Mathematica und verwandter Systeme I,* Monatshefte für Mathematik und Physik 38, 173–198.

1938 *The Consistency of the Axiom of Choice and of the Generalized Continuum Hypothesis,* Proceedings of the National Academy of Sciences, U.S.A. 24, 556–557; reprinted in [Gödel 1989], 26–27.

1947 *What Is Cantor's Continuum Problem?*, American Mathematical Monthly 54, 515–525; reprinted in [Gödel 1989], 176–187.

1989 Collected Works of Kurt Gödel, Vol. II, ed. Solomon Feferman, John W. Dawson, Jr., Stephen C. Kleene, Gregory H. Moore, Robert M. Solovay, and Jean van Heijenoort (New York: Oxford University Press).

GRATTAN-GUINNESS, IVOR

1970 *An Unpublished Paper by Georg Cantor: Principien einer Theorie der Ordnungstypen. Erste Mittheilung*, Acta Mathematica 124, 65–107.

1977 Dear Russell—Dear Jourdain (London: Duckworth).

GUTBERLET, CONSTANTIN

1885 *Das Problem des Unendlichen*, Zeitschrift für Philosophie und philosophische Kritik 88, 179–223.

HANNEQUIN, ARTHUR

1895 Essai critique sur l'hypothèse des atomes dans la science contemporaine (Paris: Masson).

HAUSDORFF, FELIX

1901 *Ueber eine gewisse Art geordneter Mengen*, Königlich Sächsische Gesellschaft der Wissenschaften zu Leipzig, Mathematisch-Physische Klasse, Sitzungsberichte 53, 460–475.

1906 *Untersuchungen über Ordnungstypen*, Königlich Sächsischen Gesellschaft der Wissenschaften zu Leipzig, Mathematisch-Physische Klasse, Sitzungsberichte 58, 106–169.

1907 *Untersuchungen über Ordnungstypen*, Königlich Sächsischen Gesellschaft der Wissenschaften zu Leipzig, Mathematisch-Physische Klasse, Sitzungsberichte 59, 84–159.

1907a *Über dichte Ordnungstypen*, Jahresbericht der Deutschen Mathematiker-Vereinigung 16, 541–546.

1908 *Grundzüge einer Theorie der geordneten Mengen*, Mathematische Annalen 65, 435–505.

1914 Grundzüge der Mengenlehre (Leipzig: de Gruyter).

1916 *Die Mächtigkeit der Borelschen Mengen*, Mathematische Annalen 77, 430–437.

HILBERT, DAVID

1900 *Mathematische Probleme. Vortrag, gehalten auf dem internationalem Mathematiker-Kongress zu Paris. 1900*, Nachrichten von der Königlichen Gesellschaft der Wissenschaften zu Göttingen, 253–297; translated in Bulletin of the American Mathematical Society (2) 8 (1902), 437–479.

1926 *Über das Unendliche*, Mathematische Annalen 95, 161–190;

reprinted in [1935], 178–191; translated in [van Heijenoort 1967], 367–392.

1928   *Die Grundlagen der Mathematik*, Abhandlungen aus dem mathematischen Seminar der Hamburgischen Universität 6, 65–85; translated in [van Heijenoort 1967], 464–479.

1935   Gesammelte Abhandlungen, vol. 3 (Berlin: Springer).

JOURDAIN, PHILIP E.B.

1915   Contributions to the Founding of the Theory of Transfinite Numbers, translation of [Cantor 1895] and [1897].

KERRY, BENNO

1885   *Ueber G. Cantor's Mannigfaltigkeitsuntersuchungen*, Vierteljahrsschrift für wissenschaftliche Philosophie 9, 191–232.

KÖNIG, JULIUS

1905   *Zum Kontinuum-Problem*, Verhandlungen des dritten internationalen Mathematiker-Kongresses in Heidelberg von 8. bis 13. August 1904 (Leipzig: Teubner), 144–147.

1905a   *Über die Grundlagen der Mengenlehre und das Kontinuumproblem*, Mathematische Annalen 61, 156–160; translated in [van Heijenoort 1967], 145–149.

KOWALEWSKI, GERHARD

1950   Bestand und Wandel (Munich: Oldenbourg).

KURATOWSKI, KAZIMIERZ

1933   Topologie I. Espaces métrisables, espaces complets (Warsaw).

LEVI, BEPPO

1900   *Sulla teoria delle funzioni e degli insiemi*, Atti della Reale Accademia dei Lincei, Rendiconti, Classe di scienze fisiche, matematiche e naturali (5) 9, 72–79.

LEVY, AZRIEL

1964   *Measurable Cardinals and the Continuum Hypothesis*, Notices of the American Mathematical Society 11, 769–770.

LINDENBAUM, ADOLF, AND ALFRED TARSKI

1926   *Communication sur les recherches de la théorie des ensembles*, Comptes Rendus des Séances de la Société des Sciences et des Lettres de Varsovie, Classe III 19, 299–330.

LUZIN, NIKOLAI NIKOLAYEVICH

1914   *Sur un problème de M. Baire*, Comptes rendus hebdomadaires des séances de l'Académie des Sciences, Paris 158, 1258–1261.

1920   *Problèmes*, Fundamenta Mathematicae 1, 224.

1929   *Sur les voies de la théorie des ensembles*, Atti del Congresso Inter-
        nazionale dei Matematici, Bologna, 3–10 settembre 1928, Vol. I,
        295–299.

1935   *Sur les ensembles analytiques nuls*, Fundamenta Mathematicae 25,
        109–131.

MARTIN, DONALD A.
1977   *Descriptive Set Theory: Projective Sets*, Handbook of Mathematical
        Logic, ed. Jon Barwise (Amsterdam: North- Holland), 783–815.

MARTIN, DONALD A., AND ROBERT M. SOLOVAY
1970   *Internal Cohen Extensions*, Annals of Mathematical Logic 2, 143–
        178.

MESCHKOWSKI, HERBERT
1965   *Aus den Briefbüchern Georg Cantors*, Archive for History of Exact
        Sciences 2, 503–519.

MYCIELSKI, JAN
1964   *On the Axiom of Determinateness*, Fundamenta Mathematicae 53,
        205–224.

MOORE, GREGORY H.
1982   Zermelo's Axiom of Choice: Its Origins, Development, and In-
        fluence (New York: Springer).

1988   *The Origins of Forcing*, Logic Colloquium '86, ed. F.R. Drake and
        J.K. Truss (Amsterdam: North-Holland), 143–173.

1989   *Introductory Note to [Gödel 1947]*, in [Gödel 1989], 154–175.

NOETHER, EMMY, AND JEAN CAVAILLÈS
1937   (eds.) Briefwechsel Cantor-Dedekind (Paris: Hermann).

PEIRCE, CHARLES S.
1897   *The Logic of Relatives*, The Monist 7, 161–217.

1933   Collected Papers of Charles Sanders Peirce, edited by Charles
        Hartshorne and Paul Weiss, volume IV: The Simplest Mathe-
        matics (Cambridge, Mass.: Harvard University Press).

1976   The New Elements of Mathematics, ed. Carolyn Eisele, Vol. 3:
        Mathematical Miscellanea (Paris: Mouton).

PURKERT, WALTER, AND HANS J. ILGAUDS
1987   Georg Cantor 1845–1918 (Basel: Birkhäuser).

RUSSELL, BERTRAND
1896   *On Some Difficulties of Continuous Quantity* (unpublished manu-
        script).

1899   *Sur les axiomes de la géométrie*, Revue de métaphysique et de
        morale 7, 684–707.

1903   The Principles of Mathematics (Cambridge, U.K.: Cambridge University Press).
1919   Introduction to Mathematical Philosophy.

SCHOENFLIES, ARTHUR
1900   *Die Entwicklung der Lehre von den Punktmannigfaltig-keiten*, Jahresbericht der Deutschen Mathematiker-Vereinigung 8, Heft 2, 1–251.
1908   *Die Entwickelung der Lehre von den Punktmannigfaltigkeiten*, Jahresbericht der Deutschen Mathematiker-Vereinigung, Ergänzungsband 2, 1–331.
1922   *Zur Erinnerung an Georg Cantor*, Jahresbericht der Deutschen Mathematiker-Vereinigung 31, 97–106.

SCHRÖDER, ERNST
1898   *Über zwei Definitionen der Endlichkeit und G. Cantor'sche Sätze*, Deutsche Akademie der Naturforscher, Nova Acta Leopoldina 71, 303–362.

SIERPIŃSKI, WACLAW
1919   *Sur un théorème équivalent à l'hypothèse du continu ($2^{\aleph_0} = \aleph_1$)*, Bulletin international de l'Académie des Sciences et des Lettres, Cracovie, classe des sciences mathématiques et naturales, Série A, sciences mathématiques (1919), 1–3.
1926   *Sur une propriété des ensembles (A)*, Fundamenta Mathematicae 8, 362–369.
1934   Hypothèse du continu (Warsaw: Garasiński).

SILVER, JACK H.
1975   *On the Singular Cardinals Problem*, Proceedings of the International Congress of Mathematicians, Vancouver 1974, 265–268.

SKOLEM, THORALF
1923   *Einige Bemerkungen zur axiomatischen Begründung der Mengenlehre*, Den femte skandinaviska matematikerkongressen, Redogörelse (Helsinki: Akademiska Bokhandeln), 217–232; translated in [van Heijenoort 1967], 290–301.

SOLOVAY, ROBERT M.
1965   *Measurable Cardinals and the Continuum Hypothesis*, Notices of the American Mathematical Society 12, 132.
1969   *On the Cardinality of Σ_2^1 Sets of Reals*, Foundations of Mathematics, ed. J. Buloff et al., 58–73.

SUSLIN, MIKHAIL
1917   *Sur une définition des ensembles mesurables (B) sans nombres transfinis*, Comptes rendus hebdomadaires des séances de l'Académie des Sciences, Paris 164, 88–91.

TANNERY, PAUL
1884   *Note sur la théorie des ensembles,* Bulletin de la Société Mathématique de France 2, 90–96.

TARSKI, ALFRED
1925   *Quelques théorèmes sur les alephs,* Fundamenta Mathematicae 7, 1–14.

WHITEHEAD, ALFRED NORTH, AND BERTRAND RUSSELL
1913   Principia Mathematica, vol. 3 (Cambridge, U.K.: Cambridge University Press).

YOUNG, WILLIAM H.
1903   *Zur Lehre der nicht abgeschlossenen Punktmengen,* Königlich Sächsischen Gesellschaft der Wissenschaften zu Leipzig, Mathematisch-Physische Klasse, Sitzungberichte 55, 287–293.

YOUNG, WILLIAM H. AND GRACE C.
1906   The Theory of Sets of Points (Cambridge, U.K.: Cambridge University Press).

ZERMELO, ERNST
1908   *Untersuchungen über die Grundlagen der Mengenlehre. I,* Mathematische Annalen 65, 261–281.

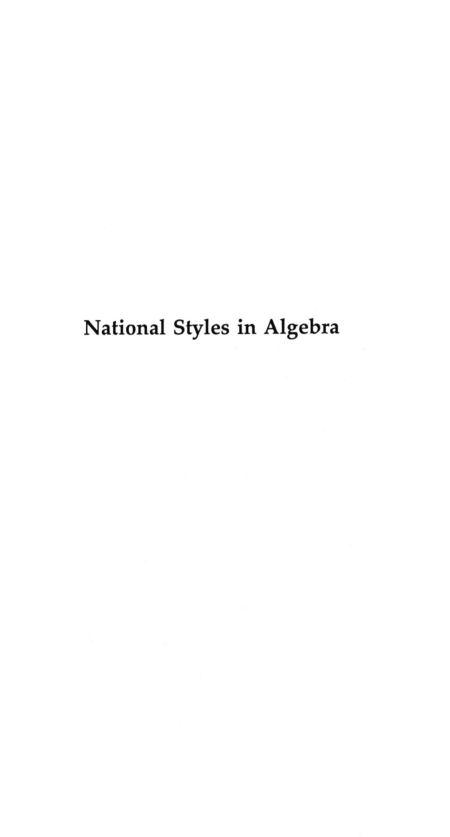

National Styles in Algebra

Benjamin Peirce (1809–1880)
Courtesy of Springer-Verlag Archives

British Synthetic vs. French Analytic Styles of Algebra In the Early American Republic

Helena M. Pycior

The mathematics of colonial America was more homogeneous than that of the early Republic. Colonial students of mathematics learned the subject from imported British textbooks.[1] Following the War of Independence, however, Americans became more familiar with French ideas and institutions. Contemporaneously they were exposed to the self-criticism of some British scientists who conceded French superiority in science and mathematics.[2] Political sympathy with France, growing acceptance of the thesis that British science had declined over the course of the eighteenth century, and even the early Napoleonic victories then turned some Americans towards French institutions as models for American higher education. The French influence helped shape the Military Academy at West Point; it also led to the adoption of new mathematical textbooks for the American classroom. According to the American historians Florian Cajori and, more recently, Robert Bruce, "French [mathematical] authors displaced the English in many of our best institutions."[3]

The present paper uses four influential American algebra textbooks as bases for a reexamination of the mathematical reform movement of the early Republic. Algebra—a required course at most American colleges of the period—is well suited as a focus of this study, since by the turn of the nineteenth century the British and the French had developed distinctly different ways of presenting the subject, and by the 1810s some Americans were already weighing the relative merits of these two algebraic styles.[4] Contrary to Cajori's basic "British, then French" model, the American educators did not universally opt for the French analytic style and French textbooks. Rather, one of the most popular American algebras of the period, Jeremiah Day's *Introduction to Algebra*, followed the British synthetic style. Even Charles Davies' *Elements of Algebra: Translated from the French of M. Bourdon*—the most successful of the so-called French "translations"—was not a faithful rendering of the original textbook. Davies took great liberties with the original, as he sought to adapt it for American students by uniting Bourdon's theory with British "practical methods." Benjamin Peirce's *Elementary Treatise on Algebra*, on the other hand, essentially ignored both styles. In fact, the only of the four

125

textbooks fitting Cajori's model was John Farrar's translation of Lacroix's *Elémens d'algèbre*.[5]

The present paper thus shows that the large, developing American nation supported mathematical heterogeneity—with some mathematical educators squarely aligned with the British, others with the French, and still others in the middle. In so doing, the study suggests that no model so passive as Cajori's is appropriate. For much of the early Republic, American educators pursued the active roles of judging, adapting, and synthesizing the British and the French mathematics, especially the rival synthetic and analytic styles of algebra.

JEREMIAH DAY's *Introduction to Algebra*

The British tradition in American mathematics continued beyond the Revolution, but took forms reflective of the new American-English relationship. Americans now prepared their own editions of British works, including Charles Hutton's *Course of Mathematics*, revised and corrected by the American Robert Adrain. More boldly, Harvard's Samuel Webber copied long sections from the mathematical writings of Hutton and a few other British authors, and published these under his own name in *Mathematics, Compiled from the Best Authors....*[6] Although Adrain's and Webber's works met the immediate needs of early-nineteenth-century American colleges, such compendia soon gave way to topical mathematical textbooks. Of the latter genre, Jeremiah Day's *Introduction to Algebra* kept the British algebraic tradition alive in the United States throughout the first half of the nineteenth century. An original textbook based on essentially British sources, the work was printed sixty-seven times between 1814 and 1850 before its author and Anthony Stanley produced a revised version, which subsequently went through sixteen printings between 1852 and 1869.[7]

Day (1773-1867) was the professor of mathematics and natural philosophy at Yale from 1801 to 1825, and the president of the college from 1817 to 1846. In the early years of his professorship, he seems to have followed the mathematical curriculum and textbooks adopted at Yale around the turn of the nineteenth century: "Freshman, Webber's Mathematics; Sophomores, Webber's Mathematics and Euclid's Elements; Juniors, Enfield's Natural Philosophy and Astronomy, and Vince's Fluxions; Seniors, natural philosophy and astronomy."[8] Eventually, however, he became dissatisfied with Webber's compilation. He considered replacing it with original British textbooks but concluded that none was perfectly suited for the American classroom. He then determined to prepare his own *Course of Mathematics* for American colleges. In the original preface to his *Algebra* (the first and most successful of the four volumes in his *Course*),

Day explained that his review of the British works had revealed two basic types:
  (1) "extended treatises, which enter into a thorough investigation of the particular departments which are the objects of their inquiry" and
  (2) "mere *text-books*, containing only the *outlines* of subjects which are to be explained and enlarged upon, by the professor in his lecture room, or by the private tutor in his chamber."

Algebras of the first type were "too voluminous for the use of the body of [American] students," and those of the second assumed more guidance from professors and tutors than available at American colleges.[9] His country needed, Day thought, textbooks that were accessible to students with minimal mathematical background and were complete in themselves—and thus useful at colleges like Yale, where recent graduates assumed professorships and taught primarily through recitations.[10]

Although Day dismissed British textbooks as inappropriate for American colleges, his explanation of the role of mathematics in the liberal arts curriculum had a distinctly British tone. If mathematics education aimed simply at the transmission of the practical rules of the subject, he observed, then British books of the second type would be sufficient for American students. "But a higher object is proposed...," he wrote,

> It is the *logic* of the mathematics which constitutes their principal value.... The time and attention devoted to them, is for the purpose of forming *sound reasoners*, rather than expert mathematicians. To accomplish this object it is necessary that the principles be clearly explained and demonstrated, and that the several parts be arranged in such a manner, as to show the dependence of one upon another...[11]

This emphasis on mathematics as a tool for training the rational powers betrayed the influence of Timothy Dwight (1752-1817), Day's mentor. The mathematical and intellectual sympathies of Dwight—who put Day in charge of his successful Greenfield Hill Academy when he was called to the presidency of Yale in 1795—were British. As a tutor at Yale in the early 1770s, Dwight had "three times or more... plowed a straight, lone course through the demonstrations in Sir Isaac's monumental *Principia*." As president of Yale he had campaigned vigorously against the ideas of the French Revolution, explicitly contrasting the "abstract declarations" of the French philosophes with the logical arguments of British thinkers.[12] In 1801 moreover he succeeded in forcing Josiah Meigs, a supporter of the French Revolution, out of the professorship of mathematics and natural philosophy at Yale and naming Day as his replacement.[13]

In his *Algebra* Day acknowledged his debt to the British algebraic tradition through specific references to the algebras of Newton, Maclaurin,
Saunderson, Simpson, and Emerson.[14] He not only wrote of mathematics
as a tool for developing the rational powers, but adopted the synthetic
style. Thus the work opened with "Introductory Observations on the
Mathematics in General" in which Day explained the importance of definitions, axioms, and deductive arguments.[15] Next came section 1 on the
"Notation, Definitions, Negative Quantities, Axioms, &c." of algebra, and
section 2 which began the presentation of algebraic rules, with accompa-
nying proofs.

Like many of his British mathematical predecessors, Day of course
knew that algebra was not quite the logical paradigm that geometry was.
"Euclid and others have given to the geometrical part [of mathematics],"
he admitted, "a degree of clearness and precision which would be very
desirable, but is hardly to be expected, in algebra." Specifically, algebra
involved "a mode of thinking so abstract" that a "plain and diffuse manner" of presentation best suited elementary treatments of the subject.[16]

Quite expectedly, the "negative quantity" provided a main example
of an abstraction to which young minds were unaccustomed and which
therefore required "plain and diffuse" explanation. The explanation—
found in articles 49 through 56—followed Day's British sources: it began
with acknowledgment of the problem of the negatives, provided a definition, and then offered illustrations from everyday life. "To one who has
just entered on the study of algebra," the discussion commenced,

> there is generally nothing more perplexing, than the
> use of what are called *negative* quantities. He supposes
> he is about to be introduced to a class of quantities
> which are entirely new; a sort of mathematical *noth-*
> *ings*, of which he can form no distinct conception.

Trying to dispel this view and following Maclaurin, Day defined "a negative quantity" simply as "one which is required to be subtracted" and
gave some of the usual examples—loss vs. gain, debt vs. credit, back-
ward motion vs. forward motion, and descent vs. ascent.[17] As Day had
explained in the preface, the definition alone should have legitimated the
negatives (as definitions legitimated geometrical objects), but in this case
illustrations were necessary to "catch...[the student's] attention, and aid
his conceptions."[18]

Day's subsequent elaboration on the negative numbers showed lin-
gering and typically British preoccupation with their definition. Such
premier mathematicians as Newton and Euler, he noted, had defined the
negatives as quantities less than nothing. "But this is an exceptionable
manner of speaking," he continued. "No quantity can be really less than

nothing. It may be diminished, till it vanishes, and gives place to an *opposite* quantity." He admitted that even his own definition required qualification, since: "The terms positive and negative, as used in the mathematics, are merely *relative*. They imply that there is, either in the nature of the quantities, or in their circumstances, or in the purposes which they are to answer in calculation, some such *opposition* as requires that one should be *subtracted* from the other."[19]

Thus British mathematicians shaped Day's algebraic work. He shared their view of mathematics as a tool for developing logical powers, their synthetic aspirations for algebra, their emphasis on clear definitions, their concern for the problem of the negative numbers, and even their frustration when the reality of algebra failed to match the synthetic ideal. As an American partisan of British algebra, he was not alone. His *Algebra* was well received not only at Yale, where it fit with Dwight's anti-French sentiment, but also at other early-nineteenth-century American colleges, where the British mathematical tradition was already entrenched and where mathematics was seen primarily as an introduction to deductive thinking.[20]

JOHN FARRAR'S TRANSLATIONS OF FOREIGN ALGEBRAS

In his translation of the *Elémens d'algèbre* by S. F. Lacroix, John Farrar (1779-1853) produced an algebra textbook that rivalled Day's in immediate influence if not in enduring popularity.[21] As Day's *Algebra* represented the survival of the British algebraic style in America, Farrar's translation of 1818 symbolized a growing appreciation for French mathematics among some American mathematical practitioners and educators, especially those associated with Harvard College. Key members of the Harvard circle argued that French mathematical superiority mandated American use of French mathematical textbooks and, moreover, that the French style of algebra was more natural and beneficial to students than the rival British style.

Harvard's early mathematical cosmopolitanism was due, first and foremost, to Nathaniel Bowditch (1773-1838)[22] and later, to John Farrar. Arguably the most talented of America's early nineteenth-century mathematical practitioners, Bowditch took the lead in the Boston area in studying the scientific and mathematical works of continental Europe, and thereby found his "principal role in science—that of a very able and learned corrector and extender of the work of others." In this role he produced a memorandum of 1805 to Lacroix containing corrections to the latter's *Traité du calcul différentiel et du calcul intégral* and eventually the translation of Laplace's *Mécanique céleste* from the original French into English.[23]

Although Bowditch served as a member of the Harvard Corporation for a dozen years and helped shape Harvard's mathematical program for nearly a half-century,[24] he never held an academic post at the college. Offered the Hollis Professorship of Mathematics at Harvard in 1807, he declined and thus set the stage for the appointment of Farrar[25]—who subsequently became the dean of the American mathematical translators. Farrar, however, had little specific preparation for the professorship. But at the turn of the nineteenth century the Hollis Professorship was neither mathematically demanding nor especially attractive. As Farrar's biographer explained,

> Webber's Course of Mathematics, beginning with Numeration, and closing with an elementary chapter on Spherical Astronomy, was the [Harvard] College textbook, and ...even the end of that manual was never reached or approached by any class. Pure mathematics was a very unwelcome subject.[26]

On the positive side, Farrar, a graduate of Harvard, brought to his professorship a knack for languages (evidenced by his earlier position as tutor of Greek at the college) as well as enthusiasm for the study of the mathematical sciences. His first objective as Hollis Professor was to raise "his own knowledge up to the European standard of the day."[27] His next objective was to introduce his students to the best of the European mathematics—primarily through translations of and selections from French mathematical textbooks. Harvard's answer to Day's *Course of Mathematics*, then, was Farrar's series of six works: *An Introduction to the Elements of Algebra* (from Euler), *An Elementary Treatise on Arithmetic* (Lacroix), *Elements of Algebra* (Lacroix), *Elements of Geometry* (Legendre), *An Elementary Treatise on Plane and Spherical Trigonometry, and on the Application of Algebra to Geometry* (Lacroix and Bézout), and *First Principles of the Differential and Integral Calculus* (Bézout).[28] Of these books, "the two first... [were] to be studied previous to admission into college,"[29] while the others were intended for use at the college level.

Harvard's adoption of the French mathematical textbooks was certainly related to the growing conviction that by the end of the eighteenth century French mathematicians and scientists had overtaken their British counterparts. George Barrell Emerson, a tutor of mathematics at Harvard under Farrar, made this point explicitly. He described the translation of Lacroix's algebra as

> the first step of a direct introduction to the excellent treatises [in French] on astronomy and the various branches of physics. ...one step, and a very considerable one, towards removing the reproach, to which,

from community of language, we have been obnoxious, together with the English, of being almost a century behind the rest of the world in all that relates to mathematical and physical science.[30]

But introduction of continental mathematics through translation was not an originally American idea. As British mathematicians themselves had taken the lead in elaborating the decline thesis, so they had evidenced some interest in translations of continental books. In fact, Farrar's early work, *An Introduction to the Elements of Algebra*, contained "selections ...from the first English edition" of Euler's *Elements of Algebra*.[31] The latter book, originally published in German, had been translated first into French by Jean Bernoulli and then (from the French) into English by the young Scotsman Francis Horner.[32] Moreover, there was a modest translation program even at Cambridge University in the mid-1810s. There young scholars—proponents of the decline thesis who hoped for scientific and mathematical reform—produced English translations of Book I of Laplace's *Mécanique céleste* (1814) and of Lacroix's *Traité du calcul différentiel et du calcul intégral* (1816).[33]

A well-established mathematical program (complete with its own textbooks) and national pride, among other factors, restrained the translation movement at Cambridge University, England. But, on the other side of the Atlantic, American educators were free to carry it to the limit. And Farrar did, once he had concluded—as stated in the advertisement to the *Elementary Treatise on Arithmetic* (the first of his French translations)—that "the first principles, as well as the most difficult parts of Mathematics, have ...been more fully and clearly explained by the French elementary writers, than by the English...."[34] In short, in Farrar's eyes, French mathematicians excelled not only in research but in pedagogy as well.

Specifically, Farrar and Emerson extolled Lacroix as one of the "very distinguished" elementary French writers,[35] and Lacroix's *Elements of Algebra* as the leading textbook on the subject. They were aware of the book's popularity within France, and, on their own, found its analytic style most appealing. Thus, the advertisement to Farrar's translation stated that "Lacroix's Algebra has been in use in the French schools for a considerable time. It has been approved by the best judges, and been generally preferred to the other elementary treatises, which abound in France."[36] The thesis that the key to the textbook's effectiveness was its analytic style was developed by Emerson in his review of the first four books of Farrar's series, in which he evidenced familiarity with Lacroix's *Essais sur l'enseignement* of 1805 as well as his mathematical textbooks. Quoting from the former work, Emerson showed that Lacroix saw "the study of the [mathematical] sciences... as the means of exercising the mind, of developing the intellectual faculties and rendering them fit for meditation and discussion."[37] Unlike most British thinkers, Lacroix thought that the analytic style could accomplish such goals. Thus, first among "the rules by which Lacroix seem[ed]... to have been guided in composing ... [mathematical] books ... [was] making use of the analytical method, to pursue, as nearly as possible, the steps of invention." This method—inspired by Condillac, used by Clairaut, and sanctioned by Laplace—followed "the very philosophical conception of introducing the artifices and making the parts succeed each other in the same order in which they might be supposed to have occurred to an original inventor."[38] This method also set Lacroix's *Elements* apart from contemporary British algebras. "The [algebra] treatises commonly in use," Emerson wrote, "are not strictly analytical. They are algebra delivered in the synthetical form. Of such we have had enough." Emerson, however, intended no blanket condemnation of synthetic methods, since he quickly described the synthetic form as an appropriate one for textbooks on nearly every other mathematical subject. But algebra was part of analysis, an instrument of discovery; its textbooks, above all, ought to stand in the "course of liberal study" as introductions to "the means of discovering new [truths]."[39]

Moreover, Emerson saw the discovery process as an especially effective way of conveying to students a clear understanding of the nature and content of algebra. Lacroix, he enthused,

> endeavors always to take the reader along with him, never to lay down a rule until it begins to be anticipated, never to give a new process, or bring forward a new principle, until its necessity is felt. We are thus enabled to keep pace with him and be present at his discoveries, to divine his reasons and partake of his power.

According to Emerson, the pedagogical merits of this approach were evident especially in the introductory section of the *Elements* and in its development of the negative quantities. "All the apparent absurdity of the usual statement of the doctrine of plus and minus both in addition and multiplication is entirely avoided," he wrote. Later: "some examples tend... to explain the nature of negative, of infinite, and of indeterminate quantities, when they occur in a result, and throw... much light on the metaphysics of algebra."[40]

Lacroix's introductory section highlighted the differences between algebras written in the synthetic and analytic styles. No core of definitions greeted the student. Rather Lacroix's *Elements* began in an almost leisurely fashion with "Preliminary Remarks upon the Transition from

Arithmetic to Algebra—Explanation and Use of Algebraic Signs." Here Lacroix reconstructed the history of algebraic symbols, using an example to explain their origin in the perceived need for a mathematical short-hand. The example was: "To divide a given number into two such parts, that the first shall exceed the second by a given difference." Arguing strictly in prose, Lacroix reached the conclusion: "The less part is equal to half the number to be divided, diminished by half the given excess."[41] He next turned to a specific case of the problem, where the number to be divided is 9 and the difference, 5. Using the rule he had just developed, he easily found the smaller number to be $9/2 - 5/2 = 4/2 = 2$; the greater number, 7. Only after the prose solution and specific example did Lacroix discuss algebra and its signs. Noting that the paths to solution of the above and similar problems appealed to such expressions as "added to," "diminished by," and "is equal to," he observed: "it is evident, that the[se] expressions might be abridged by representing each of them by a sign." Their naturalness thus established, Lacroix introduced the traditional algebraic symbols for the four basic operations and equality. Continuing to work slowly from his prose problem to algebra, he added: "These abbreviations, although very considerable, are still not sufficient, for we are obliged often to repeat *the number to be divided, the number given, the less part, the number sought*, &c. by which the process is very much retarded." He was finally ready to introduce the unknown! "With respect to the unknown quantities," he explained, "the practice has been to substitute in their stead a conventional sign $[x]$...."[42]

Illustrating the transition from prose to algebraic language and simultaneously nudging his students towards generalization, Lacroix redid his original problems. He took the specific case first; letting x stand for the smaller number, he arrived at: the greater number is $x + 5$; the sum of the two numbers is $x + 5 + x = 9$; thus $2x + 5 = 9$, $2x = 9 - 5$, $2x = 4$, and $x = 2$. He observed:

> The number 2, the result of the preceding operations, will answer only for the particular example which is selected, while the course of reasoning considered by itself, by teaching us, that *the less part is equal to half the number to be divided, minus half the given excess*, renders it evident, that the unknown number is composed of the numbers given, and furnishes a rule by the aid of which we can resolve all the particular cases comprehended in the question.

Thus the above solution suggested a method for solving all similar problems, by showing $2x$ to be the difference of the number to be divided (9) and the given excess (5). But this course of reasoning was lost in the final solution, since: "when we have come to the result, there is nothing to show how the number 2... has been formed from the given numbers 9 and 5." Now, just as the students were supposed to be feeling the need for general solutions, Lacroix introduced the use of the early letters of the alphabet to stand for indeterminate numbers, and solved anew the original problem of the *Elements*. In all its generality, the problem now read: "To divide a given number represented by a into two such parts that the greater shall have with respect to the less a given excess represented by b." Simple algebraic calculations led to: x (the smaller number) = $a/2 - b/2$.[43] The message of the long, careful introduction was thus clear: algebra developed naturally from a desire to abbreviate or shorten, and then generalize, simple reasoning.

Lacroix came to the negatives by a similarly gradual route. Unlike the synthetic approach to algebra that required early definition of the negatives, his analytic approach involved two stages, only the second of which concerned explanation—and even then the explanation took the form of an equation. The signs $+$ and $-$ originally appeared in the *Elements* exclusively as signs of operations: $+$ stood for addition and $-$ for subtraction.[44] Later, in the section on methods for performing operations on "quantities that are represented by letters," however, Lacroix wrote: "we give the name of *simple quantities* to those which consist only of one term, as $+2a$, $-3ab$, &c." This was Stage I of his development of the negatives: in which he introduced quantities bearing the sign $-$ without any discussion and essentially as terms of polynomials (not as isolated terms), avoided the names "positive" and "negative,"[45] and set about developing the rules governing operations on such quantities. For "quantities preceded by the sign $-$," the latter rules were derived indirectly through consideration of such quantities exclusively as terms of polynomials rather than as isolated terms. Thus Lacroix did not directly consider the subtraction of the term $-c$ from a, but instead took the quantity $b - c$ from a. Suppose, he began, we first wrote $a - b$. We would thus be taking "the whole quantity b" from a. "But the subtraction ought to have been performed after having first diminished b by the quantity c; we have taken therefore this last quantity too much, and it is necessary to restore it with the sign $+$, which gives for the true result $a - b + c$." From this he concluded that subtraction of algebraic quantities involved changing their signs. Lacroix approached the rule of signs for multiplication in a similar fashion. Reasonable explanations (in which quantities preceded by the sign - appeared only as terms of polynomials) came first; the rule followed. The key to the rule was the product $(a - b) \times c$. Consider the product ac, he began. Now this product is too large, since "each time that we take the entire quantity a, ... we take the quantity b too much;

the product ac therefore...exceeds the product sought by the quantity b, taken as many times as is denoted by the number c, that is by the product bc." Thus, $(a - b) \times c = ac - bc$. Eventually from this formula and $(a - b)(c - d) = ac - bc - ad + bd$, he generalized to the standard rule of signs for multiplication.[46]

Stage II of Lacroix's development of the negatives—in which isolated negatives were directly confronted—came much later in the textbook, in a separate section on "Questions Having Two Unknown Quantities, and of Negative Quantities." Beginning with the particular, he posed a problem that generated a negative solution:

> A laborer having worked for a person 12 days, and
> having with him, during the first 7 days, his wife and
> son, received 46 francs; he worked afterward with the
> same person 8 days more, during 5 of which, he had
> with him his wife and son, and he received at this time
> 30 francs; how much did he earn per day himself, and
> how much did his wife and son earn?[47]

Letting x stand for the daily wages of the man and y for those of the wife and son, he wrote: $12x + 7y = 46$ and $8x + 5y = 30$, and thus arrived at $x = 5$ and $y = -14/7$. Now he made an issue of isolated negatives, asking: "how are we to interpret the sign $-$, which affects the insulated quantity 14? We understand its import, when there are two quantities separated from each other by the sign $-,\ldots$; but how can we subtract a quantity when it is not connected with another in the member where it is found?" His first tack was to present negative solutions as the results of contradictions in the enunciations of problems. For example, the solution to the specific problem at hand showed that, rather than earning anything, the laborer's wife and son incurred expenses charged to his account. At this point, Lacroix continued, the problem could be rewritten to reconcile the contradictions. The correct statement would begin: "A laborer worked for a person 12 days, having had with him the first 7 days, his wife and son at a certain expense...." "In every case," he generalized, "where we find, for the value of the unknown quantity, a number affected with the sign $-$, we can rectify the enunciation in a manner analogous to the preceding ....."[48]

This lengthy exercise on a laboring family's wages was preliminary to a sub-section on what were finally called "negative quantities." Here the opening paragraph outlined Lacroix's ultimate strategy for the negatives. "Since negative quantities resolve in a certain sense the problems, which give rise to them," he began, "it is proper to inquire a little more particularly into the use of these quantities, and to settle once for all the manner of performing operations in which they are concerned."

Thus, having established that the negatives arose naturally as solutions of problems, Lacroix felt no need to define or otherwise legitimate them; rather he concentrated on "settling" the rules governing their manipulation. His earlier development of related rules, he now suggested, was not completely satisfactory since the rules had "not been demonstrated with reference to insulated quantities." Thus his derivation of the rule for subtraction had hinged on the example of $(b - c)$ taken from a, not $-c$ (an isolated negative) taken from a. "Strictly speaking," he thought,

> the reasoning does not depend upon the value of b; it would still apply when $b = 0$.... But the theory of negative quantities being at the same time one of the most important and most difficult in algebra, it should be established upon a sure basis. To effect this, it is necessary to go back to the origin of negatives.[49]

So Lacroix the analyst turned once again to the history of algebra: he reconstructed the moment at which the negatives first arise; worked through to "the idea...attached to a negative quantity," which he expressed in an equation; and used the equation to derive the rules governing the negatives. Beginning his discussion of the origin of the negatives, he noted that the "greatest subtraction" from a quantity is "to take away the quantity itself." In such cases, the result is zero, which is expressed $a - a = 0$. Sometimes however the subtraction of a greater quantity from a lesser is attempted. In order to indicate that "we cannot subtract... [the greater] entirely," Lacroix explained, we introduce negative quantities. Thus, in the case of $3 - 5$, the mathematician writes the answer as -2, where "the sign $-$, which precedes 2, shows what is necessary to complete the subtraction; so that, if we had added 2 to the first of the quantities, we should have had $3 + 2 - 5$, or zero." Without any further elaboration, Lacroix now stated his version of "the idea that is to be attached to a negative quantity $-a$." This idea, he wrote, could be "express[ed]..., with the help of algebraic signs... by forming the equation $a - a = 0$, or by regarding the symbols $a - a$, $b - b$, &c., as equivalent to zero." The equation $a - a = 0$, however, was a key to the rules governing negative quantities rather than an ontological statement. Using the equation Lacroix derived these rules anew. Given $b - b = 0$, $a = a + b - b$. Since "in order to subtract any one of the... quantities $[+b$ or $-b]$, it is sufficient to efface it," $a - (+b) = a - b$ and $a - (-b) = a + b$. "With respect to multiplication...," Lacroix began, "the product $a - a$ by $+b$ must be $ab - ab$, because the multiplicand being equal to zero, the product must be zero; and the first term being ab, the second must necessarily be $-ab$ to destroy the first. We infer from this, that $-a$, multiplied by $+b$, must give $-ab$." Similar arguments justified the remaining rules for multiplication.[50]

Thus Farrar's translation of Lacroix's *Elémens d'algèbre* brought to America a new style of algebra and algebra education—according to which key methods and concepts (including the negative numbers) were approached slowly through examples and the history of the subject. Americans like Bowditch, Farrar, and Emerson had confidently set themselves up as judges of the rival British and French mathematics. They had deemed the French mathematics superior, and had chosen the analytic style of algebra because of both its connection with French mathematics and its pedagogical promise. The country's other mathematical educators now had a clearcut choice between the British and French styles.

CHARLES DAVIES AND BOURDON'S *Elements of Algebra*

The most popular of the American algebra "translations" was Charles Davies' *Elements of Algebra: Translated from the French of M. Bourdon. Revised and Adapted to the Course of Mathematical Instruction in the United States.* "Davies' Bourdon" went through over forty printings from 1835 to 1901.[51] Widespread use of the work was certainly tied to the overall success of Davies' "Course of Mathematics," a series consisting of thirteen textbooks and three keys that the publisher Alfred Smith Barnes personally promoted in a two-year tour of American towns and villages.[52] Still, Davies' self-proclaimed new approach to the subject lay behind the particular popularity of his algebra. Unlike Day, Davies (1798-1876) was no consistent partisan of the British synthetic tradition; neither was he, like Farrar, primarily a translator of French textbooks. Rather, he revised and adapted Bourdon's *Elémens*, and in the process sought a synthesis of the "practical methods of the English" and the "scientific discussions of the French."[53] Although Davies' originality and mathematical judgment may be questioned, this synthesis struck a responsive chord with his fellow mathematical educators.

Davies was not the first English translator of Bourdon's *Elémens d'algèbre*; in fact, he used an existing translation of the work as the basis of his textbook. Augustus De Morgan, the English mathematician, seems to have made the original English translation of approximately one-fourth of the work. Published in 1828, this translation was intended for the "use of students in and preparing for the University of London," who were to benefit from "the care which... [Bourdon had] taken to deduce every rule from first principles, and to distinguish between the results of convention and those of demonstration."[54] Later, two Americans went beyond De Morgan's translation in independent textbooks, both of which appeared in 1831. Covering "a little more than one half of the original work," the anonymous translation published in Boston—and most likely by Farrar—gave no indication of where in the mathematics curriculum it was to be used—in particular, of whether or not Bourdon's *Elements* was intended as a replacement for Lacroix's.[55] Whatever the source and purpose of this translation, it mattered little, since Edward C. Ross's more extensive translation of the same year became the basis for Davies' algebra textbook.

Ross's and Davies' interest in translations of French textbooks seems to have been stimulated at least indirectly by Farrar. The former were graduates and then instructors of mathematics at the United States Military Academy at West Point. Established in 1802 for the scientific training of army engineers, the Academy emerged as a leader among American technical colleges in the late 1810s under the superintendency of Sylvanus Thayer (1785-1872). Before taking charge of the Academy in 1817, Thayer had spent almost two years in Europe studying military and technical developments as well as the methods of the Ecole Polytechnique.[56] As superintendent, he used the latter institution as a model in his reform of the Military Academy. Mathematical reform, however, took time. Although all cadets studied French,[57] most apparently mastered the language too slowly to use it in their mathematical studies, which stretched from the freshman through senior years. Therefore, Hutton's *Course of Mathematics* and a few other English textbooks formed the basis of the Academy's mathematical curriculum in the early years of Thayer's superintendency. In 1821, while visiting West Point, the Harvard professors Andrews Norton and George Ticknor—the latter nearly a lifelong friend of Thayer[58]—suggested that the Academy adopt Farrar's translations. Norton sent copies of Farrar's works to the Academy, which began using them (including Lacroix's *Elements of Algebra*) in 1823.[59]

An assistant professor of mathematics when Thayer assumed control of the Academy, Davies became professor of mathematics just as the French textbooks were being introduced. Although as a cadet he had cut his mathematical teeth on Hutton's *Course* and later as an assistant professor, taught from the work, "he...at once imbibed the spirit and fully sympathized with the desires of the superintendent, and labored earnestly to carry them out in building up a logical system of instruction and recitation...."[60] Davies' assistant professor of mathematics—Edward C. Ross, a member of the Academy's class of 1821—was at least equally committed to the French reforms. But by the early 1830's Ross had become dissatisfied with Lacroix's *Elements*. As recounted by General Francis H. Smith (class of 1833), Ross "gave to ...[his] class many extra discussions of the difficult points in algebra,...for he was not pleased with Farrar's translation of La Croix, our text-book in algebra, and he was preparing his translation of Bourdon."[61] The latter translation seems to have been published just once under Ross's name.[62]

Within five years of the appearance of Ross's translation, however, Davies had revised and adapted it for publication anew—now under Davies' name with a mention in the preface that the work was "an abridgment of Bourdon, from the translation of Lt. Ross...." Bourdon's *Elements*, Davies contended, was "a work of singular excellence and merit." "One of the leading text books" in France, it had already been adopted (in Ross's translation) by the Military Academy and five American colleges. Still Bourdon's original work and, as Davies implied, even Ross's translation needed to be modified for American students. The work—six hundred seventy pages in the original and nearly four hundred pages in Ross's translation—required abridgment because: "The time which is given to the study of Algebra... is too short to accomplish so voluminous a work." More importantly, its content had to be adapted to American mathematical instruction. Here Davies claimed some originality: in revising and adapting Bourdon's *Elements*, he sought "to unite... the scientific discussions of the French, with the practical methods of the English school; that theory and practice, science and art, may mutually aid and illustrate each other."[63]

A comparison of the development of the negative numbers in Davies' *Elements* with that given in Bourdon's original *Elémens* reveals some of the specific alterations by which Davies tempered the French theory with English practicality. Bourdon's two-stage development of the negatives resembled Lacroix's. Thus Bourdon originally introduced the signs + and − as signs of operation.[64] Then, in Stage I, he decomposed polynomials into two kinds of terms ("those which are preceded by the sign + [and] are called *positive* terms, and those which are preceded by the sign −, [and are called] *negative* terms") and demonstrated the naturalness of the standard rules for addition and multiplication of such terms.[65] True to Lacroix's pattern, he dealt with isolated negatives only late in his book and in a separate section—entitled "Problems Which Give Rise to Negative Results. Theory of Negative Quantities." This section—Stage II—opened with an admission that: "The use of algebraical signs in the resolution of problems often leads to results which at first view are embarrassing; on reflection, however, it will appear, not only that they are capable of explanation, but that by their means the language of algebra may be still further generalized." Then, following the analytic style, Bourdon turned to two problems that produced negative results. The first and easier involved "find[ing] a number which, added to a number b, gives for their sum a number a." He wrote the formula $x = a - b$, then set $a = 24$ and $b = 31$, and arrived at $x = -7$—which he called "a *negative solution*," and of which he asked: "what is the meaning of it?"[66] Working from the two given problems, he reached general conclusions:

(1) "a negative value... is indicative of some incorrectness in the manner of stating the question" and

(2) the "value, independently of its sign, may be regarded as the solution of a problem, which differs from the proposed problem in this, that certain quantities, which were additive in the first, are subtractive in the second, and the reverse."

Thus, in his first example, -7 showed that (1) there is no number that added to 31 gives 24, but (2) "$x = 7$... is the solution of the following problem, *To find a number which, subtracted from 31, gives 24.*" Stating that he had arrived at his general conclusions by analogies, he next offered demonstrations of them.[67]

Although thus quite similar to Lacroix's, Bourdon's development of the negatives included a dimension that set it apart from—and, in the eyes of some, above—earlier algebra textbooks. Put simply, Bourdon used the negatives to stress the role of convention in algebra. For example, as he introduced the names "positive" and "negative" in Stage I (and thus earlier than Lacroix), he quickly explained that these names were "improper" but "sanctioned" by custom.[68] More positively, in the rich "Remarks" that concluded Stage II Bourdon made convention the pillar of the explanation of the negatives. He proposed that

(1) some of the rules of algebra could not be proven rigorously but came by extension from other results;

(2) algebra concerned "subjects ... [that] are often *imaginary;*" and

(3) it was by convention that "negative expressions" were "considered" as quantities.

A review of the demonstrations of his conclusions regarding negative solutions led to these three observations. He noted that, since his demonstrations had involved multiplication of $+a$ by $-x$, division of $-b$ by $+a$, and so on, "the results were obtained by applying to simple quantities the rule of the signs established for the multiplication and division of polynomials." When faced with this charge, earlier authors (Lacroix included) had "demonstrate[d] these rules [anew] with reference to insulated simple quantities." But, Bourdon contended, "the demonstrations which they have given have only the appearance of rigour, and leave much uncertainty in the mind." Here, he argued, convention rather than demonstration should take over:

> We say then, that the rule for the signs, established
> for polynomial quantities, is extended to simple quan-
> tities, in order to interpret the peculiar results to which
> algebraical operations lead. Those who do not admit
> this extension deprive themselves of one of the princi-
> pal advantages of algebraical language, which consists

in comprehending, under one formula, the solutions of
several questions of the same nature....

This extension led to such results as the rule that the product of two negative expressions is positive, a rule that Bourdon used as a springboard
for additional remarks on the nature of algebra. "In arithmetic and geometry," he wrote, "the things reasoned on are real, and such that the
mind can form a distinct conception of them; while in algebra, the subjects of the reasoning and of the operations are often *imaginary*, or contain
symbols of operations which cannot be performed." Negative quantities
fell into the category of imaginary entities—but, as Bourdon (continuing
to stress convention) explained, "in order to interpret the peculiar results
which the algebraical solution of a problem leads to, we have agreed to
consider *negative expressions* as quantities."[69]

Davies' development of the negatives took great liberties with Bourdon's. To be sure, Davies initially presented the signs + and − as signs of
operations, then (in Stage I) sorted polynomial terms into those "preceded
by the sign +, and the others by the sign −," and next demonstrated the
rules of operations on such terms.[70] At this point, however, he departed
from Bourdon's format: he appended material from Bourdon's Stage II to
Stage I. Thus, after determining the rules of addition and subtraction of
polynomial terms, he distinguished between arithmetical and algebraic
sums as well as arithmetical and algebraic differences. In arithmetic, he
observed, addition always "convey[s] the idea of augmentation"; in algebra, however, $-b$ may be added to a with a "result [that] is numerically
less than a." "It follows from this," Davies concluded, "that an algebraic
sum may, in the numerical applications, be reduced to a *negative* number,
or a number affected with the sign −."[71] From the perspective of the analytic style, Davies had clearly jumped the gun: he had introduced isolated
negatives without carefully setting the scene for them and, unlike both
Bourdon and Lacroix, he had even defined them—although minimally as
"numbers affected with the sign −." From the same perspective, Davies'
version of Stage II of Bourdon's textbook was anomalous. His later section
on the "Theory of Negative Quantities" did not begin as Bourdon's—that
is, with admission of the embarrassment caused by the negatives and then
consideration of problems with negative solutions. Rather it opened with
some general remarks on algebraic signs that led quickly to a definition
of the negative numbers. Davies' point here was that, if a particular sign
stood for a particular operation, "that operation should always be performed on every quantity before which the sign is placed." Therefore, he
observed, "if the sign of a quantity is +, we understand that the quantity
is to be added; if it is −, we understand that it is to be subtracted. For
example, if we have −4, we understand that this 4 is to be subtracted

from some other number, or that it is the result of a subtraction in which the number to be subtracted was the greatest." Only afterwards did he turn to Bourdon's problems. Taking $x = a - b$, with $a = 24$ and $b = 31$, he concluded that $x = -7$, referred to -7 as a "negative solution," and even asked: "How is it to be interpreted?" But, given the book's earlier remarks (including those on algebraic addition), the question had lost its punch. The answer was immediately forthcoming: "We have found the value of x to be -7, and this number added, in the algebraic sense, to 31, gives 24 for the algebraic sum, and therefore satisfies both the equation and the enunciation." Bourdon's second problem, which required more discussion, gave Davies the opportunity to talk about rewriting problems to assure positive solutions—the end of his Stage II.[72]

As Davies' development of the negative numbers evidences, his "Bourdon" contained elements of both the British synthetic and the French analytic styles, but these elements fit uneasily together. On the analytic side, he avoided defining the negatives in his opening pages, began an approach to these numbers through polynomials, and, in a later section, worked through two problems producing negative solutions, so that he could ask: "How are they to be interpreted?" But, with a concern for practical algebra befitting a professor at West Point, he seemed unwilling to defer results involving isolated negatives too long. Thus, he introduced these numbers in Stage I (even minimally defining them) and opened Stage II with a full-blown definition of the numbers—one similar to that used by such mathematicians as Maclaurin and Hutton. Viewed traditionally, then, "Davies' Bourdon" was a hodgepodge of the synthetic and analytic styles. The union lacked the elegance of the synthetic textbooks that defined negative numbers at the very beginning as basic algebraic entities. But, it also lacked the pedagogical effectiveness of the analytic textbooks that worked slowly from polynomial terms to isolated negatives through examples. Furthermore, it missed entirely Bourdon's metaphysical sophistication, which De Morgan had applauded. Having defined negative numbers, Davies had no need of Bourdon's remarks of Stage II on the roles of convention and extension in algebra. Here American practicality, aided by succinct definition, had indeed replaced French "scientific discussion." In a nutshell, reared on Hutton's practical British mathematics but pushed towards the French by Thayer, Davies was satisfied with neither, struck out on his own, and produced a perhaps unhappy, but popular, union of the two.[73]

BENJAMIN PEIRCE'S *Elementary Treatise on Algebra*

America's first research mathematician, Benjamin Peirce (1809-1880), took yet a different stand on the traditional analytic vs. synthetic debate in algebra. In his *Elementary Treatise on Algebra* of 1837, he essentially ignored both styles.[74]

Peirce was a protégé of Bowditch and Farrar. Appointed as a tutor in mathematics at Harvard in 1831, he became Hollis Professor of Mathematics the next year and soon thereafter began preparation of his own series of mathematical textbooks, including the *Elementary Treatise on Algebra*. In general, these books "were so full of novelties that they never became widely popular."[75] In particular, his algebra was unlike any earlier textbook on the subject. In it, he avoided the lengthy analytic development of algebra to which he had been exposed as a Harvard undergraduate. At the same time, he stressed neither the deductive approach nor the training of men's minds, and his treatment of the negative numbers was terse and matter-of-fact. He omitted any reference to the "problem of the negatives" and merely introduced them in the opening section as terms of a polynomial: "The terms of a polynomial which are preceded by the sign + are called the *positive* terms, and those which are preceded by the sign − are called the *negative* terms." Any strong resemblance between Peirce's approach to the negatives and that of Lacroix and Bourdon, however, ended here. Peirce moved immediately to a "rule for reducing polynomials, which contain similar terms," according to which similar positive terms were to be summed by adding their coefficients, the same done for similar negative terms, and then "the difference of these sums preceded by the sign of the greater...substituted as a single term for the terms from which it is obtained." Without further ado, he gave exercises asking students to find the sums of $11x$ and $-9x$, $-11x$ and $9x$, and so on.[76] Similarly Peirce's section on the law of signs for multiplication ignored the careful distinction between negative terms of polynomials and isolated negatives that Lacroix and Bourdon had used. He proved $(A - C)(B - D) = AB - BC - AD + CD$, and immediately asked students to multiply $-a$ by b and $-a$ by $-b$.[77]

Like Lacroix and Bourdon, he included a separate section on "cases of negative value of unknown quantity." This, however, was a brief section and in no way a prelude to a general discussion of the negatives. The whole section consisted of one corollary—dealing with the rewriting of problems that produced negative solutions—and a few exercises. In the exercises students were referred to earlier problems and of each asked: "In what case would the value of the unknown quantity...be negative? why should it be so? and could the enunciation be corrected for this case?"[78]

In short, even more so than Davies' algebra textbook, Peirce's *Elementary Treatise on Algebra* represented a break with the British and French algebraic styles. At the very least, his handling of the negatives was novel. He tried neither to capture their essence through definition and analogy (as the British textbooks had) nor to make the negatives appear natural through a lengthy analytic approach (as the French had tried). Rather, he avoided theoretical issues concerning the meaning of "negative quantities" and the like; instead he matter-of-factly introduced the negatives as terms of polynomials with the sign − and then proceeded to manipulate isolated negatives. And, unlike his European predecessors, he left the matter at that—and got on with the mathematics.[79]

How are we to account for Peirce's new approach to the genre of the algebra textbook? First, although many American mathematical educators took as a major objective the development of the mental (rational or inventive) powers of the bulk of their students, Peirce did not see mathematics primarily as a tool for improving men's minds, but rather believed that his mission as a professor of mathematics was to infect a few talented young men with enthusiasm for the subject.[80] But there were more personal reasons for his eccentric textbook. Thus it is likely that his opinion of the traditional analytic style was lower than Farrar's, since even as a student the talented Peirce had argued for mathematical economy.[81] At the same time, the "conventional forms of demonstration" were not for him. According to his student and then colleague Thomas Hill, Peirce had a "habit of using simple conceptions, axioms, and forms of expression, without reference to established usage" to produce "demonstrations... [of] exceeding brevity."[82] Perhaps also he did not expect students to worry about such concepts as the negatives, since no mathematical concept seemed to pose a metaphysical problem for him. Indeed, over the years Peirce evolved a theological argument that justified the pursuit of pure mathematics in whatever direction and with whatever means the mathematician's mind dictated, be the means novel, seemingly useless, or even questionable. This argument—what man thought, God thought, and so it was reality—was reiterated again and again in Peirce's writings. It appeared as early as 1854 in his address as retiring president of the American Association for the Advancement of Science. Using conic sections as an example of pure mathematics that waited many centuries for its most fruitful applications in astronomy, he came to his main thesis:

> And so it has ever been; in the ambitious flights to
> which geometry has been impelled by its impatience
> of the restraints of observation, it has never soared
> above the Almighty presence. And so it must ever be;
> the true thought of the created mind must have had

its origin from the Creator; but with him thought is reality. It must then be that the loftiest conceptions of transcendental mathematics have been outwardly formed, in their complete expression and manifestation, in some region or other of the physical world; and that there must always be... these striking coincidences of human thought and nature's law.[83]

Peirce's eccentric algebra textbook acquired a reputation of appropriateness for "those of decided taste for mathematics, or those of mature mind who wish to pursue that science."[84] But, in general, even at Harvard where Peirce's textbooks enjoyed their most consistent use, they "were found very difficult" and their assignment remained controversial.[85]

CONCLUSION

Rather than settling the problem of algebra instruction for Americans, the French algebra textbooks brought the new nation into the longstanding European debate on the relative merits of analysis vs. synthesis in mathematics—more specifically, analytic vs. synthetic algebra. Day and Farrar belonged to the first generation of Americans exposed to the rival algebraic styles; Davies and Peirce, to the second generation. Textbook authors of the first generation understood, evaluated, and worked within the traditional synthetic and analytic styles. Some of the more independent authors of the second generation, however, twisted or ingored these styles. In Davies' case, practical concerns led to the gutting of the basic analytic style and theory of the French textbook he was supposedly translating. In Peirce's, genius spawned an eccentric textbook and one ahead of its time, foreshadowing modern textbooks in which no questions of mental training, the meaning of negative numbers, and the like would be raised. Clearly, then, this second generation included educators who were simple partisans of neither the British nor the French mathematics. No longer content to judge, translate, and select from European mathematics, these independent Americans were trying to establish an American style—if not in mathematics, at least in mathematical education.

NOTES*

1. On algebra in colonial America, see Lao Genevra Simons, *Introduction of Algebra into American Schools in the Eighteenth Century* (Washington: Government Printing Office, 1924), pp. 51–56, 72–73. Textbooks used in early American colleges included Nathaniel Hammond's *Elements*

---

* An earlier version of this paper was given at the annual meeting of the History of Science Society, October 1987, Raleigh-Durham, North Carolina.

of Algebra (London, 1742) and Thomas Simpson's *Treatise of Algebra* (London, 1745). Furthermore, among British books imported for sale were the algebras of Newton, Colin Maclaurin, and Nicholas Saunderson.

2. For a discussion of the British "decline thesis," see J. M. Dubbey, *The Mathematical Work of Charles Babbage* (Cambridge: Cambridge University Press, 1978), pp. 10–30.

3. Florian Cajori, *The Teaching and History of Mathematics in the United States* (Washington: Government Printing Office, 1890), p. 99. Without giving details, Cajori cautions that "we must guard against the impression that French authors and methods entirely displaced the English. English books continued to be used in some of our schools" (p. 100). For repetition of Cajori's basic claim, see Robert V. Bruce, *The Launching of Modern American Science, 1846–1876* (New York: Alfred A. Knopf, 1987), p. 85.

4. The two styles were the "synthetic" and "analytic." The paper uses the term "synthetic style" to characterize the loosely deductive manner in which British mathematicians tried to present algebra. Using geometry as a model, they sought clear definitions and self-evident axioms, from which the rules of algebra were supposed to follow. This ideal of presenting algebra in a deductive fashion lay behind the British difficulty with the negative numbers, which defied clear definition. The style inspired, for example, Maclaurin's *Treatise of Algebra* (1748) and Francis Maseres' *Dissertation on the Use of the Negative Sign in Algebra* (1758). The "analytic style" refers to a slow, natural development of algebra that loosely follows its history. Algebras of this style—such as Lacroix's *Elémens d'algèbre* of 1799—used examples to convince students of the need for and reasonableness of algebraic methods and concepts. Thus analytic algebras built slowly to explanations of the negative numbers rather than beginning with their definition.

5. Unlike Cajori, the present author explains the motives behind these textbooks and compares and contrasts their handling of a specific topic—the negative numbers. Still the present study is by no means complete since it includes but four of many algebras of the period. These four were chosen for various reasons: Day's algebra was the first algebra by an American and an extremely popular textbook; Farrar's and Davies' algebras are the standard examples given to support Cajori's thesis; and Peirce's algebra was the product of America's first research mathematician.

6. For details on these works and others of the same category, see Simons, *Introduction*, pp. 75–76. The American edition of Hutton's *Course* appeared in 1812; Webber's compendium, in 1801.

7. The various editions are listed in Louis C. Karpinski, *Bibliography of Mathematical Works Printed in America through 1850* (Ann Arbor: University of Michigan Press, 1940), pp. 199–202. All references in this paper are to the new edition: Jeremiah Day, *An Introduction to Algebra, Being the First Part of a Course of Mathematics, Adapted to the Method of Instruction in the American Colleges* (New Haven: Durrie and Peck, 1857). This edition includes the original preface. (A last-minute opportunity to study the thirty-second edition of 1838 suggests that there are no essential differences—on the issues discussed in this paper—between the editions of Day's *Algebra* published through and after 1850. Interestingly, the edition of 1838 cites Maseres' *Dissertation*.)

8. Cajori, *Teaching and History*, p. 63.

9. Day, *Algebra*, preface, p. iii

10. Arithmetic was the sole mathematical entrance requirement at Yale from 1745 through 1845. Once at Yale, the men studied mathematics through reading and recitation. In the late 1810s, for example, students were given daily assignments of a few pages of a textbook to memorize for presentation in the classroom. (For these and other details on mathematics at Yale, see Brooks Mather Kelley, *Yale: A History* [New Haven and London: Yale University Press, 1974], pp. 70, 157–160, 167.)

11. Day, *Algebra*, preface, pp. iii–iv.

12. Charles E. Cuningham, *Timothy Dwight, 1752-1817: A Biography* (New York: Macmillan, 1942), pp. 42, 297.

13. Kelley, *Yale*, pp. 130–131.

14. Day, *Algebra*, preface, p. vi. Here Day mentioned also the algebras of Euler and Lacroix. In a later work he referred to John Wallis's *Treatise of Algebra* (Day, *Treatise of Plane Trigonometry* [New Haven: Durrie and Peck, 1838], p. 14).

15. Ibid., pp. 1–8, 397-399 (note). The note to this chapter—on whether quantity included number or not—showed that Day was familiar with the mathematico-philosophical reflections of Isaac Barrow, John Locke, Thomas Reid, and Dugald Stewart.

16. Ibid., preface, p. v.

17. Ibid., pp. 18–21. For comparison, see Colin Maclaurin, *A Treatise of Algebra*, 2nd ed. (London, 1756), p. 4.

18. Ibid., preface, p. vi.

19. Ibid, pp. 19–21. Day's remarks on Newton's definition resemble those of Nicholas Saunderson, *The Elements of Algebra in Ten Books*, vol. 1 (Cambridge: University Press, 1741), pp. 50–51.

20. A lengthy, favorable review of Day's *Course* appeared in 1817; its author was well read in British mathematics and believed that college mathematics was primarily intended "to discipline the intellectual faculties" (Art. I, *Analectic Magazine* 9 [1817], 441–467, p. 441). According to Cajori, "Day's mathematics were....introduced in nearly all our colleges;" Dartmouth used his *Algebra* through 1839, and Yale through at least 1848. Moreover, Karpinski offered evidence of the use of the work through the late 1830s in such states as New York, Connecticut, and Pennsylvania. See Cajori, *Teaching and History*, pp. 64, 158, and Karpinski, *Bibliography*, pp. 597–600.

21. *Elements of Algebra, by S. F. Lacroix. Translated from the French for the Use of the Students of the University at Cambridge, New England*, trans. John Farrar (Cambridge: Hilliard and Metcalf, 1818). The book went through five editions, from 1818 to 1837. In this paper references are to the third edition (Boston: Hilliard, Gray, Little, and Wilkins, 1831).

22. A brief biography of Bowditch with some related letters is found in Nathan Reingold, *Science in Nineteenth Century America, a Documentary History* (New York: Hill and Wang, 1964), pp. 11–28. Still useful is Nathaniel Ingersoll Bowditch's "Memoir" of his father in Marquis de LaPlace, *Mécanique céleste*, translated, with commentary, by Nathaniel Bowditch, 4 vols. (Boston: Hilliard, Gray, Little, and Wilkins, 1829-1839), 4: 9–168.

23. Reingold, *Science in Nineteenth Century America*, p.12. Prior to Bowditch, the Englishman John Toplis and the Irishman Henry Harte had translated sections of Laplace's *Mécanique céleste*. See I. Grattan-Guinness, "Before Bowditch: Henry Harte's Translation of Books 1 and 2 of Laplace's 'Mécanique Céleste,'" *NTM* 24 (1987), 53–55.

24. As late as 1847 Louis Agassiz, in describing his new American colleagues, wrote that "the mathematicians have also their *culte*, dating back to Bowditch...." (Agassiz to Milne Edwards, 31 May 1847, in *Louis Agassiz: His Life and Correspondence*, ed. by Elizabeth Cary Agassiz [Boston and New York: Houghton Mifflin, 1885], p. 438).

25. J. G. Palfrey, "Professor Farrar," *Christian Examiner* 55 (1853), 121–136, p. 126. Farrar was the corporation's third choice for the professorship, after Bowditch and Joseph McKean.

26. Ibid.

27. Ibid., p. 127.

28. For some details on these works as well as an early discussion of "The Influence of French Mathematicians at the End of the Eighteenth Century," see Lao Genevra Simons, *"Fabre and Mathematics" and Other Essays*, Scripta Mathematica Library, no. 4 (New York: Scripta Mathematica, 1939), pp. 53–54, 68–69.

29. [George Barrell Emerson], Article XVII (review of the first four books of Farrar's series), *North American Review* 13 (1821), 363–380, p. 363.

30. Ibid., p. 374. Emerson had firsthand knowledge of Farrar's translations. He was a tutor under and close friend of Farrar, and completed Farrar's translation of Bézout's calculus. See George B. Emerson, *Reminiscences of an Old Teacher* (Boston: Alfred Mudge and Son, 1878), pp. 33–34.

31. *An Introduction to the Elements of Algebra, Designed for the Use of Those Who Are Acquainted Only with the First Principles of Arithmetic. Selected from the Algebra of Euler*, 2nd ed. (Cambridge, New England: Hilliard and Metcalf, 1821), p. vii. Although this work was published anonymously, Farrar was identified as the author of the "remarks prefixed to the volume" in "Intelligence and Remarks: *Introduction to the Elements of Algebra*," *North American Review* 6 (1818), 412–414, p. 412.

32. Horner, who had attended the University of Edinburgh from 1792 to 1795, made the translation (which was published in 1797) while studying under John Hewlett in Shackleford, Surrey. See the advertisement to *Elements of Algebra, by Leonard Euler...*, ed. by John Hewlett, 3rd ed. (London: Longman, Hurst, Rees, Orme, 1822), p. iv.

33. See Grattan-Guinness, "Before Bowditch," and Dubbey, *Mathematical Work*, pp. 33–36.

34. *An Elementary Treatise on Arithmetic, Taken Principally from the Arithmetic of S. F. Lacroix and Translated from the French with Such Alterations and Additions as Were Found Necessary in Order to Adapt It to the Use of American Students*, trans. John Farrar, 3rd ed. (Cambridge, New England: Hilliard and Metcalf, 1825), n.p.

35. Ibid.

36. *Elements of Algebra, by S. F. Lacroix. Translated from the French for the Use of the Students of the University of Cambridge, New England ,* trans. John Farrar(Cambridge: Hilliard and Metcalf, 1818), n.p. Hereinafter all references to this work will be to the fourthth edition (Boston: Hilliard, Gray, and Co., 1833), which was based on the eleventh edition (1815) of Lacroix's *Elémens.* (On significant differences between the various editions fo Lacroix's work, see Gert Schubring, "Ruptures dans le statut mathématique des nombres négatifs," *Petit x,* no. 12 [1986], 5–32, pp. 14–16.) The fairness of Farrar's remarks is supported by Jean Dhombres' recent work on French textbooks. Lacroix's *Elémens d'algèbre* was included among textbooks imposed in the French lycées of the early nineteenth century. Moreover, by 1812 the work (originally published in 1799) reached its tenth edition, with "the usual 3,000 copies" being printed in 1812. See Jean Dhombres, "French Mathematical Textbooks from Bézout to Cauchy," *Historia Scientiarum* 28 (1985), 91–137, pp. 101, 103–104, 127.

37. [Emerson], Article XVII, p. 366. (The translation is apparently Emerson's.)

38. Ibid., pp. 366–367, 373. This method is known in today's pedagogical literature as the genetic method.

39. Ibid., pp. 373–374. At this point of the review, Emerson repeated John Playfair's criticism of the synthetic style.

40. Ibid., pp. 370–371.

41. *Elements of Algebra, by S. F. Lacroix,* pp. 1–2. The rhetorical proof began as follows: "The greater part is equal to the less added to the given excess." And: "The greater added to the less forms the number to be divided." Thus, "The less part added to the given excess, added moreover to the less part, forms the number to be divided." Abridging the language, "Twice the less part, added to the given excess, forms the number to be divided."

42. Ibid., pp. 2–3.

43. Ibid., pp. 3–4.

44. Ibid., p. 2.

45. At this point, Lacroix referred to quantities such as $+a$ and $-a$ by a variety of terms, but never as "positive" and "negative" quantities. He called them "simple quantities" (ibid., p. 24), distinguished between "quantities having the sign $+$" and "quantities having the sign $-$" (p. 25), and wrote of "terms...being...subtractive" (p. 25).

46. Ibid., pp. 27, 32–34.

47. Ibid., pp. 65, 67. This problem is taken from the former page with the modifications introduced by Lacroix on the latter page.

48. Ibid., pp. 67–71.

49. Ibid., p. 72.

50. Ibid., pp. 72–73.

51. The textbook was first published in 1835 by Wiley and Long. Starting in 1838, however, the publisher was A. S. Barnes and Company. With the 1853 edition, the title of the work changed to: *Elements of Algebra: On the Basis of M. Bourdon; Embracing Sturm's and Horner's Theorems, and Practical Examples.* All citations in this paper are to the revised edition of 1838 (Hartford, CT: A. S. Barnes and Co.).

52. John Barnes Pratt, *A Century of Book Publishing, 1838-1938: Historical and Personal* (New York: A. S. Barnes and Company, 1938), pp. 3–4. Davies' textbooks covered mathematics from the *First Lessons in Arithmetic* through algebra, geometry, drawing, trigonometry, surveying, linear perspective, and the calculus.

53. Davies, *Elements of Algebra*, preface, p. iv.

54. *The Elements of Algebra; Translated from the First Three Chapters of the Algebra of M. Bourdon, and Designed for the Use of Students in and Preparing for the University of London,* trans. Augustus De Morgan, (London: John Taylor, 1828), advertisement.

55. *Elements of Algebra, by Bourdon, Translated from the French for the Use of Colleges and Schools* (Boston: Hilliard, Gray, Little, and Wilkins, 1831), advertisement. The anonymous translator–whom Simons identified as Farrar (Simons, "Influence of French Mathematicians," pp. 61–62)– acknowledged "considerable use... of a translation by Augustus De Morgan."

56. Thayer's European tour is covered in Sidney Forman, *West Point: A History of the United States Military Academy* (New York: Columbia University Press, 1950), pp. 41-42. Thayer instigated the tour because he believed that "only by study of European, particularly Napoleonic, military methods and European scientific treatises could the American officer be adequately trained" (Forman, pp. 41–42). For additional details on Thayer's study of the Ecole Polytechnique, see R. Ernest Dupuy, *Where They Have Trod: The West Point Tradition in American Life* (New York: Frederick A. Stokes, 1940), pp. 92–93.

57. Instruction in French began at the Military Academy in 1804 following the creation of a "teachership" of French. An act of Congress of 1812, written at the height of anti-British sentiment, continued to provide for this teachership in the newly organized Academy. For an outline of the history of the Academy, see Edward D. Mansfield, "The Military Academy at West Point," *American Journal of Education* 30 (1863), 17–46.

58. Thayer and Ticknor met as students at Dartmouth College, before Thayer left to study at the Military Academy (Dupuy, *Where They Have Trod*, pp. 18–19). As young men, both spent considerable time in Europe and later both played influential roles in bringing the ideas of continental Europe to America.

59. For a brief history of mathematics at the Academy from the late 1810s through 1845, see Peter Michael Molloy, "Technical Education and the Young Republic: West Point as America's Ecole Polytechnique, 1802-1833" (Ph.D. diss., Brown University, 1975), pp. 442–443.

60. Albert E. Church, as quoted in Cajori, *Teaching and History of Mathematics*, pp. 118–119. A member of the Academy's class of 1828, Church later served as a professor of mathematics at his alma mater.

61. Quoted in Cajori, *Teaching and History*, p. 122. Smith later served as a professor of mathematics at the Virginia Military Institute.

62. *Elements of Algebra, Translated from the French of M. Bourdon, for the Use of the Cadets of the U. S. Military Academy*, trans. Lieutenant Edward C. Ross (New York: E. B. Clayton, 1831). Ross's translation—which did not mention De Morgan's—omitted chapters 4, 9, and 10 of the original, except for figurate numbers and related series, as well as parts of chapters 7 and 8. Ross later taught at Kenyon College and City College of New York (Dupuy, *Where They Have Trod*, p. 370).

63. Davies, *Elements of Algebra*, preface, pp. iii–iv. Davies here probably thought of Hutton's *Course* as an example of the practical British mathematics.

64. *Elements of Algebra, by Bourdon*, p. 1 (1–2). For the comparison, I am using Farrar's English translation of Bourdon's *Elémens*. I have however checked the relevant passages from the latter against M. Bourdon, *Elémens d'algèbre*, huitième édition (Bruxelles: Alexandre de Mat, 1836). The reference to p. 1 (1–2) here is to page 1 of Farrar's work and pages 1–2 of the eighth French edition.

65. Ibid., pp. 9, 11–17 (10, 12–19). The French edition calls terms preceded by the sign + "termes additifs" as well as "termes positifs," and those

preceded by the sign −, "termes soustractifs" as well as "termes négatifs."

66. Ibid., p. 71 (79). The French edition actually asks: "Mais comment l'interpréter?"

67. Ibid., pp. 71–74 (79–83).

68. Ibid., p. 9 (10). The French description is "dénominations assez impropres que l'usage seul a consacrées."

69. Ibid., pp. 75–78 (83–87).

70. Davies, *Elements of Algebra*, pp. 1, 15, 24–30. Davies called the terms with the sign +, "*additive terms*, the others, *subtractive terms.*"

71. Ibid., p. 26. These remarks follow very closely those made by Bourdon in Stage II (*Elements of Algebra, by Bourdon*, p. 77 [86]).

72. Ibid., pp. 94–97.

73. Davies' translation was used at various times by students at Dartmouth, Cornell, and the Virginia Military Institute, as well as the universities of Michigan, Mississippi, South Carolina, Tennessee, and Transylvania (Simons, "Influence of French Mathematicians," pp. 64–65).

74. Benjamin Peirce, *An Elementary Treatise on Algebra: to Which Are Added Exponential Equations and Logarithms* (Boston: James Munroe and Company, 1837). Peirce however mentioned the two styles in his preface: "The form ... adopted in English works of instruction, of dividing the subject as much as possible into separate propositions, is probably the best adapted to the character of the English pupil. This form has, therefore, been adopted in the present treatise, while the investigation of each proposition has been conducted according to the French system of analysis" (p. iii). This preface, which minimized the originality of Peirce's work, was omitted from later editions, which appeared through 1875. All further references in this paper are to the second edition of 1842.

75. For a list of Peirce's textbooks and this evaluation by Thomas Hill, see Cajori, *Teaching and History*, p. 134.

76. Peirce, *Elementary Treatise*, pp. 1–7, especially pp. 6–7.

77. Ibid., pp. 11–12.

78. Ibid., pp. 81–85.

79. For a discussion of other eccentricities of Peirce's algebra, see [Thomas Hill], review of *Physical and Celestial Mechanics*, by Benjamin Peirce, *North American Review* 87 (1858), 1–21, pp. 3, 9.

80. Benjamin Peirce, "The Intellectual Organization of Harvard University," *Harvard Register*, 1 (1880), p. 77. See also Benjamin Peirce to Daniel C. Gilman, 4 October 1875, Gilman Collection #1, Johns Hopkins University Library.

81. Cajori, *Teaching and History*, p. 134.

82. [Hill], review, p. 9.

83. Benjamin Peirce, "Address of Professor Benjamin Peirce," *Proceedings of the American Association for the Advancement of Science* 8 (1854), 1–17, pp. 13–14. For a more detailed discussion of this belief, see Helena M. Pycior, "Benjamin Peirce's *Linear Associative Algebra*," *Isis* 70 (1979), 537–551, pp. 549–551. On Peirce's lack of concern for foundations and rigor, see J. L. Coolidge, "Three Hundred Years of Mathematics at Harvard," *American Mathematical Monthly* 50 (1943), 347–356, pp. 351–352.

84. Review of *Elements of Algebra, Being an Abridgment of Day's Algebra*, by James B. Thompson, *Christian Examiner* 35 (1843), 398.

85. Cajori, *Teaching and History*, pp. 136, 140–141.

Paul Gordan (1837–1912)
Courtesy of Springer-Verlag Archives

James Joseph Sylvester (1814–1897)
Courtesy of Springer-Verlag Archives

Toward a History of Nineteenth-Century Invariant Theory

Karen Hunger Parshall

In his influential treatise of 1939 entitled *The Classical Groups: Their Invariants and Representations*, Hermann Weyl outlined the reconstruction of the development of nineteenth-century invariant theory now most commonly accepted within the mathematical community.[1] According to Weyl, the systematic study of invariants began in the British Isles in mid-century and was developed there in the hands of mathematicians such as Arthur Cayley, James Joseph Sylvester, and George Salmon. Then, attracted by this work, other researchers like Siegfried Aronhold, Alfred Clebsch, and Paul Gordan in Germany, Francesco Brioschi and Luigi Cremona in Italy, and Charles Hermite in France enlisted in the invariant-theoretic cause. Looking back on the nineteenth century from his 1939 vantage point, Weyl categorized the work of all of these contributors as formal in character, that is, he understood their efforts as directed toward "...the development of formal processes and the actual computation of invariants..."[2] As he saw it, these emphases abruptly changed in 1890-1892 with the work of David Hilbert. By shifting his attention away from the explicit calculation of invariants and toward the development of "...suitable general notions and their general properties along such abstract lines as have lately come into fashion all over the whole field of algebra", Hilbert, in Weyl's view, "...solve[d] the main problems [of invariant theory] and almost kill[ed] the whole subject."[3]

Almost, but not quite, for in his book Weyl resurrected invariant theory in the guise of the representation theory of semisimple groups. Unfortunately, though, Weyl's reincarnation of the theory of invariants failed to result in its revitalization. While recasting old ideas in newer terms may make those ideas appear more natural in the modern context and may illustrate the far-reaching and all-encompassing features of a new theory, it does not necessarily reveal the sort of new, outstanding problems necessary to rekindle an old area.

This was precisely the shortcoming of Weyl's book. Still, invariant theory refused to die. In his book *Geometric Invariant Theory* of 1965, David Mumford translated invariant-theoretic ideas into yet another language,

THE HISTORY OF
MODERN MATHEMATICS

that of algebraic geometry, and did open up a new avenue of research into an area which had supposedly died in the 1890's.[4]

Even more recently, invariant theory infused with combinatorics has led to new results and the promise of more to come.[5] In their paper entitled "The Invariant Theory of Binary Forms", Joseph P.S. Kung and Gian-Carlo Rota took the formal, constructivist approach and the symbolic notation of the German school of nineteenth-century invariant theorists and made it rigorous. As they explained in the introduction to their study, "... what is perhaps the main novelty of the present work is a rigorous and yet manageable account of the umbral or symbolic calculus, what Hermann Weyl called 'the great warhorse of nineteenth[-]century invariant theory'."[6] Then, using their updated version of this notation, they reproved, recast, and extended many of the classical results while pointing the way to further research.

Thus, Jean Dieudonné's now oft-quoted remark that "[i]nvariant theory has already been pronounced dead several times, and like the phoenix it has been again and again rising from its ashes", remains as true now as it was when it was written in 1968.[7] The symbolic notation, which Weyl disdained in 1939 in favor of the newer axiomatic language and approach of abstract algebra, has arisen again in work like that of Kung and Rota. In so definitively closing the historical book on the work of the pre-Hilbertian nineteenth century, Weyl would seem to have underestimated the historical importance of those contributions. In this paper, then, I return to the middle of the nineteenth century to examine more closely the invariant-theoretic researches being conducted on both sides of the English Channel.

The historical picture which will emerge will be one more complicated than that suggested by Weyl. An examination of the work principally of Cayley and Sylvester on the British side and that of Aronhold, Clebsch, and Gordan on the Continental side, will reveal not one but two distinct, formal approaches to the subject. Furthermore, the correspondence between Cayley and Sylvester relative to the latter's repeated attempts to prove Gordan's finiteness theorem of 1868 will underscore just how profound the differences between those two approaches were.

Invariant Theory and the British Approach

In his Vorlesungen über die Entwicklung der Mathematik im 19. Jahrhundert, Felix Klein traced the beginnings of invariant theory back to that ultimate source, Carl Friedrich Gauss.[8] In his Disquisitiones arithmeticae of 1801, Gauss had investigated, among many other things, the question:

how is a binary quadratic form with integer coefficients affected by a linear transformation of its variables? More particularly, he considered the form

(1) $$ax^2 + 2bxy + cy^2, \qquad a, b, c \in \mathbf{Z}$$

and the linear transformation of its variables

(2) $$\begin{aligned} x &= \alpha x' + \beta y' \\ y &= \gamma x' + \delta y', \end{aligned} \qquad \alpha, \beta, \gamma, \delta \in \mathbf{Z}$$

Substituting (2) into (1), he derived a new binary form

(3) $$a'(x')^2 + 2b'x'y' + c'(y')^2$$

and explicitly exhibited that

$$\begin{aligned} a' &= a\alpha^2 + 2b\alpha\gamma + c\gamma^2 \\ b' &= a\alpha\beta + b(\alpha\delta + \beta\gamma) + c\gamma\delta \\ c' &= a\beta^2 + 2b\beta\delta + c\delta^2. \end{aligned}$$

Relative to these last three equations, Gauss then remarked that by "[m]ultiplying the second equation by itself, the first by the third, and subtracting we get

(4) $$b'b' - a'c' = (b^2 - ac)(\alpha\delta - \beta\gamma)^2.\text{"}[9]$$

He further observed that the discriminant $(b')^2 - a'c'$ of (3) is divisible by the discriminant $b^2 - ac$ of (1) and that their quotient is square. To use the term James Joseph Sylvester coined in 1851, Gauss had discovered that the discriminant $b^2 - ac$ is an "invariant" of the binary form (1).[10]

Even before Gauss considered binary forms in his *Disquisitiones Arithmeticae*, however, Joseph-Louis Lagrange had encountered and dealt with the problem of the transformation of homogeneous polynomials by linear substitutions of the variables in his two-volume *Mécanique analytique* of 1788. There, in his discussion of the rotational motion of rigid bodies, for example, Lagrange had examined the linear transformation

(5) $$\begin{aligned} p &= p'x + p''y + p'''z \\ q &= q'x + q''y + q'''z \\ r &= r'x + r''y + r'''z \end{aligned}$$

satisfying $p^2+q^2+r^2 = x^2+y^2+z^2$ which transforms the ternary quadratic form

(6) $T = (1/2)(Ap^2 + Bq^2 + Cr^2) - Fqr - Gpr - Hpq$

into the sum of squares

(7) $T = (1/2)(\alpha x^2 + \beta y^2 + \gamma z^2).$

He noticed in particular that the coefficients of the transformation (5), that is, p' ,p'', p''', q', q'', q''', r', r'', r''' satisfied certain additional conditions, namely,

$$(p')^2 + (q')^2 + (r')^2 = 1 \quad p'p'' + q'q'' + r'r'' = 0$$
$$(p'')^2 + (q'')^2 + (r'')^2 = 1 \quad p'p''' + q'q''' + r'r''' = 0$$
$$(p''')^2 + (q''')^2 + (r''')^2 = 1 \quad p''p''' + q''q''' + r''r''' = 0.^{11}$$

In his reading of the *Méchanique analytique* sometime between 1831 and 1838, the young, self-taught, British mathematician, George Boole focused on precisely this aspect of that multifaceted work and got a first glimpse at the mathematics involved in linearly transforming the variables of homogeneous polynomials.[12]

Following Lagrange's lead, Boole developed his own ideas and techniques for dealing with transformations such as the one which takes (6) to (7) and wrote up his approach in 1839 for the newly established *Cambridge Mathematical Journal*. The paper, entitled "Researches on the Theory of Analytical Transformations, with a Special Application to the Reduction of the General Equation of the Second Order", marked Boole's debut as a published mathematician, and its reception encouraged him to continue his studies in this area.[13] By 1841, in fact, he had not only extended this earlier work but also established his indisputable priority as the founder of invariant theory with his two-part article, "Exposition of a General Theory of Linear Transformations."[14]

Beginning his "Exposition" with a sketch of the intellectual lineage of the ideas he was about to present, Boole explained that

> The transformation of homogeneous functions by linear substitutions, is an important and oft-recurring problem of analysis. In the *Mécanique Analytique* of Lagrange, it occupies a very prominent place, and it has been made the subject of a special memoir by Laplace. More recently it has engaged the attention of Lebesgue

and Jacobi, the former of whom has extended his investigations to homogeneous functions of the second degree ...

The most general conclusion to which the labours of the above-mentioned writers have led, is, that it is always possible to take away the products of the variables x_1, x_2, ..., x_m, from a proposed homogeneous function of the second degree, Q, by the linear substitution of a new set of variables, y_1, y_2, ..., y_m, connected with the original ones by the relation

$$x_1^2 + x_2^2 + \ldots + x_m^2 = y_1^2 + y_2^2 + \ldots + y_m^2 \ldots^{15}$$

Given the limited nature of the results of his predecessors in this area, Boole tackled the obvious generalization of the problem. For a homogeneous polynomial of degree n in m unknowns and a linear transformation of its variables, Boole wanted "... to determine the relations by which the [coefficients of the polynomial before and after the transformation is applied] are held in mutual dependence."[16] The method he devised for solving this problem hinged on eliminating the variables from the given homogeneous polynomial by means of partial differentiation of that polynomial with respect to each of the m unknowns.

Taking the simplest case as his first example, Boole demonstrated his technique on the binary quadratic form $Q = ax^2 + 2bxy + cy^2$.[17] (Notice that this is the same form that had arisen in Gauss's number-theoretic research in the *Disquisitiones Arithmeticae* except that Boole imposed no explicit restrictions on the coefficients.) He began by calculating the partial derivatives of Q with respect to x and y and equating them to zero:

$$\frac{\partial Q}{\partial x} = 2ax + 2by = 0$$

$$\frac{\partial Q}{\partial y} = 2bx + 2cy = 0.$$

Then, the elimination of x and y from these equations yielded the expression

(8) $$\theta(Q) = b^2 - ac,$$

the desired relation between the coefficients of Q.[18] Carrying the analysis a bit further, Boole applied the linear transformation

$$x = mx' + ny' \qquad m, n, m', n' \in \mathbb{R}; \quad mn' - m'n \neq 0$$
$$y = m'x' + n'y'$$

to Q to get a new binary form $R = A(x')^2 + 2Bx'y' + C(y')^2$.[19] Obviously, calculating the partial derivatives of R relative to x' and y' produced an expression $\theta(R) = B^2 - AC$ perfectly analogous to (8).

Through his explanation of the relationship between expressions like $\theta(Q)$ and $\theta(R)$, Boole hit upon what subsequently became the defining criterion of an invariant. Presenting his theorem in the fullest generality but proving it only for binary quadratic and binary cubic forms, Boole stated that "[i]f Q_n be a homogeneous function of the nth degree, with m variables, which by the linear theorems ... is transformed into R_n, a similar homogeneous function; and if γ represents the degree of $\theta(Q_n)$ and $\theta(R_n)$, then

$$\theta(Q_n) = \frac{\theta(R_n)}{E^{\frac{\gamma n}{m}}},$$

E being the result obtained by the elimination of the variables from the second members of the linear theorems ... equated to 0."[20] In other words, $\theta(Q_n)$ and $\theta(R_n)$ were equal up to a power of the determinant of the linear transformation.

After carrying out a series of calculational examples in the second installment of the paper, examples which illustrated the effectiveness of his techniques as well as the applicability of his theorems, Boole closed his study by passing the mathematical torch which he had lit. He augured that "[t]o those who may be disposed to engage in the investigation, it will, I believe, present an ample field of research and discovery. It is almost needless to observe that any additional light which may be thrown on the general theory, and especially as respects the properties of the function $\theta(q)$, will tend to facilitate our further progress, and to extend the range of useful applications."[21]

One upstart mathematician who read and heeded this advice was the Senior Wrangler at Cambridge in 1842, Arthur Cayley. Writing to Boole on June 13, 1844, the twenty-two year old Cayley could hardly contain his enthusiasm for the area Boole had opened up for him. "Will you allow me," he asked, "to make an excuse of the pleasure afforded me by a paper of yours published some time ago in the mathematical journal 'On the theory of linear transformations' and of the interest I take in the subject, for sending you a few formulae relative to it, which were suggested to me by your very interesting paper. I should be delighted if they were to prevail upon you to resume the subject which really appears inexhaustible."[22] Although Boole effectively left the theory of invariants after his 1841 work, this letter marked the beginning of Cayley's long and fruitful engagement with the theory and hinted at just how prophetic Boole's closing remarks would prove to be. In fact, just five months later on November 11, 1844, Cayley wrote back to Boole to tell him that the "few formulae" had grown into something more substantial, namely, "...a paper on linear transformations for the next [number] of the Journal which I believe is to be printed soon."[23]

In this paper, entitled "On the Theory of Linear Transformations" and published in 1845, Cayley presented a new theoretical construct for generating invariants, the so-called hyperdeterminant.[24] Typical of Cayley's approach to mathematics he developed his new theory in the most general setting possible. As noted above, Boole worked with homogeneous polynomials in m variables of degree of homogeneity (or order) n in the variables. Characteristically, Cayley realized "...that it might be generalized by considering ... not a homogeneous function of the nth order between variables, but one of the same order, containing n sets of m variables, and the variables of each set entering linearly."[25] In other words Cayley chose to work up a process for generating invariants of multilinear forms.[26]

Although the details of his computationally nightmarish method of hyperdeterminants need not concern up here, the idea behind that method is relevant: by generating invariants of a multilinear form, he could uncover invariants of the binary form which resulted from a suitable identification of the variables in the multilinear form. To illustrate this idea, Cayley concluded his paper with a series of examples, but one of them proved particularly important, namely, that of the homogeneous function in four sets of two variables. Applying his method of hyperdeterminants to the multilinear form

$$
\begin{aligned}
(9) \quad U = &a x_1 y_1 z_1 w_1 + b x_2 y_1 z_1 w_1 + c x_1 y_2 z_1 w_1 + d x_2 y_2 z_1 w_1 + \\
&e x_1 y_1 z_2 w_1 + f x_2 y_1 z_2 w_1 + g x_1 y_2 z_2 w_1 + h x_2 y_2 z_2 w_1 + \\
&i x_1 y_1 z_1 w_2 + j x_2 y_1 z_1 w_2 + k x_1 y_2 z_1 w_2 + l x_2 y_2 z_1 w_2 + \\
&m x_1 y_1 z_2 w_2 + n x_2 y_1 z_2 w_2 + o x_1 y_2 z_2 w_2 + p x_2 y_2 z_2 w_2,
\end{aligned}
$$

Cayley produced the invariant

$$
(10) \qquad u = ap - bo - cn + dm - el + fk + gj - hi.
$$

Identifying $x_1 = y_1 = z_1 = w_1 = x$ and $x_2 = y_2 = z_2 = w_2 = y$, then, reduced (9) to

$$
\begin{aligned}
U = &a x^4 + (b + c + e + i) x^3 y + (d + f + g + j + k + m) x^2 y^2 \\
&+ (h + l + n + o) x y^3 + p y^4
\end{aligned}
$$

which, setting

$$
(11) \qquad
\begin{aligned}
&a = \alpha, \ b = c = e = i = \beta, \ d = f = g = j = k = m = \gamma, \\
&h = l = n = o = \delta, p = \epsilon
\end{aligned}
$$

yielded $U = \alpha x^4 + 4\beta x^3 y + 6\gamma x^2 y^2 + 4\delta x y^3 + \epsilon y^4$, the binary quartic form. Finally, the application to (10) of the identification equations in (11) generated

(12) $u = \alpha\epsilon - 4\beta\delta + 3\gamma^2,$

an invariant of the binary quartic.[27]

As innocent as this may seem, it represented a fundamental break-through in the theory of invariants, for Cayley's method of hyperdeterminants had begotten an invariant of the binary quartic *different* from the invariant

(13) $\alpha\gamma\epsilon - \alpha\delta^2 - \epsilon\beta^2 - \gamma^3 + 2\beta\gamma\delta$

generated by Boole's method. Furthermore, as Boole discovered shortly after receiving the letter from Cayley in which he detailed this new finding, the binary quartic had yet another invariant

(14) $(\alpha\epsilon - 4\beta\delta + 3\gamma^2)^3 - 27(\alpha\gamma\epsilon - \alpha\delta^2 - \epsilon\beta^2 - \gamma^3 + 2\beta\gamma\delta)^2,$

a polynomial function of the invariants in (12) and (13).[28] Thus, new invariants arose as special combinations of known invariants, and while this observation came to determine the course of nineteenth-century invariant theory, neither Cayley nor Boole anticipated such developments in 1845. For Cayley, especially, the most immediate concern was to generate simple invariants, and by 1846, he had devised a new method for doing just that.

Cayley viewed this fresh approach, dubbed the hyperdeterminant derivative, as one "...which, at the same time that it is much more general, has the advantage of applying directly to the only case which one can possibly hope to develope with any degree of completeness, that of functions of two variables."[29] Thus, in spite of his consistent insistence on generality, Cayley had recognized by 1846 that the sorts of computational techniques he proposed could realistically handle only the simplest case of the binary forms. With this in mind, then, he posed what would become the fundamental problem of invariant theory, namely "[t]o find all the derivatives [i.e., invariants] of any number of functions, which have the property of preserving their form unaltered after any linear transformation of the variables."[30]

Furthermore, based on Boole's observation of the dependence relation or syzygy (14) between the two known invariants (12) and (13) of the binary quartic, Cayley identified the greatest potential stumbling block of the computational approach to the subject. As he warned, "... there remains a question to be resolved, which appears to present many great difficulties, that of determining the *independent* derivatives, and the relation between these and the remaining ones."[31] Boole's method, as well as Cayley's own techniques of hyperdeterminants and hyperdeterminant derivatives, generated invariants, but they provided no insight into the possible interrelations between the invariants so generated. The development of a theory to explain and to anticipate such dependence relations as (14) thus represented yet another avenue of invariant-theoretic research, down which Cayley took the first tentative steps. As he admitted in 1846, though, he had "... only succeeded in treating a very particular case of this question, which shows however in what way the general problem is to be attacked."[32] In light of this advance and others in his 1846 paper, Cayley proved the prescience of Boole's 1842 prediction.[33] The theory of invariants did indeed have the potential of becoming a new and vital area of mathematical research. The further realization of its potential, however, would not occur for another five years or so.

In 1846, when Cayley's formulation of the main problems of the theory appeared in print, he had given up his fellowship at his old college, Trinity, at Cambridge, for the more lucrative prospects of a career in law. Studying at Lincoln's Inn in London beginning in April 1846, Cayley soon met another mathematician-turning-lawyer, James Joseph Sylvester.

A fellow Cantabrigian, but from the rival St. John's College, Sylvester had followed a rocky academic road compared to that of his new friend, Cayley. In spite of gaining the position of Second Wrangler on the Mathematical Tripos of 1837, Sylvester had been denied not only his degree but also the possibility of competing for further prizes or fellowships on the basis of his professed Jewish faith. While serving as a Professor of Natural Philosophy at dissenting University College, London from 1838 to 1841, he did earn the M.A. but from Trinity College, Dublin, another institution independent of the Church of England. With the prospects of mathematical employment in England slim, Sylvester left his homeland for the United States in 1841 to assume the Professorship of Mathematics at the University of Virginia. Unfortunately, Sylvester's stay in Charlottesville lasted less than four months, and his repeated attempts to find a new academic position in America failed.[34] Returning to England in 1844, he accepted an actuarial post at the Equity and Law Life Assurance Society, and by 1846 had also decided to take up the law in London.

The meeting of these two misplaced mathematicians, Cayley and Sylvester, marked the beginning of a personal friendship and mathematical partnership which would end only with Cayley's death in 1895. By

1850 when Sylvester was admitted to the Bar (Cayley had been called up one year earlier), the two friends were thinking about invariant theory.

Judging from the sequence of published papers and the surviving correspondence, Sylvester apparently served as the catalyst that precipitated Cayley's renewed interest in the invariant-theoretic program announced in the paper of 1846. Although Cayley had dealt with the problem of calculating invariants in several short articles between 1846 and 1850, his progress in this area had already slowed by 1846.[35] His hyperdeterminants and hyperdeterminant derivatives simply proved too daunting calculationally, especially given the primitive state of combinatorics in the 1840's.[36] Coming to invariant-theoretic ideas from the vantage point of his work on the problem of solving systems of n equations in n unknowns, however, Sylvester brought new perspectives on the theory's similarly spirited, calculational techniques.[37] By 1851, hard at work on a general theory of forms, Sylvester discovered a more general technique for generating invariants, called compound permutation, that yielded Cayley's hyperdeterminant as a bonus.[38] Sylvester's success had a highly positive effect both on his own work and on that of his friend, Cayley.

Sylvester followed the 1851 paper, "Sketch of a Memoir on Elimination, Transformation, and Canonical Forms", with two lengthy and seminal papers in 1852 and 1853. In the first of these, entitled "On the Principle of the Calculus of Forms" and published in two installments in the *Cambridge and Dublin Mathematical Journal*, Sylvester began a systematic assault on the theory of invariants by drawing together and organizing known and new facts and techniques.[39] His next massive effort, a 141-page treatise, entitled "On a Theory of Syzygetic Relations of Two Rational Integral Functions, Comprising an Application to the Theory of Sturm's Functions, and That of the Greatest Algebraical Common Measure", represented a first attempt at coming to terms with the phenomenon of syzygies like (14) between invariants such as (12) and (13). In this work, Sylvester also set down, once and for all, a glossary of the developing vocabulary of invariant theory, a vocabulary that he had coined almost singlehandedly in his extensive correspondence with Cayley.[40]

Sylvester captured well the early sense of unbounded optimism for the future of invariant theory and for its impending development into a major branch of mathematics in the quotation he chose to introduce his 1853 paper:

> How charming is divine philosophy!
> Not harsh and crabbed as dull fools suppose,
> But musical as is Apollo's lute,
> And a perpetual feast of nectar'd sweets,
> Where no crude surfeit reigns!–Comus[41]

One year later in 1854, Cayley returned to sample from the "perpetual feast of nectar'd sweets" which Sylvester had laid out for him. In the first of his ten "Memoirs on Quantics", a series of works which spanned a period of almost twenty-five years, Cayley set British invariant theory on its permanent course.[42]

In his "Introductory Memoir", Cayley abandoned his old, computationally difficult techniques of hyperdeterminant differentiation in favor of a new point of view. After 1854, Cayley understood an invariant as a special kind of covariant, that is, as "[a] function which stands in the same relation to the primitive function from which it is derived as any of its linear transforms to a similarly derived transform of its primitive."[43] In more modern terms, if

$$(15) \qquad a_0 x^m + a_1 \binom{m}{1} x^{m-1} y + a_2 \binom{m}{2} x^{m-2} y^2 + \cdots + a_m \binom{m}{m} y^m$$

is a binary quantic and if $T : x \mapsto aX + bY, y \mapsto a'X + b'Y$ is a linear transformation of nonzero determinant Δ which sends (15) to $A_0 X^m + A_1 \binom{m}{1} X^{m-1} Y + A_2 \binom{m}{2} X^{m-2} Y^2 + \cdots + A_m \binom{m}{m} Y^m$, then a homogeneous polynomial $K(a_0, a_1, \ldots, a_m; x, y)$ in the variables and non-binomial coefficients of (15) is a covariant provided that $K(A_0, A_1, \ldots, A_m; X, Y) = \Delta^k K(a_0, a_1, \ldots, a_m; x, y)$ for some positive integer k.[44] In this context, Cayley recognized a covariant as a homogeneous polynomial K annihilated, to use Sylvester's term, by the following partial differential operators:

$$(16) \qquad \begin{aligned} & a_0 \partial_{a_1} + 2a_1 \partial_{a_2} + 3a_2 \partial_{a_3} + \ldots + m a_{m-1} \partial_{a_m} - y \partial_x, \text{ and} \\ & a_m \partial_{a_{m-1}} + 2a_{m-1} \partial_{a_{m-2}} + 3a_{m-2} \partial_{a_{m-3}} + \cdots + m a_1 \partial_{a_0} - x \partial_y. \end{aligned}$$

Calling this pair of operators the "entire system" for this case, Cayley stated that "...besides the quantic itself, there are a variety of other functions which are reduced to zero by each of the operations of the entire system; any such function is said to be a covariant of the quantic, and in the particular case in which it contains only the elements [i.e., the coefficients a_i in (15)], an invariant."[45]

While this method of generating co- and invariants by means of differential operators immediately supplanted both of Cayley's hyperdeterminant techniques, it was not entirely new. As early as the 1845 paper "On the Theory of Linear Transformations", Cayley had alluded to such a method without really putting it to use. Then later in a letter to Sylvester on December 5, 1851, he resurrected the idea and opined that "[t]his will constitute the foundation of a new theory of Invariants."[46] As Tony Crilly has pointed out, this "new synthesis" provided a calculationally simpler way of uncovering covariants.[47] Furthermore, since invariants were just special types of covariants, the goals of the new theory became to determine the fundamental system of covariants for the binary forms and to determine the dependence relations or syzygies between the covariants of the various degrees in the coefficients and orders in the variables.[48] Hand-in-hand with these goals went two distinct counting questions, namely, for a given binary quantic 1) how many independent (or asyzygetic) covariants of a given degree and order were in the fundamental system?, and 2) was the total number of independent covariants always finite? Cayley provided answers to both of these questions in his "Second Memoir on Quantics" published in 1856.[49]

Actually, though, judging from the postscript dated October 7, 1854 to his "Introductory Memoir" and from a jubilant but undated letter to Sylvester, Cayley had discovered the solution to the first problem long before it finally appeared in print in 1856.[50] To the 1854 article, he had appended the seductive remark that he had "... since the preceeding memoir was written, ... been ... led to the discovery of the law for the number of asyzygetic covariants of a given order and degree in the coefficients."[51] He left the mathematical world in suspense vis à vis this discovery until he finally published a "Second Memoir" in 1856. There, Cayley began by laying the foundation upon which he built his new law. To understand his version of the result, however, we must introduce several more of the concepts Sylvester had developed in 1853.

Consider a covariant K of (15) as above. It is not hard to show that K has a constant degree of homogeneity in $a_0, a_1, \ldots, a_n$, denoted by θ and called the degree, and a constant degree of homogeneity in x and y, called the order and denoted μ. Furthermore, taking x to have weight 1, y have weight 0, and a_i to have weight i by convention, each monomial of K has a computable weight. Thus, if $a_1^2 a_2^3 a_3 x^2 y$ is a monomial of K, it has weight $1 \cdot 2 + 2 \cdot 3 + 3 \cdot 1 + 1 \cdot 2 + 0 \cdot 1 = 13$. Using these notions, the proof that each monomial of K has the same weight follows easily. In fact, for a covariant K of degree θ and order μ, each monomial has weight $1/2 \cdot (m\theta + \mu)$ follows easily.[52]

With this theory at his disposal, then, Cayley gave an explicit construction of co- and invariants for a binary quantic. Relative to the quantic (15), he defined two differential operators, namely,

$$X = a_0 \partial_{a_1} + 2a_1 \partial_{a_2} + 3a_2 \partial_{a_3} + \ldots + ma_m \partial_{a_{m-1}} \text{ and}$$
$$Y = ma_1 \partial_{a_0} + (m-1)a_2 \partial_{a_1} + (m-2)a_3 \partial_{a_2} + \ldots a_m \partial_{a_{m-1}}.$$

He next took a polynomial A_0 in the coefficients of (15) of degree θ and of weight $1/2 \cdot (m\theta - \mu)$ which was annihilated by X, that is, $X A_0 = 0$. Using

A_0 and the other operator Y, Cayley recursively constructed a series of new coefficients $A_1 = YA_0$, $A_2 = 1/2 \cdot YA_1$, ..., $A_\mu = 1/\mu \cdot YA_{\mu-1}$, and showed that the form

$$(17) \qquad A_0 x^\mu + A_1 x^{\mu-1} y + \ldots + A_{\mu-1} x y^{\mu-1} + A_\mu y^\mu$$

was always a covariant of (15). Furthermore, "...a function A of the degree θ and of weight $1/2 \cdot m\theta$, satisfying the condition $XA = 0$ will (also satisfy the equation $YA = 0$ and will) be an invariant."[53] Given these theoretically explicit forms (17) of the covariants, Cayley was also able to solve problem 1) above. Taking ". . . for $[A_0]$ the most general function of the coefficients, of the degree θ and of the weight $1/2 \cdot (m\theta - \mu)$; then $[XA_0]$ is a function of the degree θ and of the weight $1/2 \cdot (m\theta - \mu) - 1$, and the arbitrary coefficients in the function A are to be determined so that $[XA_0] = 0$ ... hence ... the number of asyzygetic covariants of the order μ and the degree θ is equal to the number of terms [of A_0] of the degree θ and weight $1/2 \cdot (m\theta - \mu)$ less the number of terms [of XA_0] of the degree θ and weight $1/2 \cdot (m\theta - \mu) - 1$."[54]

In his proof of this result, however, Cayley glossed over one key point which, had it been false, would have vitiated most of the subsequent results of the British invariant theorists. Cayley had based his result on the assumption that the system of linear equations resulting from setting $XA_0 = 0$ was linearly independent In fact, in the same undated letter in which he had announced his new theorem to Sylvester, Cayley had confidently stated that "...the coefficients of $[A_0]$ satisfy a certain number of linear equations [and] there is no reason for doubting that these equations are independent ..."[55] While he may have been morally certain of the linear independence of the system, Cayley nevertheless left a gaping mathematical hole in a result which lay at the heart of his and his countrymen's approach to invariant theory. As we shall see, almost twenty-five years would pass before Sylvester would finally prove that there indeed had been "no reason for doubting" that linear independence.[56]

This one oversight aside, Cayley had provided an algorithm for determining the number of independent covariants of a given degree and order in the fundamental system of a binary quantic in his "Second Memoir Upon Quantics" of 1856. Likewise in that paper, he thought he had also answered that second open question, namely, was the number of covariants in the fundamental system always finite? In his treatment of this matter, however, Cayley was not as lucky. Again, an assumption about linear independence, this time a false assumption, led him badly astray. As Cayley himself later realized, "...certain linear relations, which I had assumed to be independent, are really not independent, but, on the contrary, linearly connected together. ..."[57] His failure at this juncture in the argument to recognize the possibility of a linear dependence relation resulted in his incorrect application of the only technique available at the time for handling the combinatorial question "how many?", namely, Euler's theory of generating functions.[58] Since the dependence relation or "...interconnexion in question does not occur in regard to the quadric, cubic, or quartic, [however,] for these cases respectively the theory is true as it stands."[59] Thus, Cayley successfully calculated that the binary quadratic had two irreducible covariants, the binary cubic had four, and the binary quartic had five. Relative to the binary quintic, though, his luck ran out. The system of equations which arose from his theory relative to the quintic was not linearly independent, and so the argument by generating functions predicted infinitely many irreducible covariants in this case.[60]

In the face of this result, which he believed to be true, Cayley proceeded to calculate eleven different covariants and three invariants for the binary quintic. So great was his zest for calculation that he plunged headlong and unrelentingly into a calculational problem of infinite proportions. Only in 1868, when the German invariant theorist, Paul Gordan proved that the number of irreducible covariants of any binary quantic was, in fact, always finite, did Cayley's research problem assume more human dimensions.[61]

THE GERMAN APPROACH TO INVARIANT THEORY

Whereas Boole, and through him what would become a British school, followed an initially algebraic path to the theory of invariants, the rival German school had its origins not in algebra but rather in number theory and geometry. As indicated earlier, Gauss had recognized and discussed the invariance of the discriminant of the binary quadratic within the context of the number-theoretic researches he presented in the *Disquisitiones Arithmeticae* of 1801. Like many of the far-reaching ideas in this work, however, the notion of invariance attracted little, if any, attention in the years immediately following the book's publication. Yet, between 1837 and 1842, the Gymnasium student and mathematical prodigy, Gotthold Eisenstein, engaged in a careful study of Gauss's work. By the summer of 1842, he had become intimately acquainted with the *Disquisitiones Arithmeticae* and had embarked upon a research program aimed at extending Gauss's theory of quadratic forms to those of the third degree.[62]

In his first paper, published in *Crelle's Journal* in 1844 but written in December 1843, Eisenstein proposed to study cubic forms by looking at a certain "corresponding" quadratic form. Given a cubic

(18) $$ax^3 + 3bx^2y + 3cxy^2 + dy^3$$

and a linear substitution of the variables $x = \alpha x' + \beta y'$, $y = \gamma x' + \delta y'$ where $\alpha \delta - \beta \gamma = 1$, Eisenstein noted that

Each cubic form has a corresponding quadratic form

[19] $\qquad Ax^2 + 2Bxy + Cy^2 = F = (A, B, C)$

the coefficients of which are linked to those of the cubic form by the very simple equations

$$A = b^2 - ac, 2B = bc - ad, C = c^2 - bd,$$

and if one applies an arbitrary substitution ... to the cubic form as well as to the quadratic form, the same relation will exist between the coefficients of the two new forms which one gets by this double transformation ... [Thus,] a whole class of *cubic* forms will always correspond to a complete class of *quadratic* forms ... a remarkable property which sheds new light on the nature of cubic forms.[63]

In the case of the binary cubic, Eisenstein had, in effect, anticipated Ludwig Otto Hesse's work on what Sylvester would later dub "the Hessian"[64] Eisenstein's "corresponding" quadratic form was, as we shall soon see, precisely the Hessian of the original cubic. In fact, in a more definitive paper on cubic forms also written in December 1843, Eisenstein verified that (19) satisfied the invariantive property.[65]

At essentially the same time that covariants arose in Eisenstein's number-theoretic researches in Berlin, they cropped up in Otto Hesse's geometrically inspired work in Königsberg. In the second of a two-part paper published in *Crelle's Journal* in 1844, Hesse analyzed the theory of third-order plane curves.[66] He based his analysis, however, on the paper's first installment dated earlier that same year on January 16.[67] There, Hesse exploited the theory of determinants (perfected by Jacobi in 1841) to develop a computational device for determining critical points of curves in three-space.[68] More generally, though, Hesse considered a homogeneous function f of degree m in n unknowns $x_1, x_2, \ldots, x_n$, and he defined the functional determinant

(20) $\qquad \left| \dfrac{\partial^2 f}{\partial x_i \partial x_j} \right|, \qquad 1 \leq i, j, \leq n.$

Furthermore, given a linear transformation

(21) $\qquad x_i \mapsto y_i = a_{i1}x_1 + a_{i2}x_2 + \ldots + a_{in}x_n,$

Hesse proved that if "[t]he determinants [as in (20)] of the function f be denoted ϕ or ϕ' according to whether one considers $x_1, x_2, \ldots, x_n$ or $y_1, y_2, \ldots, y_n$ as the variables [of f and if o]ne denotes by r the determinant formed by the coefficients of the variables in the given linear equations [21], then

[22] $\phi = r^2 \phi'$."[69]

Thus, in studying the properties of his new functional determinant (20), Hesse proved in general what Eisenstein had found in the special case of a binary cubic, namely, that (20) satisfies the invariantive property (22).[70]

Whereas Hesse's interest in (20) lay in its utility as a tool for answering certain geometrical questions, by 1849 Siegfried Aronhold had investigated some of the purely algebraic ramifications of his teacher's work.[71] As Aronhold explained in the introduction to his first paper on this subject:

> In the twenty-eighth volume of this journal [*Crelle*], Hesse exposed a series of problems which were as important for the theory of third degree homogeneous functions in three variables as they were interesting for algebra, in that they furnish the first example of a new kind of higher equation. ...
>
> I have undertaken researches on third degree homogeneous functions in three variables so as to ascertain the most important algebraic relations of their coefficients and the functions which are coordinate and subordinate to them. These researches have led to new characteristics of these functions which I shall take the liberty to share in a detailed discussion.[72]

Thus, in 1849 Aronhold consciously launched the first systematic German attack on invariant theory inspired by Hesse's algebraic analysis of a geometrical problem.[73]

In the course of his work, Hesse had proven, in a highly calculational, determinant-theoretic way, that if f denotes a ternary cubic form

$$
\begin{aligned}
(23) \quad \Sigma a_{\chi,\lambda,\mu} x_1 x_2 x_3 = {} & a_{1,1,1} x_1^3 + a_{2,2,2} x_2^3 + a_{3,3,3} x_3^3 + \\
& 6 a_{1,2,3} x_1 x_2 x_3 + 3 a_{2,2,3} x_2^2 x_3 + 3 a_{2,3,3} x_2 x_3^2 + \\
& 3 a_{1,3,3} x_1 x_3^2 + 3 a_{1,1,3} x_1^2 x_3 + 3 a_{1,2,2} x_1 x_2^2 + \\
& 3 a_{1,1,2} x_1^2 x_2
\end{aligned}
$$

and if $\Delta f = \Sigma b_{\chi,\lambda,\mu} x_1 x_2 x_3$ denotes its Hessian, then for four distinct values λ, $\lambda f + \Delta f$ decomposes into the product of three factors linear in x_1,

x_2, x_3.[74] Theoretically at least, Hesse was able to produce the four values λ by recognizing them as the roots of a certain auxiliary fourth-degree equation. To construct this biquadratic, Hesse began by taking f and Δf as above and letting a and b be two indeterminants over the complex numbers. He then showed that the relation $\Delta(af + b\Delta f) = Af + B\Delta f$ always holds, where A and B are homogeneous functions of the third degree in a and b and in the coefficients $a_{i,j,k}$ of f. Using this relation to get a homogeneous twelfth-degree equation in a and b (which he then had to solve!), Hesse finally came up with the equation $G = Ba - Ab = 0$, a fourth-degree homogeneous equation in a and b. Solving $G = 0$ yielded the four solutions λ of the decomposition problem.[75]

From Aronhold's point of view, this construction of G, which was auxiliary to Hesse's *geometric* goals, served as the springboard for an *algebraic* study. As Aronhold put it: "Hesse's presentation, which has a different point of departure, brings out the algebraic problem rather as a corollary to the other developments. Therefore, I have considered it necessary to prove systematically the theorems he gives, but with a slight change of notation, ... [so that] they give the solution of the problem directly",[76] that is, so that G falls out naturally, thereby avoiding the twelfth-degree equation, etc. In order to effect this more intrinsic development of G, Aronhold began by analyzing the expression $af + b\Delta f$ in a new way.

Letting $F = af + b\Delta f$ and keeping the above notation, the equation

(24)
$$\Delta F = Af + B\Delta f$$

holds. If $F = \Sigma A_{\chi,\lambda,\mu} x_\chi x_\lambda x_\mu$ and $\Delta F = \Sigma B_{\chi,\lambda,\mu} x_\chi x_\lambda x_\mu$, then it follows immediately from the definition of F and (24) that

(25)
$$A_{\chi,\lambda,\mu} = a \cdot a_{\chi,\lambda,\mu} + b \cdot b_{\chi,\lambda,\mu}$$
$$B_{\chi,\lambda,\mu} = A \cdot a_{\chi,\lambda,\mu} + B \cdot b_{\chi,\lambda,\mu}.$$

Now using these coefficients, define the determinants

(26)
$$A_{\chi,\lambda,\mu} B_{\chi_1,\lambda_1,\mu_1} - A_{\chi_1,\lambda_1,\mu_1} B_{\chi,\lambda,\mu},$$

where χ, λ, μ, and χ_1, λ_1, μ_1 take on all possible permutations of 1, 2, 3.

Aronhold showed that F decomposes into the product of three linear factors if and only if all of these determinants vanish. Substituting (25) into (26), this decomposability criterion becomes $(a \cdot a_{\chi,\lambda,\mu} + b \cdot b_{\chi,\lambda,\mu})(A \cdot a_{\chi_1,\lambda_1,\mu_1} + B \cdot b_{\chi_1,\lambda_1,\mu_1}) - (a \cdot a_{\chi_1,\lambda_1,\mu_1} + b \cdot b_{\chi_1,\lambda_1,\mu_1})(A \cdot a_{\chi,\lambda,\mu} + B \cdot b_{\chi,\lambda,\mu}) = 0$
or

(27)
$$(Ba - Ab)(a_{\chi,\lambda,\mu} b_{\chi_1,\lambda_1,\mu_1} - a_{\chi_1,\lambda_1,\mu_1} b_{\chi,\lambda,\mu}) = 0.$$

Thus, it follows that

> ...all the conditional equations have $G = Ba - Ab$ as a factor and will be satisfied whenever the single equation $G = Ba - Ab = 0$ is satisfied. ...
>
> One sees from [27] that if $G = Ba - Ab$ does not vanish, then the equations $a_{\chi,\lambda,\mu}b_{\chi_1,\lambda_1,\mu_1} - a_{\chi_1,\lambda_1,\mu_1}b_{\chi,\lambda,\mu} = 0$ must be satisfied, which are the conditions for f to be decomposable into linear factors. It follows from this that if f is not decomposable, the decomposability [of F] can be effected through the solution of the equation $G = 0$.[77]

Based on this observation, Aronhold realized that the crux of the decomposability problem lay in understanding the functions A, B, and G. Since A and B are each homogeneous third-degree polynomials in a and b, they look like:

$$(28) \qquad \begin{aligned} A &= \alpha_1 a^3 + \alpha_2 a^2 b + \alpha_3 ab^2 + \alpha_4 b^3 \text{ and} \\ B &= \beta_1 a^3 + \beta_2 a^2 b + \beta_3 ab^2 + \beta_4 b^3. \end{aligned}$$

Understanding them, and so G, thus hinged on determining the coefficients α_i and β_j. In his efforts to express A and B explicitly and using intrinsic, algebraic methods, Aronhold developed not only a first approximation of the symbolic notation, which would later characterize the German school of invariant theory, but also demonstrated its utility in interpreting invariants.

Consider the two, second-degree, homogeneous functions $\Sigma a_{\chi,\lambda} x_\chi x_\lambda$ and $\Sigma b_{\chi,\lambda} x_\chi x_\lambda$ in x_1, x_2, x_3. The first of these (as well as the second *mutatis mutandis*) has monomials with coefficients

$$(29) \qquad \begin{matrix} a_{1,1} & a_{1,2} & a_{1,3} \\ a_{2,1} & a_{2,2} & a_{2,3} \\ a_{3,1} & a_{3,2} & a_{3,3} \end{matrix}$$

where the binomial coefficients are disregarded and where it is assumed that $a_{i,j} = a_{j,i}$. Using this array of coefficients, Aronhold defined a new 3×3 array $(aa)^{i,j}$ where $(aa)^{i,j}$ is the signed minor formed by striking the ith row and j^{th} column in (29). So, for example, $(aa)^{1,1} = (-1)^{1+1}(a_{2,2}a_{3,3} - a_{3,2}a_{2,3}) = a_{2,2}a_{3,3} - a_{2,3}a_{3,2}$. Based on the new systems (29), Aronhold gave an algorithm for forming a third system common to the first two, "namely:

$$\begin{matrix} (ab)^{1,1} & (ab)^{1,2} & (ab)^{1,3} \\ (ab)^{2,1} & (ab)^{2,2} & (ab)^{2,3} \\ (ab)^{3,1} & (ab)^{3,2} & (ab)^{3,3} \end{matrix}$$

which one gets from [29] if one differentiates the functions of this system by every value a and substitutes, in place of the increments, the corresponding b ... "[78]

An example best illustrates what Aronhold had in mind here. To compute $(ab)^{1,1}$, for instance, take $(aa)^{1,1}$ and differentiate it relative to each of the $a_{i,j}$'s present. Then form the differential of $(aa)^{1,1}$, that is, $a_{3,3}\Delta a_{2,2} + a_{2,2}\Delta a_{3,3} - 2a_{2,3}\Delta a_{2,3}$. Finally, replace $\Delta a_{i,j}$ by $b_{i,j}$ to get

$$(ab)^{1,1} := a_{3,3}b_{2,2} + a_{2,2}b_{3,3} - 2a_{2,3}b_{2,3}.$$

Never losing sight of the fact that the problem at hand involved ternary cubic forms and so coefficients of the form $a_{i,j,k}$ not $a_{i,j}$, Aronhold grafted his new symbolism onto the old. Thus, he defined $(a_r, a_\rho)^{1,1} := a_{r,3,3}b_{\rho,2,2} + a_{r,2,2}b_{\rho,3,3} - 2a_{r,2,3}b_{\rho,2,3}$.[79] Having adapted the new notation to the ternary cubic, Aronhold proceeded to use it in his quest for the coefficients in (28). This search began with an invariant-theoretic observation.

If $S = (a_1a_1)^{1,1}(a_1a_1)^{1,1} + (a_1a_1)^{2,2}(a_2a_2)^{1,1} + (a_1a_1)^{3,3}(a_3a_3)^{1,1} + 2(a_1a_1)^{2,3}(a_2a_3)^{1,1} + 2(a_1a_1)^{1,3}(a_1a_3)^{1,1} + 2(a_1a_1)^{1,2}(a_1a_2)^{1,1}$, then "...$S$ is homogeneous and of the fourth degree in the coefficients $a_{\chi,\lambda,\mu}$ [of f]. Furthermore, [S] has the property that its form does not change if it is constructed out of the coefficients of that function into which f changes when one transforms [f] by a linear substitution."[80] In particular, if r denotes the determinant of the linear transformation and if S_1 denotes the transformed version of S, then $S_1 = r^4S$. In other words, S satisfies the invariantive property.[81]

Continuing in this vein, Aronhold constructed another new invariant T which satisfied $T_1 = r^6T$. He completed this phase of his analysis by expressing the coefficients of (28) in terms of the invariants S and T. Explicitly, $A = 3Sa^2b + 6Tab^2 + 3S^2b^3$ and $B = a^3 - 3Sab^2 - 2Tb^3$, which resulted in $G = a^4 - 6Sa^2b^2 - 8Tab^3 - 3S^2b^4$. Since his original problem collapsed to solving $G = 0$, Aronhold simplified this equation by first dividing through by b^4 and then setting $\lambda = a/b$. This left him with the following biquadratic equation in one variable: $G = \lambda^4 - 6S\lambda^2 - 8T\lambda - 3S^2 = 0$. Given, as we noted above, that Eisenstein had provided an invariant-theoretic solution to the general biquadratic in 1844, Aronhold could use those findings to determine the four roots of $G = 0$, thereby solving Hesse's decomposition problem intrinsically.[82]

By 1858, Aronhold had developed both the ideas and the symbolic notation of his 1849 paper into a single, unifying theory, which explained the behavior and properties of ternary cubics from an invariant-theoretic point of view. Also by 1858, he had familiarized himself with some of the work of the British school of invariant theory and had adopted some of their terminology, most importantly the terms "invariant" and "covariant".[83] Not content with this special theory, however, Aronhold, using a key result proved by Alfred Clebsch, managed to extend it to the general case of m-ary n-ics, that is, homogeneous forms of degree n in m variables. In his 1863 paper entitled, "Ueber eine fundamentale Begründung der Invariantentheorie", Aronhold provided the developing German school with its invariant-theoretic manifesto.[84]

Simply stated, Aronhold's goal in this work was not unlike that of his British contemporaries, namely, to encompass, in a single theory, all invariants and covariants of all orders and degrees associated to any m-ary n-ic and to establish, in so doing, not only their interrelationships but also a justification of their existence.[85] Essential to his program, however, were both the symbolic notation, a notation of sufficient generality to handle and to describe invariant-theoretic relationships succinctly, and its associated symbolic method. Before examining Gordan's symbolic proof of the finiteness theorem, a result which caught the British school totally offguard, we must look briefly at the development of the symbolic approach in the hands of Aronhold and Clebsch between 1858 and 1863.

As the name implied, the notation and method developed by the German school was symbolic rather than explicit in nature. Whereas the British invariant theorists sought explicit expressions (or Cartesian expressions, to use Tony Crilly's terminology [86]) of the invariants and covariants of a given quantic in terms of the coefficients of that quantic, the Germans preferred a notation which allowed them to operate at a somewhat higher, somewhat more abstract level. For example, rather than writing the mth degree binary form explicitly in terms of its coefficients as in (15), the German school wrote it symbolically as

$$(30) \qquad\qquad (a_1 x_1 + a_2 x_2)^m$$

or more simply as a_x^m or even as a^m.[87] They recovered (15) by expanding the binomial in (30) formally and equating corresponding coefficients. In this way, the coefficient a_i in (15) equaled the coefficient $a_1^{m-1} a_2^i$ in (30), which the German school then compressed into $a_{\underbrace{11\cdots1}_{m-i}\underbrace{22\cdots2}_{i}}$. Obviously, since the a_1 and a_2 represented true symbols in the German scheme, the mth degree binary form could just as well be written as $b^m = b_x^m = (b_1 x_1 + b_1 x_1)^m$, or as c^m, or as d^m, and so on. Based on these various symbolic representations of the original form, they defined the 2×2 determinants $(ab) = a_1 b_2 - a_2 b_1$, $(ac) = a_1 c_2 - a_2 c_1$, $(bd) = b_1 d_2 - b_2 d_1$, etc.[88] Relative to the binary m-ic, the German approach to invariant theory hinged on the theorem which stated that any covariant, and hence invariant, of (30) could be expressed in the form

$$(31) \qquad (ab)^{\alpha}(ac)^{\beta}(bd)^{\gamma} \cdots a^p b^q c^r \cdots$$

where the number of a's, b's, c's, etc. appearing in the symbolic product equaled m or an integral multiple of m. Conversely, everything of the form (31) was a covariant.[89]

As noted above, Aronhold had introduced symbolic determinant forms into his analysis of the ternary cubic form as early as 1849. More importantly, by 1858 and still relative to the ternary cubic, he had noted the relationship between symbolic products analogous to (31) and covariants. Somewhat later still, Clebsch completely generalized Aronhold's observation of 1858, and in so doing provided the German school with its guiding theorem.

As Clebsch explained in his 1861 paper entitled "Ueber symbolische Darstellung algebraischer Formen",

> In the 55th volume of this journal [*Crelle*], Mr. Aron-hold, relative to the groundforms of homogeneous functions of degree three in three variables, proceeded from a very remarkable symbolic expression which seems especially adapted to exhibit the true properties of such forms. In the 58th volume of this journal (p. 117), I briefly indicated a general proof of this fact, namely, that the given symbolic form [Gestalt] applies not only to [the ternary cubic] but also to all invariants [and] covariants ... however big the degree and the number of variables in the original function may be. It is even appropriate, perhaps, to take this symbolic representation generally as the definition of such forms ...[90]

By setting up the theory from precisely this point of view, Clebsch determined the course of pre-Hilbertian, German invariant theory. He showed that, theoretically at least, invariants and covariants could be analyzed and categorized via symbolic expressions like (31). In practice, though, this presented a serious problem. In 1861, the German school lacked an effective calculus for manipulating and identifying symbols such as (31). In his paper of 1863, however, Aronhold overcame this final obstacle. The method he put forth hinged largely on the clever application of symbolic identities to reduce an arbitrary covariant to known ones. An example perhaps best conveys the spirit of this method and its implementation.

As above, let $a = a_1 x_1 + a_2 x_2$, $b = b_1 x_1 + b_2 x_2$, and $c = c_1 x_1 + c_2 x_2$. It is easy to verify formally that the identity

$$(32) \qquad a(bc) + b(ca) + c(ab) = 0$$

holds.[91] We can then use (32) to generate a new identity

$$(33) \qquad b^2(ca)^2 + c^2(ab)^2 - a^2(bc)^2 = 2bc(ab)(ac)$$

by shifting the summand $a(bc)$ to the righthand side of (32) and squaring. Now, recalling that any symbolic expression of the form (31) is a covariant of a binary m-ic, we can generate two equal covariants from (33) by multiplying each side of the identity by the appropriate powers of a, b, and c, that is, by $a^{m-2}b^{m-2}c^{m-2}$. This yields $2a^{m-2}b^{m-1}c^{m-1}(ab)(ac) = b^m a^{m-2} c^{m-2}(ca)^2 + c^m a^{m-2} b^{m-2}(ab)^2 - a^m b^{m-2} c^{m-2}(bc)^2$. Recalling that a^m, b^m, and c^m are all equivalent notations for the binary m-ic, we see that, up to sign, each term on the righthand side of this equation is identical, and so the equation collapses to

$$(34) \qquad a^{m-2}b^{m-1}c^{m-1}(ab)(ac) = (1/2)a^m b^{m-2} c^{m-2}(bc)^2.$$

In symbolic notation, the Hessian of a binary m-ic is given by the expression $b^{m-2}c^{m-2}(bc)^2$; we finally see that the covariant on the lefthand side of (34) is actually equivalent to half the product of the original binary m-ic, a^m, and its Hessian.[92]

Using identities such as (32), Aronhold and the German school established such general reduction rules as: "every symbol having a pair of factors with a common letter $(ab)(ac)$ may be reduced to a more compact form by substituting for this pair their value from equation [33], and so expressing the symbol by others in which this pair of factors is replaced by a single square factor."[93] Fundamentally, it was this ability to reduce and simplify covariants symbolically which allowed Paul Gordan to prove his celebrated finiteness theorem in 1868.

In presenting his "Beweis, dass jede Covariante und Invariante einer binären Form eine ganze Function mit numerischen Coefficienten einer endlichen Anzahl solcher Formen ist", Gordan realized that he was correcting a longstanding error in the invariant-theoretic literature. He explained that

> In the 146th volume of the Philosophical Transactions p. 101, Mr. Cayley dealt with the question whether all covariants and invariants originating from a binary form are representable as a function of a bounded number of forms with numerical coefficients.

He showed that forms of the second, third, and fourth
degrees are expressible in this way. In what follows, I
specify for binary forms of the nth degree a finite system of covariants and invariants from which I show
that, and how, all forms derived from the [original]
form are representable as rational functions of the [fi-
nite system] with numerical coefficients. ... I have also
carried out this reduction for forms of the fifth and
sixth degrees and have produced a smallest possible
system of groundforms [in each of these cases].[94]

Thus, with the utmost diplomacy, Gordan informed his readers that he
had not only corrected Cayley's erroneous conclusions from 1856 about
binary forms of degrees greater than four but also explicitly exhibited
the fundamental system of groundforms for the quintic and sextic. At
the heart of his success lay the process of Über(einander)schiebung or
transvection.

In exploiting their symbolic notation, the German school of invariant
theorists had found a successful procedure for generating new covariants
from given ones.[95] Taking two covariants ϕ and ψ of a binary form f,
where ϕ and ψ have symbolic representations $(\phi_1 x_1 + \phi_2 x_2)^\mu$ and $(\psi_1 x_1 + \psi_2 x_2)^\nu$, respectively, Gordan "... use[s] in particular the known procedure
by which to build the simplest new covariants and invariants from the
two known covariants ϕ and ψ of f ..., namely, the forms:

$$(\phi\psi)^0 = \phi \cdot \psi$$
$$(\phi\psi)^1 = \phi_x^{\mu-1} \psi_x^{\nu-1} (\phi\psi)$$
$$(\phi\psi)^2 = \phi_x^{\mu-2} \psi_x^{\nu-2} (\phi\psi)^2, \ldots$$

[He] will say that these forms arise through the '0th, 1st, 2nd, ... transvection of ϕ and ψ'"[96] Notice, then, that the kth transvectant is of the form
$(\phi\psi)^k = \phi_x^{\mu-k} \psi_x^{\nu-k} (\phi\psi)^k$, where $1 \leq k \leq \min\{\mu, \nu\}$.[97]

As the first step in his argument toward the proof of the finiteness
theorem, Gordan showed that *any* covariant of the form (31) could actually be expressed as a linear combination of specially generated transvectants. More specifically, he proved "... that all symbolic products of order
m and therefore also all [covariants] of this order are linear combinations of forms built by means of transvection from f and forms of order
$m - 1$."[98] His conclusion here turned on a careful analysis of the component transvectants via the symbolic method, that is, using identities
like (32) and (33) to dissect them into transvectants of the appropriate
orders.[99] Focusing next on the transvectants so obtained, he proved a
technical lemma concerning their precise structure by induction on their

order. This allowed him to show that "...the forms for a quantic of the nth degree consist either of the forms which had occurred already for a quantic of the degree $n-1$, or else of the mutual transvectant between such forms and the series of two lettered forms $(ab)^2$, $(ab)^4$, &c."[100] With this additional structure at his disposal, he set up an algorithm for constructing a generating system from the transvectants and proved its finiteness by induction on the order of the transvectants.[101] In this constructive, but symbolic, tour de force, Gordan gave nineteenth-century invariant theory its deepest pre-Hilbertian result.

A BRITISH PROOF OF GORDAN'S THEOREM

Gordan's proof of the finiteness theorem caught the English school of invariant theorists looking somewhat foolish. Since 1856 when Cayley published his "Second Memoir on Quantics", the British had been laboring under the mistaken assumption that the number of irreducible covariants of a binary quantic of degree five or greater was infinite. Nevertheless, they had continued undaunted to calculate invariants and covariants in spite of the seeming futility of this presumably endless task. In 1868, however, Gordan's theorem vindicated their efforts, but in some sense, he had beaten them at their own game. Given this embarrassing state of affairs, the British had little choice: they had to show that their theoretical framework for invariant theory also yielded a proof of Gordan's theorem. Only then would they have demonstrated the completeness and, to their minds at least, the superiority of their approach.

Cayley stated this new goal explicitly in 1871 in "A Ninth Memoir on Quantics", his first paper on the subject after the publication of Gordan's result. He began by acknowledging that "[i]t was shown not long ago by Professor Gordan that the number of the irreducible covariants of a binary quantic of any order is finite ..., and in particular that for a binary quintic the number of irreducible covariants (including the quintic and the invariants) is = 23, and that for a binary sextic the number is = 26."[102] He then had to admit that "[f]rom the theory given in my 'Second Memoir on Quantics' ..., I derived the conclusion, which, as it now appears, was erroneous, that for a binary quintic the number of irreducible covariants was infinite."[103] Yet an even more difficult admission followed. Although he had tried to fix up his erroneous proof of 1856 and thereby prove Gordan's theorem using British methods, Cayley was "...not able to make this correction in a general manner so as to show from the theory that the number of the irreducible covariants is finite, and so to present the theory in a complete form ..."[104] Nevertheless, he did succeed in coaxing Gordan's findings on the binary quintic out of his own theory in the "Ninth Memoir".

While this exercise certainly did not give Gordan's theorem in its full generality, it provided some indication that that goal lay within reach of the British methods. As Cayley rather wistfully concluded, "I cannot but hope that a more simple proof of Professor Gordan's theorem will be obtained—a theorem the importance of which, in reference to the whole theory of forms, it is impossible to estimate too highly."[105]

In the years after the appearance of the "Ninth Memoir", their correspondence reveals that both Cayley and Sylvester kept this idea of reproving Gordan's theorem in the backs of their minds. For Sylvester, however, the early seventies marked an uncharacteristic period of mathematical inactivity due to his premature retirement from his teaching position at the Royal Military Academy in 1870 and to his subsequent unemployment. In 1876, however, this sorry situation changed dramatically. At the age of sixty-one, Sylvester accepted the professorship of mathematics at the newly-formed Johns Hopkins University in Baltimore. Buoyed by this supreme vote of confidence in his mathematical abilities, Sylvester returned to his researches with renewed vigor, and it soon became evident that his mathematical powers had not diminished. Invariant theory in general, and the long-sought-after proof of Gordan's theorem in particular, had rekindled his mathematical flame.

Writing to Cayley midway through the spring term of his first year at Hopkins, Sylvester could hardly disguise his pride in what he viewed as a major victory for the British approach to invariant theory. "I may now announce with moral certainty", he declared, "that my method completely solves the problem of finding the *grundformen* [sic] for binary forms and systems of binary forms (without mixture of superfluous forms) in all cases—I have sent an account of the method to the *Comptes Rendus* ... my method [gives] a sure means of proceeding *step by step* from the lowest forms to higher ones until all [are] exhausted—but it w[oul]d take too long to go into this in a letter."[106]

Thus, in the two or three weeks prior to writing this letter, Sylvester had come up with a method for actually calculating the elements in the fundamental system of invariants and covariants for a given quantic or system of quantics, a collection of elements the finiteness of which was guaranteed by Gordan's theorem.[107] Given the special emphasis the British school placed upon calculation, Sylvester's algorithm would have indeed represented a major victory for the British techniques. It would have explicitly produced the elements in the fundamental system, but even more, its argument by generating functions would have yielded Gordan's finiteness theorem as a bonus. Yet as Sylvester himself sensed in the letter to Cayley, his algorithm potentially undercalculated the fundamental system. He had to admit "...that *anterior to all* verification this method could not give *superfluous* forms—but it is metaphysically conceivable that it might give *too few* grundformen [sic]."[108]

Nevertheless, confident in his ability to work out the details, he sent a brief communication of the new finding to the *Comptes rendus* of the Académie des Sciences. There he boldly claimed that he had "...completely resolved the great problem of finding the complete system of fundamental invariants and covariants ...by a purely algebraic method based on the partial differential equation [16]."[109] The method he sketched for accomplishing this feat hinged on using (16) to obtain a generating function which, when expanded out in series form, had coefficients corresponding to the desired numbers of linearly independent invariants and covariants.

His interest immediately piqued by this announcement, the French mathematician, Camille Jordan wrote to Sylvester on May 13, 1877 congratulating him on his discovery and asking him to fill in the details for the mathematical public at his earliest possible convenience. To tide him over until he had the opportunity to see the complete proof in print, though, Jordan pressed Sylvester to divulge the secret of the method. "Is it founded", he asked, "like Gordan's on considerations of the symbolic form of the covariants? I hardly think so, given the difficulties which stopped him [Gordan] at the ternary forms and at which I have just tried my hand. Another question: Does your method give the system of Grundformen without those superfluous forms which are so troublesome to eliminate?"[110]

In principle, the answers to both of these questions were in the affirmative, but, as his subsequent communications to the *Comptes rendus* testified, his intuition in that letter to Cayley had been right on the mark. Whereas Gordan's symbolic method tended to overestimate the size of the fundamental system, Sylvester's non-symbolic method tended to underestimate it.[111] Gordan's theorem gave the finiteness of the system, but neither camp could come up with a method which categorically generated *the* finite number of elements in that system. Still, thanks to Sylvester's work at Johns Hopkins in the spring of 1877, the British invariant theorists had a competitive algorithm to use in their calculational program.

The British approach to invariant theory received yet another boost from Sylvester later that same year. As noted above, Cayley had given a formula for the number of linearly independent covariants of a given degree and order in 1856. Although correct, this formula hinged on the presumed, but unverified, linear independence of a certain system of equations. Early in November 1877, Sylvester successfully provided the proof of the linear independence, thereby justifying a fact "...as certain as any fact in nature could be ..." but which "...transcended, [he had] thought, the powers of human understanding."[112] In a letter to Cayley on November 6, 1877, Sylvester put it somewhat more modestly, explaining that he had "...been fortunate enough in the last day or two to discover a *rigorous proof* of the ...Independence of the Equations given by your fundamental theorem in the second Memoir on Quantics."[113] Moreover, in the course of this work he thought he had "...[caught] a glimpse of the possibility of obtaining a simple proof of Gordan's theorem..."[114] This apparition, however, proved phantasmic. Gordan's theorem did not yield to the force of Sylvester's mathematics in 1877.

Still at work on the problem in 1878, Sylvester became convinced once again that he had finally worked the bugs out of that very general result announced in 1877 in the *Comptes rendus*. Sending yet another announcement to the French Academy, he declared that he had "...a new proof of Gordan's theorem ..." and that the finite number of elements in the fundamental system assured by this theorem was "...the coefficient of $t^j u^\epsilon$ in the development of the generating function

$$\frac{1}{(1 - tu^{2i})(1 - tu^{2i-2}) \cdots (1 - tu^{-2i+2})(1 - tu^{-2i})}$$

where j is the order and ϵ is the degree of the given covariant."[115] This time around, Gordan himself entreated Sylvester for the full particulars of his method.

In response to a letter from Sylvester sketching this new result, Gordan admitted that

> ...I did not understand the end of your letter. ...[Y]ou want to find not only the linearly independent, asyzygetic forms but generally the number of independents, i.e., also all relations between the groundforms of a system. And your theorems extend not only to the binary forms but also to those with arbitrarily many variables.
>
> That infinitely exceeds my knowledge, since I can prove the finiteness of the system only for binary forms and for the ternary forms of third and fourth degrees. That is little in itself for these forms, if one wants the statement of the number of forms of the system.... Given the respect which I have for you and the importance of the matter, I long for the chance to understand your methods; would it not be desirable for you to put it all together and send it to me for our journal ... the *Mathematische Annalen* ...? You would place me, in so

doing, under the greatest obligation, and I would take
the greatest pains to penetrate your methods.

In the hopes that you will grant me this wish, I
remain ... [116]

Unfortunately, Sylvester was unable to oblige, for the generating function
of 1878 also failed to deliver completely on its discoverer's promises.
Once again when pushed on the details, Sylvester had come up short, but
Gordan's letter revealed much more than Sylvester's inability to prove
this general theorem. It betrayed the fact that the German school of
invariant theorists found the methods of their English counterparts just
as obscure and impenetrable as the British found the German work. Be
this as it may, given Gordan's triumph for the German side, the British still
needed their own proof of the finiteness theorem to prove the equality, if
not the superiority, of their methods.

Setting Gordan's theorem aside for other invariant-theoretic concerns
in 1878, Sylvester found himself involved with it afresh in 1881. His letter
to Cayley of March 23 opened with renewed hopefulness. "I believe", he
wrote, "that I have proved Gordan's theorem and moreover can assign
a superior limit to the number of fundamental invariants. By the same
kind of processes as I applied to prove the fundamental theorem for the
number of linearly independent covariants of a given deg[ree]-order in
Crelle and the Phil[osphical] Mag[azine] I can show that the number of
fundamental in- and covariants to $(x, y)^n$ is the number of basic solutions
to the system of numerical equations in positive integers

$$x_0 + x_1 + x_2 + \cdots + x_n = y$$
$$2z + 2x_1 + 4x_2 + \cdots + 2nx_n = ny.\text{"}[117]$$

Three days later on March 26, though, Sylvester had to recant, explaining
that "[t]he theorem which I sent you the day before yesterday *is wrong*—it
proceeded from a certain new principle which is correct—but the application made of the principle was erroneous."[118]

Following another eighteen months of related research, Sylvester was
again in hot pursuit of a proof of Gordan's theorem. On September 6,
1882, he reported to Cayley that

> [t]he *lowest* weights of the ground forms for the de-
> grees 2, 3, 4, 5, ... are 2, 3, 7, 18, ... and assuming
> that the progression 2/2, 3/3, 7/4, 18/5, ... increases
> beyond any assignable limit, I can prove Gordan's the-
> orem. Of the law of progression holding I have little
> or no doubt and the proof (however difficult it may

be) belongs to the province of ordinary algebra or the theory of partitions.

Supposing it established, have I *proved* Gordan's theorem [?] Yes ..."[119]

Sylvester reaffirmed his position on September 14, declaring that "... there is no difficulty in proving Gordans [sic] theorem. I hope to have my demonstration out in the next number of the [American] Journal..."[120] By October 6, though, he had to admit once more that "[m]y supposed proof of Gordan's theorem was a *Delusion*–but I have considerable hope of being able to found one upon the method of Deduction aided by the actual application of this method to the Quintic (as a Diagram)."[121]

These hopes, too, went unrealized. After another year of periodic attempts to prove this elusive result, Sylvester's optimism seemed to wane. He finally wrote to Cayley on May 26, 1883, saying that "[i]t seems to me that it would be a good work accomplished, if some one were to translate all Gordon's and Jordan's methods or conclusions out of the language of uberschiebungen [sic] into that of alliances and so get rid for good and all of the symbolic algorithm–although perhaps it occurs to me that this might not appear to be quite a respectful remark to make to the founder of the method."[122] Clearly frustrated, Sylvester was forced to admit defeat. Neither he nor Cayley nor any of the invariant theorists of the British school ever succeeded in establishing Gordan's theorem using their methods.[123]

CONCLUSIONS

Although hinted at by Gauss in the *Disquisitiones Arithmeticae* in 1801, invariant theory owed its inception to George Boole's work of 1841. As we have seen, Cayley, and later Sylvester, developed Boole's embryonic idea into a full-blown mathematical theory during the 1850's. Their work attracted the attention of the Irish mathematician, George Salmon, as well as that of Charles Hermite and Camille Jordan in France and Francesco Brioschi in Italy.[124] Each of these mathematicians made fundamental contributions to the developing theory, and each of them was in correspondence with Sylvester and Cayley, detailing their discoveries. Thus, a truly international spirit of cooperation developed. Yet, as Wilhelm-Franz Meyer explained in his report on the state of invariant theory to the Deutsche Mathematiker-Vereinigung: "Whereas English, French, and Italian mathematicians actively exchanged ideas, thereby continually giving new thrusts to this branch of science, the situation was not the same in Germany where things were long at a standstill. ... The transition came only in 1858 with the work of Siegfried Aronhold."[125]

Aronhold had come to the study of invariants as a result of his purely algebraic study of a phenomenon which had arisen in Hesse's geometrical researches. In presenting his findings, Aronhold not only propelled the Germans into invariant theory but also set them on a trajectory widely divergent from that of their British contemporaries. In fact, as Meyer explained, the symbolic notation and method which Aronhold, and later Clebsch, developed "... [gave] birth to a new era, recognizable in that our Science [took] on a systematic development which [had], from that moment on, its center in Germany."[126] No result better typified this "new era", this eclipse of the British school, than Gordan's 1868 proof of the finiteness theorem and the subsequent, unsuccessful British attempts to prove it non-symbolically.

What caused this British failure? From the start, the British school emphasized the calculation of in- and covariants over the establishment of a firm theoretical basis for their approach.[127] The correspondence as well as the published works of Cayley and Sylvester overflowed with new computational algorithms, but theorems followed by complete and rigorous proofs appeared much less frequently. Several devotés of their point of view tried at various times to fill in these foundational gaps, to organize, and to systematize a theory which had developed so fast and furiously that few, other than its founders, could comprehend it. As early as 1859 in the first edition of his text, *Modern Higher Algebra*, George Salmon tried to pull the subject together by unquestionably establishing its theorems.[128] In 1876, the Italian mathematician, Francesco Faà de Bruno, attempted the same thing, calling into doubt some of the accepted techniques.[129] Then again in 1895, Sylvester's student at New College, Oxford, Edwin Bailey Elliott also took a stab at the problem in his *An Introduction to the Algebra of Quantics*.[130] These attempts only met with limited success because, fundamentally, the British approach to invariant theory was algorithmic. Each of these texts read, in a real sense, like a cookbook for the proper preparation of in- and covariants. In the absence both of the necessary theoretical underpinnings and of a sufficiently general notation, the British school's techniques did not lend themselves to proving existence theorems like that of Gordan.

This is not to say that the non-symbolic approach of Cayley and Sylvester did not have spectacular successes. It enabled its adherents to calculate the in- and covariants for binary forms up to the eighth degree and to determine the syzygies, or dependences, between them. They catalogued their results in massive tables, the very construction of which generated important discoveries in combinatorics and in the theory of symmetric forms. This is also not to say that the symbolic approach of

the German school surmounted every difficulty. An exceedingly complicated and technical theory, the symbolic method, like its non-symbolic
counterpart, aimed at calculating in- and covariants. After all, Gordon
followed up his proof of the finiteness theorem with an explicit calculation of the fundamental systems for the binary quintic and sextic. The
difference lay in the slightly higher degree of abstraction of the German
technique which enabled them, as in the case of the finiteness theorem,
to skim just above the hands-on level of calculation. Furthermore, the
Germans more than the British worked from firmly established theorems
rather than from principles based on moral, if not mathematical, certainty.
(Recall Cayley's very typical remark, "...there is no reason for doubting
...", relative to the linear independence question in Cayley's theorem.
See note 55.) Still, their real success came only in the case of the *binary*
quantics. Although Gordan and others had some luck in dealing with
ternary quantics symbolically, the method was, at its foundations, too
computationally oriented to embrace a general theory of *n-ary* forms.

In 1890, David Hilbert injected the new and much more abstract idea
of what has since become known as the noetherianness of polynomial algebras to reprove Gordan's theorem for binary forms.[131] By 1893, Hilbert
had used this same property to generalize the theorem completely to *n*-
ary forms.[132] Thus, rather than killing invariant theory, Hilbert boosted
it to a more abstract, less computationally oriented plane.[133] In this new
light, the old questions and research programs, which had been directed
toward calculating complete systems of in- and covariants, were gradually supplanted by new questions such as the fourteenth problem Hilbert
posed in his renowned lecture before the International Congress of Mathematicians in 1900. There Hilbert asked "...whether it is always possible
to find a finite system of relatively integral functions $X_1, \ldots, X_m$ by which
every other relatively integral function of $X_1, \ldots, X_m$ may be expressed
rationally and integrally."[134] In the twentieth century, researches directed
at answering this and other, related questions came to characterize what
we now know as modern algebra. Thus, it becomes clear that in order to
trace the evolution of algebra in the twentieth century, we need detailed
investigations of the development of invariant theory in the nineteenth
century, investigations for which the present study may hopefully serve
as a guide.

NOTES*

1. Hermann Weyl, *The Classical Groups: Their Invariants and Representations* (Princeton: University Press, 1939). For his outline of the history of invariant theory, see pp. 27–29.

2. *Ibid.*, p. 27.

3. *Ibid.* Also, on the supposed death of invariant theory, see Charles S. Fisher, "The Death of a Mathematical Theory: A Study in the Sociology of Knowledge," *Archive for History of Exact Sciences* 3 (1966):137–159.

4. David Mumford, *Geometric Invariant Theory*, Ergebnisse der Mathematik und Ihrer Grenzgebiete, n.s., vol. 34 (New York: Academic Press, Inc., 1965).

5. That combinatorics should reinfuse invariant theory with life simply repays a debt dating from the 1880's. On the role of invariant theory in the development of combinatorics, see Karen Hunger Parshall, "America's First School of Mathematical Research: James Joseph Sylvester at the Johns Hopkins University 1876–1883," *Archive for History of Exact Sciences* 38 (1988):153–196.

6. Joseph P. S. Kung and Gian-Carlo Rota, "The Invariant Theory of Binary Forms", *Bulletin of the American Mathematical Society*, n.s., 10 (1984):27–85, on p. 28.

7. Jean A. Dieudonné and James B. Carrell, *Invariant Theory, Old and New* (New York: Academic Press, 1971), p. vii.

8. Felix Klein, *Development of Mathematics in the 19th Century*, trans. M. Ackerman, Lie Groups: History, Frontiers and Applications, vol 9 (Brookline: Math Sci Press, 1979). p. 143.

9. Carl Friedrich Gauss, *Disquisitiones arithmeticae*, trans. Arthur A. Clarke (New Haven: Yale University Press, 1966), pp. 111–112.

10. James Joseph Sylvester, "On a Remarkable Discovery in the Theory of Canonical Forms and of Hyperdeterminants", *Philosophical Magazine* 2 (1851):391–410, or James Joseph Sylvester, *The Collected Mathematical Papers of James Joseph Sylvester*, ed. H.F. Baker, 4 vols. (Cambridge

* This research was partially supported by a National Science Foundation Scholars Award #SES-8509795 and was conducted primarily during my year as Visiting Assistant Professor of Mathematics at the University of Illinois at Urbana-Champaign. I would like to thank Edward Cline of Clark University and Eric Friedlander of Northwestern University for helping me secure several papers critical to the present study. Thanks are also due to the Master and Fellows of St. John's College, Cambridge for permission to quote from their archives.

University Press, 1904–1912; reprint ed., New York: Chelsea Publishing Co., 1973), 1:265–283, on p. 273 (hereinafter cited as *Math. Papers J.J.S.*).

11. Joseph-Louis Lagrange, *Méchanique Analytique*, 2 vols. (Paris: Librarie scientifique et technique Albert Blanchard, 1965), 2:183–243. The type of transformation Lagrange isolated in this part of his work is now commonly called an orthogonal transformation. Thomas Hawkins discussed the significance of Lagrange's findings and approach to the development of the spectral theory of matrices in "Cauchy and the Spectral Theory of Matrices", *Historia Mathematica* 2 (1975): 1–29 on pp. 16–18.

12. Desmond MacHale, *George Boole: His Life and Work* (Dublin: Boole Press, 1985), pp. 44–58 on Boole's early studies and efforts in mathematics.

13. George Boole, "Researches on the Theory of Analytical Transformations, with a Special Application to the Reduction of the General Equation of the Second Order", *Cambridge Mathematical Journal* 2 (1841):64–73. Although Boole first drafted this paper in 1839, it did not appear in print until 1841 owing to revisions suggested by the journal's editor, Duncan F. Gregory. See MacHale, p. 52.

14. George Boole, "Exposition of a General Theory of Linear Transformations", *Cambridge Mathematical Journal* 3 (1841–1842):1–20, 106–119.

15. *Ibid.*, p. 1.

16. *Ibid.*, p. 3.

17. Here the word "binary" refers to the number of variables involved in the form, and the term "quadratic" refers to the degree of homogeneity of the variables of the form. The British school used the word "order" for the degree of homogeneity of the variables.

18. Boole, "Exposition of a General Theory of Linear Transformations", p. 6. I have altered Boole's notation slightly here to make it conform with the earlier example from Gauss's writings. Where I have uesd a, b, and c, Boole used the corresponding uppercase letters.

19. Note that Boole did not explicitly require $mn' - m'n \neq 0$, but he tacitly assumed this restriction in his calculations.

20. Boole, "Exposition of a General Theory of Linear Transformations", p. 19. By "linear theorems" here, Boole meant the defining equations of the underlying linear transformation.

21. *Ibid.*, p. 119.

22. Arthur Cayley to George Boole, June 13, 1844, as quoted in Anthony James Crilly, "The mathematics of Arthur Cayley with particular reference to linear algebra", (Ph.D. dissertation, Middlesex Polytechnic, June 1981), p. 23, and in MacHale, p. 56.

23. Arthur Cayley to George Boole, November 14, 1844, as quoted in Crilly, "The mathematics of Arthur Cayley", p. 29 and in MacHale, p. 56.

24. Arthur Cayley, "On the Theory of Linear Transformations", *Cambridge Mathematical Journal* 4 (1845):193–209, or Arthur Cayley and A. R. Forsyth, ed., *The Collected Mathematical Papers of Arthur Cayley*, 14 vols. (Cambridge: University Press, 1889–1898), 1:80–94 (hereinafter cited as *Math. Papers A.C.* All subsequent page references will refer to this collection.)

25. *Ibid.*, p. 80.

26. For a discussion of Cayley's approach to multilinear forms, see Tony Crilly, "The Rise of Cayley's Invariant Theory (1841-1862)", *Historia Mathematica* 13 (1986):241–254, on pp. 243–244.

27. Cayley, "On the Theory of Linear Transformations", pp. 88–89.

28. *Ibid.*, pp. 93–94.

29. Arthur Cayley, "On Linear Transformations", *Cambridge and Dublin Mathematical Journal* 1 (1846):104–122, or *Math. Papers A.C.* 1:95–112, on p. 95.

30. *Ibid.*

31. *Ibid.*

32. *Ibid.*

33. That Boole's remark was indeed prescient was further born out by developments on the Continent. Approaching invariant theory essentially from the point of view of the hyperdeterminant derivative, the Germans started a distinct and rival school, which actively competed with the British during the last half-century. (See below.) For a discussion of Cayley's definition of the hyperdeterminant derivative, see Crilly, "The Rise of Cayley's Invariant Theory", p. 245.

34. On Sylvester's first American sojourn, see Lewis S. Feuer, "America's First Jewish Professor: James Joseph Sylvester at the University of Virginia", *American Jewish Archives* 36 (1984):151–201, and Lewis S. Feuer, "Sylvester in Virginia", *The Mathematical Intelligencer* 9 (1987):13–19.

35. Between 1846 and 1851, Cayley published only the following notes touching on invariant theory: Arthur Cayley, "Recherches sur l'Elimination, et sur la Théorie des Courbes", *Journal für die reine und angewandte Mathematik* 34 (1847):30–45, or *Math. Papers A.C.*, 1:337–351; "Note sur les Hyperdéterminants", *Journal für die reine und angewandte Mathematik* 34 (1847):148–152, or *Math. Papers A.C.*, 1:352–355; and "Sur les Déterminants gauches", *Journal für die reine und angewandte Mathematik* 35 (1848):93–96, or *Math. Papers A.C.*, 1:410–413. It is interesting to note that this work appeared in *Crelle's Journal* as a British response to the invariant-theoretic work of Gotthold Eisenstein, Otto Hesse, and others.

36. The situation in combinatorics began to improve by the end of the 1850's when Cayley and Sylvester turned their attentions to combinatorial questions, but major improvements did not come until the Sylvester school at Johns Hopkins set the subject on a new track. On these developments, see Parshall, "America's First School of Mathematical Research", pp. 172–188.

37. Sylvester published his first paper on this so-called theory of elimination in 1839 and brought out articles on it regularly through 1851. Most notable among these researches are: James Joseph Sylvester, "On rational derivation from equations of coexistence, that is to say, a new and extended theory of elimination, Part 1", *Philosophical Magazine* 15 (1839):428–435, or *Math. Papers J.J.S.*, 1:40–46; "Memoir on the dialytic method of elimination, Part 1", *Philosophical Magazine* 21 (1842):534–539, or *Math. Papers J.J.S.*, 1:86–90; and "On a new class of theorems in elimination between quadratic functions", *Philosophical Magazine* 37 (1850):213–218, or *Math. Papers J.J.S.*, 1:139–144.

38. James Joseph Sylvester, "Sketch of a Memoir on Elimination, Transformation, and Canonical Forms", *Cambridge and Dublin Mathematical Journal* 6 (1851):186–200, or *Math. Papers J.J.S.*, 1:184–197.

39. James Joseph Sylvester, "On the Principles of the Calculus of Forms", *Cambridge and Dublin Mathematical Journal* 7 (1852):52–97 and 179–217, or *Math. Papers J.J.S.*, 1:284–363. Sylvester came out with two addenda to this paper in 1853. See James Joseph Sylvester, "Note on the Calculus of Forms", *Cambridge and Dublin Mathematical Journal* 8 (1853):62–64, and "On the Calculus of Forms, Otherwise the Theory of Invariants", *op. cit.*, pp. 256–269, or *Math. Papers J.J.S.*, 1:402–403 and 411–422.

40. James Joseph Sylvester, "On a Theory of the Syzygetic Relations of Two Rational Integral Functions, Comprising an Application to the Theory of Sturm's Functions, and That of the Greatest Algebraical

Common Measure", *Philosophical Transactions of the Royal Society of London* 143 (1853):407–548, or *Math. Papers J.J.S.*, 1:429–586. In the *Math. Papers J.J.S.*, the glossary of invariant-theoretic terms appeared on pp. 580–586.

41. *Ibid.*, p. 407 or 429.

42. Arthur Cayley, "An Introductory Memoir on Quantics", *Philosophical Transactions of the Royal Society of London* 144 (1854):244–258, or *Math. Papers A.C.*, 2:221–234.

43. Sylvester, "On a Theory of Syzygetic Relations", *Math. Papers J.J.S.*, 1:581.

44. For a unified treatment of invariant theory based on the published papers of Cayley, Sylvester, and the Irish mathematician, George Salmon, see Edwin Bailey Elliott, *An Introduction to the Algebra of Quantics*, 2d ed. (Oxford: University Press, 1913; reprint ed., New York: Chelsea Publishing Company, 1964). This work was first published in 1895. Elliott studied invariant theory under Sylvester at New College, Oxford during Sylvester's tenure there as Savilian Professor of Mathematics.

45. Cayley, "An Introductory Memoir on Quantics", *Math. Papers A.C.*, 2:224. Keeping this notation, $I(a_0, a_1, \ldots, a_m)$ is an invariant, or satisfies the "invariantive property", provided $I(A_0, A_1, \ldots, A_m) = \Delta^k I(a_0, a_1, \ldots, a_m)$ for some positive integer k. Sylvester hit upon the differential operator approach to invariant theory at roughly the same time Cayley focused on it. Sylvester came at it from a different point of view, however, and published his results before Cayley's had appeared in print. See Sylvester, "On the Principles of the Calculus of Forms", *Math. Papers J.J.S.*, 1:353.

46. Arthur Cayley to J. J. Sylvester, December 5, 1851, Sylvester Papers, St. John's College, Cambridge, Box 2 (hereinafter cited as Sylvester Papers SJC). This passage is also quoted in Crilly, "The Rise of Cayley's Invariant Theory", p. 246.

47. See Crilly, "The Rise of Cayley's Invariant Theory", pp. 245–247.

48. For the definitions of degree and order, see below.

49. Arthur Cayley, "A Second Memoir Upon Quantics", *Philosophical Transactions of the Royal Society of London* 146 (1856):101–126, or *Math. Papers A.C.*, 2:250–281. Tony Crilly also makes this point in "The mathematics of Arthur Cayley", p. 75.

50. Arthur Cayley to J. J. Sylvester, undated, Sylvester Papers SJC, Box 2. This letter is quoted at length in Crilly, "The Rise of Cayley's Invariant Theory", pp. 246–247.

51. Cayley, "An Introductory Memoir on Quantics", *Math. Papers A.C.*, 2:234.

52. For Cayley's discussion of this fact, see *ibid.*, p. 233.

53. Cayley, "A Second Memoir Upon Quantics", *Math. Papers A.C.*, 2:256. Here, I have simplified Cayley's notation somewhat. Using the notation for quantics which had developed during the course of his correspondence with Sylvester, Cayley would have written (15) as $(a_0, a_1, \ldots, a_{m-1}, a_m \langle x, y)^m$. He would have denoted (17) with a similar notation to emphasize that the binomial coefficients were not included in this covariant of order μ. However, in the "Second Memoir", Cayley did not use the subscript notation for the various coefficients. Finally, since X and Y act on a function A_0 in the coefficients only of the quantic, X and Y are the analogues for invariants as opposed to covariants, of the two operators in (16).

54. *Ibid.*, pp. 256–257.

55. Cayley to Sylvester, undated, Sylvester Papers SJC, Box 2, and Crilly, "The Rise of Cayley's Invariant Theory", p. 247.

56. James Joseph Sylvester, "Proof of the Hitherto Undemonstrated Fundamental Theorem of Invariants," *Philosophical Magazine* 5 (1878):178–188, or *Math. Papers J.J.S.*, 3:117–126.

57. Arthur Cayley, "A Ninth Memoir on Quantics", *Philosophical Transactions of the Royal Society of London* 161 (1871):17–50, or *Math. Papers A.C.*, 7:334–353 on p. 334.

58. For a general discussion of this method as it related to invariant theory, see Parshall, "America's First School of Mathematical Research", pp. 175–178. For its subsequent development at the hands of Sylvester and his students at the Johns Hopkins University, see *ibid.*, pp. 184–188.

59. Cayley, "A Ninth Memoir on Quantics", *Math. Papers A.C.*, 7:334.

60. Cayley, "A Second Memoir Upon Quantics", *Math. Papers A.C.*, 2:269–270. For a discussion of Cayley's theorem and its relation to developments in the representation theory of Lie algebras, see Thomas Hawkins, "Cayley's Counting Problem and the Representation of Lie Algebras", *Proceedings of the International Congress of Mathematicians-Berkeley*, 2 vols. (n.p.: American Mathematical Society, 1987), 2:1642–1656.

61. Paul Gordan, "Beweis, dass jede Covariante und Invariante einer binären Form eine ganze Function mit numerische Coefficienten einer endlichen Anzahl solchen Formen ist", *Journal für die reine und angewandte Mathematik* 69 (1868):323–354. Tony Crilly discusses Cayley's

insistence on calculation in "The Rise of Cayley's Invariant Theory", pp. 247–249.

62. Kurt-R. Biermann, "Gotthold Eisenstein: Die wichtigsten Daten seines Lebens und Wirkens," in Gotthold Eisenstein, *Mathematische Werke*, 2 vols. (New York: Chelsea Publishing Co., 1975), 2:919–929 on p. 921 (hereinafter cited as *Math. Works Eisenstein.*) See also Charles C. Gillispie, ed., *The Dictionary of Scientific Biography*, 16 vols., 1 supp. (New York: Charles S. Scribner's Sons, 1970–1980), s.v. "Eisenstein, Ferdinand Gotthold", by Kurt-R. Biermann (hereinafter cited as *DSB*).

63. Gotthold Eisenstein, "Théorèmes sur les formes cubiques et solution d'une équation du quatrième degré à quatre indéterminées," *Journal für die reine und angewandte Mathematik* 27 (1844):75–79 or *Math. Works Eisenstein*, 1:1–5 on p. 1. The translation from the French is mine.

64. See Sylvester, "Sketch of a Memoir on Elimination, Transformation, and Canonical Forms", *Math. Papers J.J.S.*, 1:184–197 on p. 195 for an early, if not the first, printed use of the term "Hessian".

65. Gotthold Eisenstein, "Untersuchungen über die cubischen Formen mit zwei Variabeln", *Journal für die reine und angewandte Mathematik* 27 (1844):89–104, or *Math. Works Eisenstein*, 1:10–25 on pp. 12–13. For the definition of the "invariantive property", see note 45 above. Keep in mind that, for Eisenstein, the linear transformation was assumed to have determinant 1.

66. Otto Hesse, "Über die Wendepunkte der Curven dritter Ordnung", *Journal für die reine und angewandte Mathematik* 28 (1844):97–107.

67. Otto Hesse, "Über die Elimination der Variabeln aus drei algebraischen Gleichungen von zweiten Grade mit zwei Variablen", *Journal für die reine und angewandte Mathematik* 28 (1844):68–96.

68. For Jacobi's work on indeterminants, see, in particular, C. G. J. Jacobi, "De Formatione et Proprietatibus Determinantium", *Journal für die reine und angewandte Mathematik* 22 (1841):285–318, and C. G. J. Jacobi, "De Determinantibus Functionalibus", *ibid.*, 22 (1841):319–359, or C. G. J. Jacobi, *C. G. J. Jacobi's Gesammelte Werke*, ed. Karl Weierstrass, 8 vols. (Berlin: Königlich Preussischen Akademie der Wissenschaften, 1881–1891; reprint ed., New York: Chelsea Publishing Co., 1969), 3:354–392 and 393–438. In the second of these papers, Jacobi introduced and extensively studied the functional determinant which now bears his name, that is, the Jacobian. If u, v, w, ... are n-ary n-ics, or forms of degree n in n variables x, y, z, ..., then the Jacobian is the functional determinant given by

$$\begin{vmatrix} \dfrac{\partial u}{\partial x} & \dfrac{\partial u}{\partial y} & \dfrac{\partial u}{\partial z} & \cdots \\[6pt] \dfrac{\partial v}{\partial x} & \dfrac{\partial v}{\partial y} & \dfrac{\partial v}{\partial z} & \cdots \\[6pt] \dfrac{\partial w}{\partial x} & \dfrac{\partial w}{\partial y} & \dfrac{\partial w}{\partial z} & \cdots \\[6pt] \vdots & \vdots & \vdots & \end{vmatrix}$$

Taking his lead from Jacobi's work, Hesse defined and exploited yet another functional determinant in his work of 1844. (See (20) below.)

69. Hesse, "Über die Elimination der Variablen aus drei algebraischen Gleichungen von zweiten Grade mit zwei Variabeln", p. 89. The translation from the German is my own. The functional determinants developed by Jacobi and Hesse were similar in spirit to Cayley's hyperdeterminant derivative, a construct which he essentially abandoned after 1854. As Tony Crilly pointed out in "The Rise of Cayley's Invariant Theory", "[a] special case of [the hyperdeterminant derivative method] was the precursor of the transvection operation on which the German symbolic process was based" (p. 245). Thus, the Germans pushed through an approach which the English had abandoned as too computationally difficult.

70. Letting (18) be f in (20), it is easy to see that (19) results. Checking that (19) satisfies (22) is an easy, if somewhat tedious, exercise.

71. Aronhold had studied under both Jacobi and Hesse at Königsberg before leaving, after his graduation in 1845, to follow Jacobi to Berlin. For more on Aronhold's life and mathematical contributions, see Gillispie, ed., *DSB*, s.v. "Aronhold, Siegfried Heinrich", by Herbert Oettel.

72. Siegfried Aronhold, "Zur Theorie der homogenen Functionen dritten Grades von drei Variabeln", *Journal für die reine und angewandte Mathematik* 39 (1849):140–159 on p. 140. The translation from German is my own.

73. Aronhold's journey to the threshold of invariant theory may not have been totally unguided. Twice during the course of his 1849 paper, he cited Cayley's invariant-theoretic work, thereby betraying a familiarity with the ideas Cayley expressed in "Note sur deux Formules données par M. M. Eisenstein et Hesse", *Journal für die reine und angewandte Mathematik* 29 (1845):54–57, or *Math. Papers A.C.*, 1:113–116; and "Mémoire sur les Hyperdétermiants", *Journal für die reine und angewandte Mathematik* 30 (1846):1–37. (Since the latter paper is merely a translation into French of Cayley's two papers "On the Theory of Linear Transformations" and "On Linear Transformations", it was not reproduced in *Math. Papers A.C.*.)

74. See Hesse, p. 92, and Aronhold, "Zur Theorie der homogenen Functionen dritten Grades von drei Variabeln", p. 143. The notation adopted here is Aronhold's and not Hesse's.

75. See *ibid.*, pp. 91–92, and *ibid.*, pp. 143–144.

76. Aronhold, "Zur Theorie der homogenen Functionen dritten Grades von drei Variabeln", p. 144. The translation from German is mine.

77. *Ibid.*, p. 147. My translation.

78. *Ibid.*, p. 150. My translation.

79. *Ibid.*, p. 151. My translation. Notice that in Aronhold's notation $(aa)^{i,j}$, $(ab)^{i,j}$, $(a_r b_\rho)^{i,j}$, the symbols a, b, a_r, b_ρ have no meaning relative to the given underlying forms. They are not particular coefficients but rather are representative of any and all coefficients. In terminology which the British developed later, they were called "umbrae", shadows of specifiable coefficients, and they embodied the generality of the German symbolic notation. See below.

80. *Ibid.*, p. 152. My translation.

81. See note 45 above. Here S plays the role of $I(a_0, a_1, \ldots, a_m)$ and S_1 plays the role of $I(A_0, A_1, \ldots, A_m)$, making the necessary notational changes.

82. Aronhold was obviously well aware of Eisenstein's number-theoretic work and consciously tied it in with his own researches. He referred explicitly to Gotthold Eisenstein, "Allgemeine Auflösung der Gleichungen von den ersten vier Graden", *Journal für die reine und angewandte Mathematik* 27 (1844):81–83, or *Math. Works Eisenstein*, 1:7–9. See Aronhold, "Zur Theorie der homogenen Functionen dritten Grades von drei Variabeln", p. 156. For Aronhold's solution of the problem based on Eisenstein's work, see *ibid.*, pp. 156–159.

83. Siegfried Aronhold, "Theorie der homogenen Functionen dritten Grades von drei Veränderlichen", *Journal für die reine und angewandte Mathematik* 55 (1858):97–191 on p. 98.

84. Siegfried Aronhold, "Ueber eine fundamentale Begründung der Invariantentheorie", *Journal für die reine und angewandte Mathematik* 62 (1863):281–345.

85. As Tony Crilly pointed out, however, the motives of the two schools were not identical. The Germans sought theoretical results from which explicit calculations of invariants or covariants might (and did) follow. The British looked for effective and efficient calculational algorithms for explicitly exhibiting invariants and covariants. See Tony Crilly, "The Decline of Cayley's Invariant Theory (1863–1895)", *Historia Mathematica* 15 (1988):332–347.

86. *Ibid.* Crilly introduced this term in this context in his dissertation (see note 22 above) and developed its significance further in "The Rise of Cayley's Invariant Theory" (see note 26 above).

87. This symbolic notation actually represents an interesting shorthand for the explicit Cartesian expression (15). For the Germans, a_1 and a_2 reflect the underlying homogeneous polynomials of degree m in a_1 and a_2 which correspond to the non-binomial coefficients in (15). In modern terms, they generate a basis

$$(*) \qquad a_1^m, a_1^{m-1} \cdot a_2, a_1^{m-2} \cdot a_2^2, \ldots, a_2^m$$

for the $(m+1)$-dimensional vector space inside which the coefficients of (15) live. Now, given a linear transformation T of the variables x_1 and x_2, T also takes a_1 and a_2 in (30) to linear combinations of themselves. Likewise, focusing on (15), T takes those coefficients into linear combinations of themselves. Hence, the $a_0, \ldots, a_m$ of (15) transform in exactly the same way as the monomials in $(*)$ under the action of T.

88. Note that in the case of a ternary form the symbolic determinants would be 3×3 determinants and $n \times n$ determinants in the general case of an n-ary form. From this observation it becomes clear why neither the early German nor the British school of invariant theory had much luck beyond the case of ternary forms.

89. For an explanation of the German notations and methods aimed at the British (and later American) audience, see George Salmon, *Lessons Introductory to the Modern Higher Algebra*, 5th ed., (Dublin: Hodges, Figgis, & Co., 1885; reprint ed., Bronx: Chelsea Publishing Co., 1964). Both Clebsch and Gordan presented systematized versions for the German audience. See Alfred Clebsch, *Theorie der binären algebraische Formen* (Leipzig: B. G. Teibner, 1872); and Paul Gordan, *Vorlesungen über Invariantentheorie*, 2 vols., (Leipzig: B. G. Teubner, 1885–1887).

90. Alfred Clebsch, "Ueber symbolische Darstellung algebraische Formen", *Journal für die reine und angewandte Mathematik* 59 (1861):1–62 on p. 1. The translation from German is mine.

91. In his book on invariant theory, Edwin Bailey Elliott called this the "Jacobian identity". See Elliott, p. 69.

92. Salmon, p. 316. The relation between the Hessian in its symbolic form and its partial derivative form (20) is not at all obvious here. However, it is not hard to see that the symbolic notation $a = a_1 x_1 + a_2 x_2$ is equivalent to the differential operator $x_1 \cdot \partial/\partial x_1 + x_2 \cdot \partial/\partial x_2$, and

so a_1 is equivalent to $\xi_1 = \partial/\partial x_1$ and a_2 is equivalent to $\eta_1 = \partial/\partial x_2$. For the complete details of the equivalence, see Salmon, pp. 148–149.

93. Salmon, p. 318.

94. Gordan, "Beweis", p. 323. My translation.

95. Clebsch, "Ueber die symbolische Darstellung algebraische Formen", pp. 26–28.

96. Gordan, "Beweis", p. 325. My translation.

97. Based on their study of the kth transvectant, Gordan and Clebsch proved the pieces of what has since become known as the Clebsch-Gordan formula. In modern terms, the group $SL_2(\mathbb{C})$ has a unique irreducible (rational) representation R_d of dimension $d+1$, which can be realized as a space of homogeneous polynomials of degree d in x and y. The Clebsch-Gordan formula shows how the $SL_2(\mathbb{C})$-module $R_d \otimes R_e$ decomposes into its irreducible pieces, that is, for $d \geq e \geq 0$, $R_d \otimes R_e \cong R_{d+e} \oplus R_{d+e-2} \oplus \cdots \oplus R_{d-e}$. For the modern details, see T. A. Springer, *Invariant Theory* (New York: Springer-Verlag, 1977), pp. 49–51.

98. Gordan, "Beweis", p. 327. My translation.

99. This decomposition of the original transvectant amounted to explicitly exhibiting it according to the Clebsch-Gordan formula. For a glimpse at transvection from the modern point of view, see Springer, pp. 55–58.

100. Salmon, p. 324. Gordan carried this out in "Beweis", pp. 328–332.

101. Gordan, "Beweis", pp. 341–343. His algorithm for calculating the fundamental system gave a finite system but not necessarily the smallest possible one. Thus, it generated a system which, potentially at least, was not syzygy-free. For definitions of the terms "fundamental system" and "syzygy", see note 107 below.

102. Arthur Cayley, "A Ninth Memoir on Quantics", *Math. Papers A.C.*, 7: 334.

103. *Ibid.*

104. *Ibid.*, p. 335.

105. *Ibid.*, p. 353.

106. J. J. Sylvester to Arthur Cayley, April 23, 1877, Sylvester Papers SJC, Box 11. Sylvester's emphasis.

107. For the nineteenth-century invariant theorists, the fundamental system of a quantic or system of quantics was the smallest collection of invariants or covariants which generated all invariants or covariants.

To give an analogy to a more modern concept, the fundamental system of a system of quantics was like the basis of a vector space, except that no two elements in the fundamental system could be connected by any polynomial dependence relation. The elements in the fundamental system were called the "Grundformen" by the Germans and the "ground forms" by the British. Polynomial dependence relations among invariants or covariants were called syzygies.

108. Sylvester to Cayley, April 23, 1877. Sylvester's emphasis.

109. James Joseph Sylvester, "Sur une Méthode algébrique pour Obtenir l'Ensemble des Invariants et des Covariants fondamentaux d'une Forme binaire et d'une Combinaison quelconque de Formes binaires", *Comptes rendus* 84 (1877):1113–1116, 1211–1213, or *Math. Papers J.J.S.*, 3:58–62 on p. 58. The translation is my own.

110. Camille Jordan to J. J. Sylvester, May 13, 1877, Sylvester Papers SJC, Box 2. The translation from French is mine.

111. See James Joseph Sylvester, "Sur les Invariants", *Comptes rendus* 85 (1877):992–995, 1035–1039, 1091–1092, or *Math. Papers J.J.S.*, 3:93–100 on p. 95. Crilly also makes this point in his thesis. See Crilly, "The mathematics of Arthur Cayley", p. 140.

112. Sylvester, "Proof of the Hitherto Undemonstrated Fundamental Theorem of Invariants", *Math. Papers J.J.S.*, 3:117.

113. J. J. Sylvester to Arthur Cayley, November 6, 1877, Sylvester Papers SJC, Box 11. Sylvester's emphasis.

114. Sylvester, "Proof of the Hitherto Undemonstrated Fundamental Theorem of Invariants", *Math. Papers J.J.S.*, 3:123.

115. James Joseph Sylvester, "Détermination d'une Limite supérieure au Nombre total des Invariants et Covariants irréductibles des Formes binaires", *Comptes rendus* 86 (1878):1437–1441, 1491, 1492, 1519–1522, or *Math. Papers J.J.S.*, 3:110–116 on p. 112. The translation is my own.

116. Paul Gordan to J. J. Sylvester, October 6, 1878, Sylvester Papers SJC, Box 2. The translation from German is mine.

117. J. J. Sylvester to Arthur Cayley, March 23, 1881, Sylvester Papers SJC, Box 11. As is often the case in Sylvester's correspondence, the notation he used was not defined. Thus, the precise meaning of the system of equations he gave in this letter remains obscure. He did define his notion of a basic solution, however. As he explained in the same letter, "[a] solution that cannot be made up by the addition of any two others I term *basic* and accordingly any other solution is an *omnipositive integer linear* function of the basic ones (Sylvester's emphasis)." The papers to which Sylvester referred in this letter

are: James Joseph Sylvester, "Sur les Actions mutuelles des Formes invariantives dérivées", *Journal für die reine und angewandte Mathematik* 85 (1878):89–114, or *Math. Papers J.J.S.*, 3:218–240; and "Proof of the Hitherto Undemonstrated Fundamental Theorem of Invariants", *Math. Papers J.J.S.*, 3:117–126.

118. J. J. Sylvester to Arthur Cayley, March 26, 1881, Sylvester Papers SJC, Box 11. Sylvester's emphasis.

119. J. J. Sylvester to Arthur Cayley, September 6, 1882, Sylvester Papers SJC, Box 11. Sylvester's emphasis.

120. J. J. Sylvester to Arthur Cayley, September 14, 1882, Sylvester Papers SJC, Box 11.

121. J. J. Sylvester to Arthur Cayley, October 6, 1882, Sylvester Papers SJC, Box 11. Sylvester's emphasis.

122. J. J. Sylvester to Arthur Cayley, May 26, 1883, Sylvester Papers SJC, Box 11.

123. See Crilly, "The Decline of Cayley's Invariant Theory", pp. 340–341 for evidence of attempts on the British side to prove Gordan's theorem as late as 1886.

124. In this study, we have focused on the British and German contributions to invariant theory. Obviously, any complete history would have to deal with developments elsewhere on the Continent and, later, in the United States.

125. Wilhelm-Franz Meyer, "Bericht über den gegenwärtigen Stand der Invariantentheorie", *Jahresbericht der Deutschen Mathematiker-Vereinigung* 1 (1892):79–292. In the absence of the German text, I have used the contemporaneous French translation by H. Fehr, *Sur les Progrès de la Théorie des Invariants projectifs* (Paris: Gauthier-Villars et Fils, 1897), p. 14 The translation from the French is my own.

126. *Ibid.*, p. 18. My translation.

127. Tony Crilly draws this same conclusion in his thesis as well as in his two articles "The Rise ..." and "The Decline of Cayley's Invariant Theory".

128. See note 89 above.

129. Francesco Faà de Bruno, *Théorie des Formes binaires* (Turin: P. Marietti, 1876).

130. See note 44 above.

131. David Hilbert, "Ueber die Theorie der algebraischen Formen", *Mathematische Annalen* 36 (1890):473–534. For a modern treatment of these

developments, see Springer, pp. 15–42. He gives a sketch of the sequence of mathematical events following Hilbert's work in note 2, pp. 36–38.

132. David Hilbert, "Ueber die vollen Invariantensysteme", *Mathematische Annalen* 42 (1893):313–373.

133. Here, I take issue particularly with Fisher's argument in "The Death of a Mathematical Theory". See note 3 above.

134. David Hilbert, "Mathematical Problems: Lecture Delivered Before the International Congress of Mathematicians at Paris in 1900", *Bulletin of the American Mathematical Society* 8 (1902):437–479, in Felix Browder, ed., *Mathematical Developments Arising from Hilbert's Problems*, Proceedings of Symposia in Pure Mathematics, vol. 28, pts. 1–2 (Providence: American Mathematical Society, 1976), 1:22.

BIBLIOGRAPHY

Aronhold, Siegfried. "Theorie der homogenen Functionen dritten Grades von drei Veränderlichen." *Journal für die reine und angewandte Mathematik* 55 (1858):97–191.

_____. "Ueber eine fundamentale Begründung der Invariantentheorie." *Journal für die reine und angewandte Mathematik* 62 (1863):281–345.

_____. "Zur Theorie der homogenen Functionen dritten Grades von drei Variabeln." *Journal für die reine und angewandte Mathematik* 39 (1849):140–159.

Biermann, Kurt-R. "Gotthold Eisenstein: Die wichtigsten Daten seines Lebens und Wirkens." In Gotthold Eisenstein. *Mathematische Werke*. 2 vols., 2:919–929. New York: Chelsea Publishing Co., 1975.

Boole, George. "Exposition of a General Theory of Linear Transformations." *Cambridge Mathematical Journal* 3 (1841–1842):1–20.

_____. "Researches on the Theory of Analytical Transformations, with a Special Application to the Reduction of the General Equation of the Second Order." *Cambridge Mathematical Journal* 2 (1841):64–73.

Browder, Felix, ed. *Mathematical Developments Arising from Hilbert's Problems*. Proceedings of Symposia in Pure Mathematics. vol. 28, pts. 1–2. Providence: American Mathematical Society, 1976.

Cambridge, England. St. John's College. The Library. James Joseph Sylvester Papers.

Cayley, Arthur. "An Introductory Memoir on Quantics." *Philosophical Transactions of the Royal Society of London* 144 (1854):244–258.

_____. "Mémoire sur les Hyperdéterminants." *Journal für die reine und angewandte Mathematik* 30 (1846):1–37.

_____. "A Ninth Memoir on Quantics." *Philosophical Transactions of the Royal Society of London* 161 (1871):17–50.

_____. "Note sur deux Formules données par M. M. Eisenstein et Hesse." *Journal für die reine und angewandte Mathematik* 29 (1845):54–57.

_____. "Note sur les Hyperdéterminants." *Journal für die reine und angewandte Mathematik* 34 (1847):148–152.

_____. "On Linear Transformations." *Cambridge and Dublin Mathematical Journal* 1 (1846):104–122.

_____. "On the Theory of Linear Transformations." *Cambridge Mathematical Journal* 4 (1845):193–209.

_____. "Recherches sur l'Elimination, et sur la Théorie des Courbes." *Journal für die reine und angewandte Mathematik* 34 (1847):30–45.

_____. "A Second Memoir Upon Quantics." *Philosophical Transactions of the Royal Society of London* 146 (1856):101–126.

_____. "Sur les Déterminants gauches." *Journal für die reine und angewandte Mathematik* 35 (1848):93–96.

Cayley, Arthur and Forsyth, A. R., ed. *The Collected Mathematical Papers of Arthur Cayley.* 14 vols. Cambridge: University Press, 1889–1898.

Clebsch, Alfred. *Theorie der binären algebraische Formen.* Leipzig: B. G. Teubner, 1872.

_____. "Ueber symbolische Darstellung algebraische Formen." *Journal für die reine und angewandte Mathematik* 59 (1861):1–62.

Crilly, Anthony James. "The Decline of Cayley's Invariant Theory (1863–1895)." *Historia Mathematica* 15 (1988):332–347.

_____. "The mathematics of Arthur Cayley with particular reference to linear algebra." Ph. D. Dissertation. Middlesex Polytechnic, June 1981.

_____. "The Rise of Cayley's Invariant Theory (1841–1862)." *Historia Mathematica* 13 (1986):241–254.

Dieudonné, Jean and Carrell, James B. *Invariant Theory, Old and New.* New York: Academic Press, 1971.

Eisenstein, Gotthold. "Allgemeine Auflösung der Gleichungen von den ersten vier Graden." *Journal für die reine und angewandte Mathematik* 27 (1844):81–83.

_____. *Mathematische Werke.* 2 vols. New York: Chelsea Publishing Co., 1975.

_____. "Théorèmes sur les Formes cubiques et Solution d'une Equation du quatrième Degré à quatre Indéterminées." *Journal für die reine und angewandte Mathematik* 27 (1844):75–79.

_____. "Untersuchungen über die cubischen Formen mit zwei Variabeln." *Journal für die reine und angewandte Mathematik* 27 (1844):89–104.

Elliott, Edwin Bailey. *An Introduction to the Algebra of Quantics.* 2d ed. Oxford: University Press, 1913; reprint ed., New York: Chelsea Publishing Co., 1964.

Faà de Bruno, Francesco. *Théorie des Formes binaires.* Turin: P. Marietti, 1876.

Feuer, Lewis S. "America's First Jewish Professor: James Joseph Sylvester at the University of Virginia." *American Jewish Archives* 36 (1984):151–201.

_____. "Sylvester in Virginia." *The Mathematical Intelligencer* 9 (1987):13–19.

Fisher, Charles S. "The Death of a Mathematical Theory: A Study in the Sociology of Knowledge." *Archive for History of Exact Sciences* 3 (1966):137–159.

Gauss, Carl Friedrich. *Disquisitiones arithmeticae.* Trans. Arthur A. Clarke. New Haven: Yale University Press, 1966.

Gillispie, Charles C. ed. *The Dictionary of Scientific Biography.* 16 vols., 1 supp. New York: Charles S. Scribner's Sons, 1970–1980.

Gordan, Paul. "Beweis, dass jede Covariante und Invariante einer binären Form eine ganze Function mit numerische Coefficienten einer endlichen Anzahl solchen Formen ist." *Journal für die reine und angewandte Mathematik* 69 (1868):323–354.

_____. *Vorlesungen über Invariantentheorie.* 2 vols. Leipzig: B. G. Teubner, 1885–1887.

Hawkins, Thomas. "Cauchy and the Spectral Theory of Matrices." *Historia Mathematica* 2 (1975):1–29.

_____. "Cayley's Counting Problem and the Representation of Lie Algebras." *Proceedings of the International Congress of Mathematicians– Berkeley*. 2 vols. N.P.: American Mathematical Society, 1987.

Hesse Otto. "Über die Elimination der Variabeln aus drei algebraischen Gleichungen von zweiten Grade mit zwei Variabeln." *Journal für die reine und angewandte Mathematik* 28 (1844):68–96.

_____. "Über die Wendepunkte der Curven dritter Ordnung." *Journal für die reine und angewandte Mathematik* 28 (1844):97–107.

Hilbert, David. "Mathematical Problems: Lectures Delivered Before the International Congress of Mathematicians at Paris in 1900." *Bulletin of the American Mathematical Society* 8 (1902):437–479.

_____. "Ueber die Theorie der algebraischen Formen." *Mathematische Annalen* 36 (1890):473–534.

_____. "Ueber die vollen Invariantensysteme." *Mathematische Annalen* 42 (1893):313–373.

Jacobi, Carl G. J. *C. G. J. Jacobi's Gesammelte Werke*. Ed. Karl Weierstrass, 8 vols. Berlin: Königlich Preussischen Akademie der Wissenschaften, 1881–1891; reprint ed., New York: Chelsea Publishing Co., 1969.

_____. "De Determinantibus Functionalibus." *Journal für die reine und angewandte Mathematik* 22 (1841):319–359.

_____. "De Formatione et Proprietatibus Determinantium." *Journal für die reine und angewandte Mathematik* 22 (1841):285–318.

Klein, Felix. *Development of Mathematics in the 19ᵗʰ Century*. Trans. M. Ackerman. Lie Groups: History, Frontiers and Applications, vol. 9, Brookline: Math Sci Press, 1979.

Kung, Joseph P. S. and Rota, Gian-Carlo. "The Invariant Theory of Binary Forms." *Bulletin of the American Mathematical Society*, n.s., 10 (1984):27–85.

Lagrange, Joseph-Louis. *Méchanique analytique*. 2 vols. Paris: Librairie scientifique et technique Albert Blanchard, 1965.

MacHale, Desmond. *George Boole: His Life and Work*. Dublin: Boole Press, 1985.

Meyer, Wilhelm-Franz. "Bericht über den gegenwärtigen Stand der Invariantentheorie." *Jahresbericht der Deutschen Mathematiker-Vereinigung* 1 (1892):79–292.

_____. *Sur les Progrès de la Théorie des Invariants projectifs*. Trans. H. Fehr. Paris: Gauthier-Villars et Fils, 1897.

Mumford, David. *Geometric Invariant Theory*. Ergebnisse der Mathematik und Ihrer Grenzgebiete, n.s., vol. 34. New York: Academic Press, Inc., 1965.

Parshall, Karen Hunger. "America's First School of Mathematical Research: James Joseph Sylvester at the Johns Hopkins University 1876–1883." *Archive for History of Exact Sciences* 38 (1988):153–196.

Salmon, George. *Lessons Introductory to the Modern Higher Algebra*. 5th ed. Dublin: Hodges, Figgis, & Co., 1885; reprint ed., Bronx: Chelsea Publishing Co., 1964.

Springer, T. A. *Invariant Theory*. New York: Springer-Verlag, 1977.

Sylvester, James Joseph. *The Collected Mathematical Papers of James Joseph Sylvester*. Edited by H. F. Baker. 4 vols. Cambridge: University Press, 1904–1912; reprint ed., New York: Chelsea Publishing Co., 1973.

_____. "Détermination d'une Limite supérieure au Nombre total des Invariants et Covariants irréductibles des Formes binaires." *Comptes rendus* 86 (1878):1437–1441, 1491, 1492, 1519–1522.

_____. "Memoir on the Dialytic Method of Elimination, Part 1." *Philosophical Magazine* 21 (1842):534–539.

_____. "Note on the Calculus of Forms." *Cambridge and Dublin Mathematical Journal* 8 (1853):62–64.

_____. "On the Calculus of Forms, Otherwise the Theory of Invariants." *Cambridge and Dublin Mathematical Journal* 8 (1853):256–269.

_____. "On a New Class of Theorems in Elimination Between Quadratic Functions." *Philosophical Magazine* 37 (1850):213–218.

_____. "On the Principles of the Calculus of Forms." *Cambridge and Dublin Mathematical Journal* 7 (1852):52–97.

_____. "On rational derivation from equations of coexistence, that is to say, a new and extended theory of elimination, Part 1." *Philosophical Magazine* 15 (1839):428–435.

_____. "On a Remarkable Discovery in the Theory of Canonical Forms and of Hyperdeterminants." *Philosophical Magazine* 2 (1851):391–410.

_____. "On a Theory of Syzygetic Relations of Two Rational Integral Functions, Comprising an Application to the Theory of Sturm's Functions, and That of the Greatest Algebraical Common Measure." *Philosophical Transactions of the Royal Society of London* 143 (1853):407–548.

_____. "Proof of a Hitherto Undemonstrated Fundamental Theorem of Invariants." *Philosophical Magazine* 5 (1878):178–188.

_____. "Sketch of a Memoir on Elimination, Transformation, and Canonical Forms." *Cambridge and Dublin Mathematical Journal* 6 (1851):186–200.

_____. "Sur les Actions mutuelles des Formes invariantives dérivées." *Journal für die reine und angewandte Mathematik* 85 (1878):89–114.

_____. "Sur les Invariants." *Comptes rendus* 85 (1877):992–995, 1035–1039, 1091–1092.

_____. "Sur une Méthode algébrique pour Obtenir l'Ensemble des Invariants et des Covariants fondamentaux d'une Forme binaire et d'une Combinaison quelconque de Formes binaires." *Comptes rendus* 84 (1877):1113–1116, 1211–1213.

Weyl, Hermann. *The Classical Groups: Their Invariants and Representations.* Princeton: University Press, 1939.

Geometry and the Emergence
Of Transformation Groups

Julius Plücker (1801–1868)
Courtesy of Springer-Verlag Archives

The Early Geometrical Works
of Sophus Lie and Felix Klein

David E. Rowe

INTRODUCTION

Sophus Lie (1842-1899) and Felix Klein (1849-1925) first met in the fall of 1869 at a meeting of the Berlin Mathematics Club. It did not take long for them to hit it off. Both were steeped in line geometry, which Klein had learned at the side of Julius Plücker in Bonn and Lie by reading Plücker's works as a student in Christiania (present day Oslo). But they were also drawn to one another for a second reason, namely their common aversion to the scientific atmosphere in Berlin. In particular, they felt like outsiders in the presence of Karl Weierstrass, Leopold Kronecker, and their followers. Almost immediately after his arrival in Berlin, Klein had an inkling that he was going to feel like a fish out of water in the Prussian capital. "Everybody here is working on function theory à la Weierstrass," he wrote Max Noether, "and that almost exclusively; only sporadically does one now and then run into a *Synthetiker*."[1] The trend toward what Klein later dubbed the "arithmetization of mathematics"[2] in the methodology of Weierstrass and Kronecker effectively made these two young upstarts feel as if they had been cast out into the cold. In a letter to his mother, dated 31 October, 1869, Klein set forth these revealing remarks:

> Among the younger mathematicians I have made the acquaintance of someone who very much appeals to me. He is Lie, a Norwegian, whose name I already knew from an article he had published in Christiania. We have especially busied ourselves with similar things, so there is no lack of material for conversation. Yet we are not only united by this common love, but also a certain repulsion to the art and manner in which mathematics here asserts itself over against the accomplishments of others, particularly foreigners.[3]

The following semester Lie and Klein met again in Paris, an environment that was much more to their liking. They took neighboring rooms at a Parisian hotel, enjoyed the charms of the city in springtime, and

209

gradually became acquainted with a number of leading French mathematicians, such as Michel Chasles, Camille Jordan, and Gaston Darboux. Jordan had just published his classic *Traité des substitutions et des équations algébriques,* a work that was instrumental in promoting the development of Galois theory for the next two decades. It was Darboux, however, who provided the major stimulus to the two young foreigners, as it was through him that they became conversant with the rich, and still lively legacy of French geometry in the tradition of Monge and Chasles.

During this time Lie and Klein co-authored two notes that were published in the *Comptes rendus* of the Paris Academy, and they began work on a longer paper devoted to the same subject, namely the theory of W–curves and their associated transformations. Unfortunately, this sunny and memorable liaison in Paris turned out to be all too brief, as in July 1870 the Franco-Prussian War broke out, forcing Klein to return to Germany. Only about a week before his departure Lie made his famous discovery that the line-to-sphere transformation has the property that it carries the asymptotic curves of one surface onto the lines of curvature of another. This realization, which went hand in hand with some of Klein's previous findings, occasioned their joint paper on the asymptotic curves of the Kummer surface published in the *Monatsberichte* of the Berlin Academy.

By the time this article appeared Lie had already returned to Norway following a short sojourn to Italy. In May 1871 he completed the requirements for his doctoral degree, and a month later he was formally awarded the title, which carried with it the right to hold lectures at the university, but, of course, without remuneration. His dissertation was written in Norwegian, but by November he had refined and expanded its contents into an 111-page article entitled "Über Complexe, insbesondere Linien- und Kugel-Complexe, mit Anwendung auf die Theorie partieller Differential-Gleichungen," which appeared the next year in *Mathematische Annalen.* This monumental publication contains in a nascent state many of the essential ideas that motivated Lie's subsequent work throughout the remainder of his career. During this period, which marks the climax of their mathematical relationship, Lie and Klein carried on an intense correspondence.[4] The principal aim of this paper is to reconsider the largely forgotten constellation of ideas that dominated their attention during these early years of collaboration.

When the names Lie and Klein are mentioned together, one almost involuntarily thinks of their mutual contributions to the development of group theory, the former through the theory of Lie groups and the latter through the application of group theory to geometry as expounded

in Klein's "Erlanger Programm." Hans Wussing has, in fact, emphasized how the work of Lie and Klein forms one of the three founding
pillars of early group theory; the other two being, of course, Gauss's theory of binary quadratic forms and the work of Lagrange, Cauchy, Abel,
and Galois on permutation groups and their application to the theory
of equations.[5] The early geometrical investigations of Lie and Klein thus
constitute one of the three principal branches of mathematical research
wherein the group concept was consciously exploited, prior to its emergence as an independent structure considered worthy of investigation for
its own sake. In his article for this volume, "Line Geometry, Differential
Equations, and the Birth of Lie's Theory of Groups," Thomas Hawkins
has in fact shown that a rich variety of group-theoretic structures were
implicitly contained in Lie's earliest geometrical investigations. Without
wishing to minimize the significance of these developments, the present
study attempts to see Lie's and Klein's work in a very different light that
stresses the paths they came from rather than those that emerged in the
wake of their achievements. From this perspective, it cannot be overlooked that Lie and Klein both considered themselves representatives of
that complex and multi-faceted discipline called geometry, as practiced
by such diverse figures as Plücker, Hesse, Grassmann, and Möbius in
Germany, or the followers of Monge and Chasles in France.

One of the standard renditions of the Klein-Lie story is that they
learned about group theory from Jordan and his *Traité* during their sojourn in Paris. Thereafter, so the story goes, they divided the subject
of group theory between them, Lie taking the continuous groups and
Klein those that are discontinuous. However appealing this quaint fable
may be, no one who has studied the early work of Klein and Lie closely
would accept this account as anything better than a first-order approximation of the truth. Indeed, Lie had studied Galois theory for a semester
under Sylow back in 1862.[6] Moreover, as Hawkins has indicated in his
essay, Klein was certainly aware of Jordan's applications of group theory
to geometry well before his arrival in Paris, and there is no reason to
believe that either he or Lie actually studied Jordan's *Traité*, that "book
with seven seals,"[7] in detail. In his article on the implicit use of group-
theoretic concepts in crystallography in volume 2 of these proceedings,
Erhard Scholz has suggested that Jordan's "Mémoire sur les groupes de
mouvements" of 1869 was the primary catalyst that inspired Klein and
Lie to pursue the connection between group theory and geometry. This
interpretation is plausible, but in view of the scant documentary evidence
available, I believe it must be regarded as inconclusive. My own view is
that by early 1870 Klein and Lie were well aware of certain connections
between geometry and group theory, but that it was not until late 1871

that they attached any great significance to the analogy between discrete groups and continuous groups of transformations. Only in the light of the mathematical events that dominated Lie's and Klein's attention during the intervening period can one properly appreciate why this analogy began to play such a prominent role in their thinking.

In order to pursue this goal, however, it is first necessary to consider briefly some of the fundamental background ideas from projective geometry and the theory of line-complexes that motivated the work of Lie and Klein. This will be followed by a summary of their mathematical achievements prior to the semester they spent together in Berlin. A short discussion of the Berlin milieu then leads to a discussion of Lie's research on the geometrical properties of tetrahedral line-complexes. This precedes a lengthy account of Lie's line-to-sphere transformation, stressing its origins and the influence it exerted on Klein. The most significant expression of this influence was Klein's proof of the line-geometric analogue to Dupin's Theorem and his construction of an isomorphism between line geometry and a certain 4-dimensional submanifold in $\mathbf{P}^5(\mathbf{C})$, complex projective 5-space. These ideas were elaborated in a paper of 1871, "Über Liniengeometrie und metrische Geometrie," a work which anticipates many of the central ideas published a year later in Klein's "Vergleichende Betrachtungen über neuere geometrische Forschungen," better known as the "Erlanger Programm."

1. LINE GEOMETRY CIRCA 1870

In 1822 Poncelet developed a purely synthetic approach to projective geometry which he unveiled in his *Traité des propriétés projectives des figures*. This famous work introduced many of the concepts that were to become the basis for modern projective geometry, e.g., perspectivity, projectivity, cross-ratio, involution, and the circular points at infinity. It was Michel Chasles and his followers, however, who first fashioned the analytic apparatus that made it possible to employ these ideas in a systematic fashion. One of the distinct advantages of the analytic treatment of projective geometry is the clarity it brings to the otherwise obscure notion of imaginary points. The quasi-mystical status such entitles seemed to have for synthetic geometers in the mid-nineteenth century is brought to mind by Jacob Steiner's reference to them as "ghosts in the shadowy kingdom of geometry."[8] The following example illustrates both the utility of imaginary points and the ease with which they can be handled analytically.

In the plane each point is represented in homogeneous coordinates by an equivalence class of ordered-triples (x, y, z), where $(ax, ay, az) = (x, y, z)$ and $(0, 0, 0)$ is excluded from the system. To introduce imaginary elements, one simply lets the coordinates take on complex values. In this setting, consider the homogeneous equations associated with an arbitrary conic section and a line, respectively:

$$Ax^2 + 2Bxy + Cy^2 + 2Dxz + 2Eyz + Fz^2 = 0$$

and

$$ax + by + cz = 0.$$

In general, that is assuming no degeneracies are involved, these two equations will always intersect in two points. The points of intersection may be either real or imaginary, finite or infinite, separate or coincident. In the case of two non-degenerate conic sections, the equations will have four points in common, and similar properties hold, of course, for higher order curves. Indeed, Bezout's theorem says that two curves of degree m and n will have mn points of intersection in common. Thus, by introducing complex values and homogeneous coordinates, the manner in which two geometric figures (defined by algebraic equations) intersect one another is independent of their location, i.e., this property is a positional invariant.[9]

A particular instance of the above arises by considering the intersection of an arbitrary circle, given by the homogeneous equation

$$(x - Az)^2 + (y - Bz)^2 - r^2z^2 = 0,$$

with the line at infinity, $z = 0$. Here the equations reduce to $x^2 + y^2 = 0$ and $z = 0$, or the two points $(\pm i, 1, 0)$ on the line at infinity. These two points are known as the circular points at infinity, since, as we have just seen, every circle in the plane passes through them. A similar phenomenon occurs in three-space, where every sphere meets the plane at infinity, $w = 0$, in the conic

$$x^2 + y^2 + z^2 = 0, \quad w = 0,$$

known as the spherical-circle at infinity. The converse property also holds, namely, that every second degree curve that passes through the circular points at infinity is a circle, and every second degree surface that contains the spherical-circle at infinity is a sphere. This property therefore characterizes the three-parameter family, or three-fold infinity, of circles among the ∞^5 conics in the plane as well as the ∞^4 spheres among the ∞^9 quadric surfaces in three-space.

To explore the geometric relations that exist between various objects that intersect at points outside real space, the French geometers often applied an imaginary transformation that would bring these relations "into view." A typical example of this is given by the mapping:

$$x' = x, \quad y' = iy, \quad z' = z$$

which transforms the equation $x^2 + y^2 = 0$ into $x'^2 - y'^2 = 0$, thereby moving the circular points at infinity $(1, \pm i, 0)$ to the real infinitely distant points $(1, \pm 1, 0)$. Under this transformation the family of circles centered at $(a, b, 1)$ will be mapped to the family of rectangular hyperbolas whose asymptotes meet the coordinate axes at a 45 degree angle and which intersect at $(a, b, 1)$. A much-studied object that was handled in an analogous manner was the line-complex formed by the collection of all lines in space that intersect the spherical-circle at infinity. Often called the complex of minimal lines, Lie liked to refer to these as the "verrückten Geraden"[10] (crazy lines), as they seem to have the paradoxical property of lying perpendicular to themselves and having length zero. Actually these are only deceptions caused by the fact that their metric relations are indeterminate with respect to figures in real 3-space.

The subdiscipline of projective geometry known as line geometry may be said to have commenced with the publication in 1846 of Plücker's *System der Geometrie des Raumes in neuer analytischer Behandlungsweise, insbesondere die Theorie der Flächen zweiter Ordnung und Classe enthaltend*. In this work Plücker noted that one can regard the quantities (r, s, ρ, σ) as line-coordinates describing the 4-parameter family of all lines in $\mathbf{P}^3(\mathbb{C})$:

$$x = rz + \rho, \qquad y = sz + \sigma.$$

An algebraic equation, $F(r, s, \rho, \sigma) = 0$, then determines a 3-parameter subfamily of lines in 3-space that Plücker called a line-complex. This naive approach failed to lead to a satisfactory theory, however, owing to the fact that a linear transformation of the point coordinates

$$T: (x, y, z) \rightarrow (x', y', z')$$

will, in general, alter the degree of $F'(r', s', \rho', \sigma') = 0$, the transformed equation of F. In 1865 Plücker found a way around this difficulty: he introduced a fifth coordinate $\eta \equiv r\sigma - s\rho$ dependent on the other four.[11] Using these five coordinates, an nth-degree equation $F(r, s, \rho, \sigma, \eta) = 0$ will indeed transform into an equation of the same degree under a projective transformation of the point coordinates. This meant that one could meaningfully investigate the geometry of an nth-degree line-complex.

First-degree line-complexes, or linear complexes, arose fairly naturally in mechanics; indeed, they had been studied earlier by Möbius who referred to them as null-systems.[12] Geometrically they may be characterized as follows: given a point p, the lines of the complex that pass through p form a plane E. Möbius called E the null-plane associated with the point p, and since this is a dual relationship, it makes sense to refer to p as the null-point associated with the plane E. An important special case occurs when the complex consists of all lines that intersect a given line L; here the line L determines a so-called *special linear complex*. One obtains higher-order versions of this special type of complex by considering the family of all lines that pass through a given nth-order curve in 3-space.

The local geometry of a general nth-degree line-complex C (i.e., a complex determined by $F(r, s, \rho, \sigma, \eta) = 0$, a homogeneous nth-degree equation) has the following structure: given a non-singular point $p \in \mathbf{P}^3(C)$, the lines of C that pass through p form an nth-order cone, a cone whose plane cross-sections are nth-order algebraic curves. Such curves, as we have observed above, intersect an arbitrary line in n points. Dually, one may begin with an arbitrary (non-singular) plane E, in which case the lines of C that lie in E form a curve of the nth-class, i.e. these lines envelope a curve with the property that, given any point, there will be n tangent lines to the curve passing through it. For a 2nd-degree line-complex, the singular points p are those for which the cone of lines through p degenerates into two intersecting planes. Similarily, a singular plane E is one in which the class curve enveloped by the complex lines in E collapses to the set of lines that pass through two fixed points. The locus of the singular points (singular planes) of a line-complex is, in general, very complicated, but it is also of considerable interest in certain particular cases, as we shall see shortly.

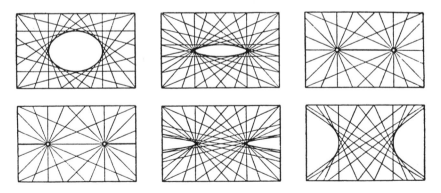

Fig. 1: The Class Curves in Various Planes of a 2nd-Degree Line-Complex

During the mid-1860s, Plücker was preparing a two-volume study entitled *Neue Geometrie des Raumes, gegründet auf die Betrachtung der geraden*

Linie als Raumelement. This novel treatment of geometry was based, as the title suggests, on the notion of regarding lines rather than points as the primary building blocks in the system. For this purpose, Plücker utilized homogeneous complex coordinates (p_i), $i = 1, 2, \ldots, 6$ subject to the restriction that $\sum p_i p_{i+3} = 0$. Following Grassmann, one can obtain such coordinates by letting (x_1, x_2, x_3, x_4) and (y_1, y_2, y_3, y_4) be any two points on a line. Then the six possible 2×2 determinants formed from the matrix having (x_i) and (y_i) as rows yield the coordinates:

$$p_1 = x_1 y_2 - x_2 y_1, \qquad p_2 = x_1 y_3 - x_3 y_1$$
$$p_3 = x_1 y_4 - x_4 y_1, \qquad p_4 = x_3 y_4 - x_4 y_3$$
$$p_5 = x_4 y_2 - x_2 y_4, \qquad p_6 = x_2 y_3 - x_3 y_2.$$

The fact that only four parameters are required to specify a line in 3-space, whereas Plücker's system requires six, is due to the circumstance that his coordinates are homogeneous (hence only five are essential) and subject to the constraint $\sum p_i p_{i+3} = 0$.

The six equations

$$L_i = \begin{cases} p_i = 0, & \text{if } i \neq j, \\ p_i \neq 0, & \text{if } i = j, \end{cases}$$

determine the six edges of a coordinate tetrahedron, often called the fundamental tetrahedron, which gives the system its frame of reference. This method of coordinization is somewhat analagous to that utilized by Möbius in his 1827 study, *Der baryzentrische Calcul*, wherein barycentric coordinates for the points in space are determined by the "weights" assigned to each of the four vertices of a tetrahedron. In his *Neue Geometrie* Plücker concentrated on those complexes of lines that arise from 1st and 2nd-degree homogeneous equations in the line-coordinates.

Families of lines had been studied by numerous mathematicians prior to Plücker, most of these researches having been motivated by the theory of optics. A well-known property of parabolic surfaces is that they reflect light rays parallel to the parabaloid's axis of symmetry into rays passing through the focal point (or *Brennpunkt* in German). In most cases, however, the light rays will not come together at a single focus after reflection or refraction, but instead they will overlap one another enveloping a so-called caustic surface (*Brennfläche*). This effect can sometimes be observed when light rays are reflected by the inside rim of a liquid container producing a caustic curve on the surface of the liquid. Drawing on their knowledge of phenomena such as these, mathematicians like Monge, Malus, and Hamilton began to create the general theory of rays, or two-parameter systems of lines.

The paradigmatic case was the family of light rays emanating from a point, which may be viewed as the collection of normal lines to a sphere. Other elementary examples of ray systems can then be obtained by considering the family of normal lines to any familiar surface. The notions of caustic curves and caustic surfaces which arise in conjunction with ray systems were first introduced by Malus.[13] In general, the caustic surface has two parts to which the system of rays form a family of double tangent lines. In 1866, Kummer showed that the *Brennfläche* for a ray system of the second order and class is, in general, a surface of the fourth order and class with 16 nodal points and 16 double tangent planes.[14] The properties of these so-called Kummer surfaces were of considerable interest to Klein, Lie, and their contemporaries, as they contain all other fourth-order algebraic surfaces as special cases.

The fundamental objects in Plücker's line geometry, on the other hand—algebraic line-complexes—were families of lines that required three parameters to be represented analytically. The lines common to two such complexes form a 2-parameter subfamily, and thus a ray system. Plücker called such systems *congruences*, and showed that if two given line-complexes have degrees m and n, then the order of the congruence they determine is mn, the number of lines that pass through an arbitrary (non-singular) point. This number need not agree with the class of the congruence, which is given by the number of its lines in an arbitrary plane. The same is true of the *Brennfläche* associated with a given congruence; in fact the difference between the order and class of the *Brennfläche* will be precisely twice that of the congruence. In case the congruence happens to belong to a linear complex, however, the order and class of the congruence will be equal. In particular, the congruence determined by the intersection of two linear complexes will be of the first order and class. Furthermore, for any congruence, its lines intersect two fixed lines, known as the *directrices* of the congruence, which may be either real or imaginary. The intersection of three line-complexes, on the other hand, yields a one-parameter family of lines that form a ruled surface. If the complexes have degrees m, n, and p, then the order and class of this surface is $2mnp$. Thus three linear complexes determine a quadric surface, for which the two lines that pass through an arbitrary point on the surface will be generators.

2. KLEIN'S FIRST PUBLICATIONS

Unfortunately Plücker was unable to bring all these ideas to their full fruition, for he died rather suddenly in May 1868. At the time of his death he had managed to complete the first 100 pages or so of the second volume of his *Neue Geometrie*. Responsibility for completing the work then fell to Alfred Clebsch, and it was this circumstance that brought Klein into close contact with the man who was to become the second great formative influence on his life as a mathematician. After completing his dissertation and passing his doctoral examination on December 12, 1868 in Bonn, Klein spent eight months in Göttingen where he became a close associate of the Clebsch school, befriending Max Noether and Ferdinand Lindemann among others. Drawing on his intimate knowledge of Plücker's theory, he was able to bring the second volume of his teacher's culminating work to a satisfactory completion during his stay there.

Yet even before Klein came to Göttingen in early 1869, Clebsch was already making his influence felt on him. While still in Bonn, Klein busied himself studying a recently published paper by Battaglini that Clebsch had urged him to read. Battaglini's work dealt with the representation of first and second-degree line-complexes by canonical algebraic forms. What Klein soon discovered, however, was that, contrary to appearances, the quadratic form Battaglini had derived was not completely general. This gave him the idea he eventually turned into his doctoral dissertation: namely, to exhibit a truly general canonical form for every second-degree line-complex.

Although the flavor of this problem appears to be highly algebraic, its main appeal for Klein lay in its implications for the geometry of line-complexes. In the 1840s, Plücker had already undertaken an analogous study in which he showed that for any surface F given by a homogeneous equation $\sum a_{ij}x_ix_j = 0$ it is always possible to find a coordinate tetrahedron such that the equation is transformed into another in which $a_{ij} = 0$ for $i \neq j$. By means of this approach, it was a simple matter to classify the possible surfaces that can arise from a second-degree equation. These are: a nondegenerate second-degree surface, a cone, a pair of planes, and a double plane. It should be emphasized here that this classification scheme utilizes transformations of 3-space that, in general, will be imaginary. The classification of all real second-degree surfaces by means of real transformations leads to many more types, as the canonical form in this case will have quadratic terms whose coefficients may be negative as well as positive.[15]

Klein's dissertation was motivated by concerns similar to these; in particular, he hoped to obtain a related form of classification for 2nd-degree line-complexes. To produce this, he sought to find a linear transformation that would map the original coordinate tetrahedron to another, with respect to which the new quadratic form would be as simple as possible. This problem was complicated, however, by the fact that the transformation also had to preserve the second form $P = \sum p_ip_{i+3} = 0$.

Battaglini had assumed that the original form X could always be transformed to $X' = \sum k_i p_i^2 = 0$, the situation analogous to Plücker's treatment of surfaces outlined above. After noticing that in fact this was not the case, Klein made a number of unsuccessful attempts to determine a truly canonical form. This problem was closely linked to the purely algebraic theory for the simultaneous diagonalization of two quadratic forms.

Klein's first inclination was to perform a series of transformations of X by means of which P would maintain its original form. Eventually he hit upon the "right" strategy: one could first diagonalize X and P to obtain X' and P', and then follow this by another transformation taking (X', P') to (X'', P''), where P'' has the same form as P. X'' will then be the desired canonical form for the line-complex. Once he saw this, he promptly wrote up his solution and sent it on to Rudolf Lipschitz, his dissertation advisor in Bonn. By sheer coincidence, the latter had just received page proofs for an article by Weierstrass which presented for the first time his theory of elementary divisors. Thus Weierstrass's paper, "Zur Theorie der quadratischen und bilinearen Formen," dealt systematically with the simultaneous transformation of two quadratic forms.[16] This was a particularly propitious circumstance, since Klein's original analysis, as was customary among mathematicians of his day, dealt only with the general case where the six roots of the characteristic polynomial given by $\det(X + kP) = 0$ are all distinct. Weierstrass's approach, on the other hand, gave the first comprehensive presentation of a theory that dealt with all possible cases. Lipschitz therefore thought it a good idea for Klein to make use of this new theory by including consideration of the ten degenerate cases that arise in the problem. This accounts for the unusually strong algebraic flavor of Klein's dissertation.

Even so, it was clearly the general case and the geometry associated with it that compelled Klein to analyze this problem in detail. After noting that it is in this case alone (i.e., when the six roots of $\det(X + kP) = 0$ are distinct) that (X, P) can be transformed into (X', P') where

$$X' = \sum k_i X_i^2 = 0$$

and

$$P' = \sum X_i^2 = 0,$$

he pointed out that this representation is by no means unique. In fact, there will be a 15-parameter family of transformations that produces the same result. These mappings are associated with fifteen distinguished tetrahedra, of which, respectively, six, two, one, or one will be real, depending on whether, respectively, zero, two, four, or six of the roots of the polynomial in k are imaginary. By utilizing such a real tetrahedron,

X' can be transformed linearly so as to take on the following canonical form:

$$X'' = a_{14}(p_1^2 + p_4^2) + a_{25}(p_2^2 + p_5^2) + a_{36}(p_3^2 + p_6^2) + 2b_{14}p_1p_4 + 2b_{25}p_2p_5 + 2b_{36}p_3p_6$$

where, as before, the indices $(i, i + 3)$ are associated with opposite edges of the coordinate tetrahedron.

In Klein's subsequent analysis he takes

$$(*) \qquad\qquad X = \sum k_i X_i^2 \quad \text{and} \quad P = \sum X_i^2$$

as his starting point—thus (X, P) is to be thought of as already in this transformed form. In these coordinates, the equations $X_i = 0$ represent six linear complexes—called the fundamental complexes—which are intimately related to the fifteen distinguished tetrahedra associated with (X, P). Taking any pair of these, such as the congruence $(1, 2)$ consisting of those lines common to $X_1 = 0$ and $X_2 = 0$, one obtains as directrices the two lines $L_{12} = (k, \pm ki, 0, 0, 0, 0)$; and these from two of the opposite edges for one of the 15 distinguished tetrahedra. Moreover, the symmetry in the equation $P = 0$ ensures that the directrices L_{12} will belong to the other four fundamental complexes, X_i $i = 3, 4, 5, 6$. It follows that for each of the 15 distinguished tetrahedra each pair of opposite edges will be shared by two other tetrahedra; hence there are only 30 edges in all, or 15 groups of opposite pairs. To obtain these tetrahedra, one therefore simply runs through the various directrices associated with each of the $C(6, 2) = 15$ congruences (i, j) determined by the possible combinations of pairs of the $X_i = 0$.

The lines common to a triple of linear complexes, as remarked earlier, form a surface of the second order and class. From this, two facts follow: first, for the fundamental complexes $X_i = 0$, the triple $(1, 2, 3)$ consisting of all lines common to $X_1 = 0$, $X_2 = 0$, and $X_3 = 0$ is such a surface. Second, this surface has two sets of generators given by the two sets of six lines: L_{12}, L_{23}, L_{31} and L_{45}, L_{56}, L_{64}, where L_{jk} is the pair of lines with ratio $\pm i$ in the jth and kth coordinates and all other entries equal to zero. It follows that the triple $(4, 5, 6)$ determines the same surface as does $(1, 2, 3)$, so that altogether there are $C(6, 3)/2 = 10$ such *fundamental surfaces*.

Returning to the form $(*)$ in the case where all the elementary divisors are real and distinct, Klein noted that an arbitrary line $(X_1, X_2, X_3, X_4, X_5, X_6)$ belonging to this complex is automatically associated with 31 other lines of the complex. This is due to the circumstance that $X = 0$ remains invariant when X_i is replaced by $-X_i$ or $-X_{i+3}$, for $i = 1, 2, 3$ (these choices then determine the value of X_{i+3} in $X = 0$). Thus, there are altogether $4^3/2 = 32$ different possible lines so obtained, since changing the sign of *all* the coordinates yields the same line. These facts led Klein to the following interesting observation. Given an arbitrary line $L = (X_i)$ of the complex, the corresponding polar line $L_j, j = 1, 2, \ldots, 10$, with respect to each of the 10 fundamental surfaces will be among the 32 lines associated with L. From this it follows that the complex $X = 0$ has the property that it is self-reciprocal with respect to each of these 10 surfaces.

Now Plücker had shown that there is a general polar relationship between the lines in space with respect to an arbitrary 2nd-degree line-complex X.[17] Given a line L, one obtains the polar L' with respect to X by considering the pencil of planes E that pass through L. Each such plane determines a conic enveloped by the lines of X, and Plücker showed that as the planes E run through the pencil the poles of L with respect to each of the corresponding conics will sweep out the line L'. In general, this relationship is not reciprocal; that is, if L' is the polar line corresponding to L, then its polar L'' will ordinarily *not* be the original line L. Klein observed, however, that if L is an edge of one of the 15 fundamental tetrahedra T associated with the complex (X, P), then its corresponding polar L' will be the opposite edge of T. This means that for the 30 edges of the system of fundamental tetrahedra the polar relationship is, in fact, reciprocal. Furthermore, these turn out to be the *only* lines for which this is the case. This key property, the culminating result of Klein's dissertation, reveals that the system of distinguished tetrahedra he had defined algebraically actually reflects the natural geometry carried by the 2nd-degree line-complex. Indeed, their role is fully analogous to that of the tetrahedra utilized by Plücker in representing an arbitrary 2nd-degree surface by an equation in the squared terms alone. In his subsequent work, Klein had occasion to explore this nexus of ideas even further, and in so doing he discovered a link between the theory of line-complexes and the Kummer surface.

These further results were presented in his article "Zur Theorie der Linienkomplexe des ersten und zweiten Grades," written in Göttingen and published in the second volume of *Mathematische Annalen*.[18] The setting here is the same as that in his dissertation of less than a year earlier, namely the investigation of the properties associated with a 2nd-degree line-complex that has been transformed into the canonical form (*). But whereas the algebra palyed a major role in his dissertation, here the underlying geometry comes fully into play, and in a startlingly unexpected manner.

In his *Neue Geometrie* Plücker discussed so-called singular surfaces associated with 2nd-degree line-complexes.[18] These arise as the locus of all points p in space where the cone of lines through p has degenerated into a pair of planes, or, dually, as the surface enveloped by the family of planes for which the correponding class curves have degenerated into two plane pencils of lines. In the latter case, the *singular lines* joining these point pairs and enveloping the singular surface form a congruence, which Plücker showed was of the fourth order and class. The locus of singular points for a 2nd-degree complex was thus one of the caustic surfaces discovered by Kummer in 1866 in the course of his work on ray systems. In the general case, these surfaces possess 16 nodal points and 16 double tangents, the maximum number possible for a 4th-degree surface. The Fresnel wave surface, among others, is a special case of the Kummer surface. Klein now went on to show that for any given Kummer surface S there is a one-parameter family of 2nd-degree line-complexes each of which has S as its singular surface. He also showed that practically all properties then known about Kummer surfaces could be derived with great ease by means of Plücker's techniques. Indeed, his presentation was a convincing argument for the viewpoint that line geometry is the natural setting for studying the Kummer surface. What follows is a brief summary of a few highlights from this paper.

Again, Klein makes considerable use of the system of 6 fundamental complexes, 10 fundamental surfaces, and the various reciprocal pole-polar and polar-polar relationships that exist with respect to both of them. Beginning with an arbitrary plane E, he notes that the 6 poles determined by the fundamental complexes $X_i = 0$ all lie on a conic section, and, dually, that the 6 polar planes determined by a given point and these 6 complexes will envelope a cone of the second class. In the first case, let $p_1, p_2, p_3, p_4, p_5, p_6$ denote the points so obtained. Then these will determine 15 new planes, 5 of which are the polar plane associated with a given point, say p_1. These 5 will be the polar planes associated with p_1 and each of the 5 complexes $X_i = 0, i = 2, 3, 4, 5, 6$. Since the polar plane determined by p_2 with respect to $X_1 = 0$ is identical with the polar of p_1 with respect to $X_2 = 0$, it follows that there are altogether $C(6, 2) = 15$ planes possible in addition to the original plane E.

Moreover, the 3 planes determined by the point pairs

$$(p_1, p_2), (p_2, p_3), (p_3, p_1)$$

will intersect in the pole of E with respect to the fundamental surface determined by the triple $(1, 2, 3)$, as will the 3 planes determined by $(p_4, p_5), (p_5, p_6)$, and (p_6, p_4), since the triple $(4, 5, 6)$ produces the same fundamental surface as $(1, 2, 3)$. These 6 planes then circumscribe a cone of the second degree. Klein concluded from these deliberations that, given a system of fundamental complexes, the points and planes of space are naturally partitioned into families of 16 each. Thus, beginning with an arbitrary plane E, one obtains 15 further planes as above. Associated with these are 6 co-planar points given by the poles of E with respect to each of the fundamental complexes, as well as 10 other poles of E with respect to each of the fundamental surfaces. Since all of these tightly-knit interrelationships are symmetrical, i.e., one could just as well have started with any of the 15 other planes or one of the 16 points for that matter, it must be the case that 6 of these 16 points lie in each of the 16 planes, and that 6 of the 16 planes pass through each of these 16 points.

This set of circumstances led Klein to a key insight: for these are exactly the same interrelations that exist between the 16 nodal points (*Doppelpunkte*) and 16 double tangents of the Kummer surface. Thus one could regard a given Kummer surface as the singularity surface associated with a one-parameter family of 2nd-degree line-complexes. If, in addition to the Kummer surface K, a tangent line L to it is given, then there will be a unique 2nd-degree complex having K as its singularity surface and L as a singular line. Should L happen to be a double tangent to K, then the corresponding complex degenerates into one of the 6 fundamental linear complexes, namely, the one that happens to contain L. In this special case, the singular lines of the complex will be the double tangents to the surface. Analytically, the one-parameter family of 2nd-degree line-complexes associated with a given Kummer surface has the form

$$\sum_{k=1}^{6} \frac{y_k^2}{a_k - \lambda} = 0.$$

As we shall see in §6, this formula, which bears a striking resemblance to the formula for a one-parameter family of confocal quadric surfaces, turned out to have a surprising and much deeper connection with French sphere geometry as developed by Darboux and others.

A number of other consequences follow from this fairly immediately. One of these concerns the points on the surface K where the singular lines are asymptotic, so-called *Haupttangenten*. One can define these asymptotic directions by means of the Dupin indicatrix, which is the curve obtained at each point p of a given surface S by intersecting S with a plane E sufficiently close to and parallel with T_p, the tangent plane to S at p. If the Gaussian curvature $K = \kappa_1 \kappa_2$ at p is positive, the Dupin indicatrix will be an ellipse, and the asymptotic directions will be imaginary. If, however, K happens to be negative, the indicatrix will be a hyperbola whose asymptotes are the asymptotic directions at p. A curve C on S whose tangent lines all point in an asymptotic direction is called an *asymptotic curve (Haupttangentenkurve)*. Such a curve may also be characterized by the property that the osculating plane and tangent plane coincide at each of its points.

What Klein observed in connection with the Kummer surface K was that the locus of points Q whose singular lines lie in an asymptotic direction form algebraic curves of degree sixteen. His proof ran as follows. Consider the fixed 2nd-degree line-complex C associated with K and the singular line L passing through point p in Q. The line L lies in the tangent plane T_p, and since p is a singular point, the remaining lines of T_p that pass through p must also belong to C. If now E ia an arbitrary plane containing p, then the intersection of T_p with E will be a common tangent line to the surface K as well as the 2nd-order curve F_1 determined by the complex C and the plane E. Thus for any plane E, the point p has the characterisitc property of being the common intersection of the tangent to F_1 with the 4th-order curve F_2 determined by the intersection of E with K. By symmetry considerations, this property also characterizes any other points of Q. Klein then noted that the curves F_1 and F_2 are tangent to one another in four points. Since F_1 and F_2 are of the 2nd and 12th class, respectively, he concluded that there must be $2(12) - 2(4) = 16$ other double tangents to both in E. Thus he has shown that, given an arbitrary plane E, one always obtains 16 points of F_2 as points of contact with these double tangents. Furthermore, all have the property that the singular lines through them lie in an asymptotic direction. It follows that each point p in Q lies on a curve of the 16th order. To obtain the totality of all such curves, one simple varies the line-complexes in the one-parameter family associated with K. Yet what Klein failed to notice at this time, although it surely must have crossed his mind, was that the curves he was studying were, in fact, the asymptotic curves of the Kummer surface. This discovery–that the asymptotic curves of a Kummer surface are algebraic and of degree 16–was made by Sophus Lie following an entirely different line of inquiry that will be taken up later.

3. LIE'S THEORY OF THE IMAGINARY

Sophus Lie was a highly original thinker whose work bore a very individualistic stamp right from the beginning. The starting point for his mathematical research began with a transformation that enabled him to study the behavior of imaginary curves in the $\mathbf{P}^2(\mathbb{C})$-plane by way of "striped" surfaces in projective 3-space, $\mathbf{P}^3(\mathbb{R})$.[20] To accomplish this, he considered the mapping $T\colon \mathbf{P}^2(\mathbb{C}) \to \mathbf{P}^3(\mathbb{R})$ which takes the point $(X, Z) = (x + yi, z + pi)$ to the point (x, y, z) with "weight" p. The "stripes" are then curves on an image surface with a given weight, among which the curve with $p = 0$ is singled out as the "null-stripe." Viewed in this way, the image of the complex line $BZ = X - A$ will be a plane E whose stripes
</caption>

comprise the family of parallel lines of maximum inclination orthogonal to the intersection of E with a horizontal plane. This property of maximal inclination also holds in the general case where the stripes are nth-order curves on an algebraic surface.

Lie gave these deliberations a Plückerian twist by fixing (X, Z) and allowing A and B to vary, so that $BZ = X - A$ represents the family of lines through the point $(X, Z) = (x + iy, z + pi)$. Under the above transformation, one obtains the collection of striped planes passing through (x, y, z). The "null-lines" among them then determine a line congruence, with the vertical line through (x, y, z) acting as its axis, and possessing the key property that it determines similar figures when cut by a horizontal plane. Moreover, following a result of Plücker, its directrices pass through the circular points at infinity associated with the (x, y)-plane. Lie's inspiration led him to associate this line congruence with the point $P = (X, Z)$ in $\mathbf{P}^2(\mathbf{C})$ (if P is a null-point, then the congruence merely reduces to the pencil of lines through its image point in $\mathbf{P}^3(\mathbf{R})$). By so doing, he managed to dispense with the notion of "weighted" points for the most part, utilizing this correspondence to establish a new link between the geometry of the complex projective plane and congruences of lines in real projective 3-space. Lie was not the first to give a geometric interpretation of imaginary points in real space, as some ten years earlier Christian von Staudt developed such a theory by utilizing involutionary mappings on the points of a line. But Lie's theory was apparently the first to forge a connection between the complex plane and line geometry.

For Gauss, the discovery that the 17-gon is constructible proved to be the decisive event that led him to become a mathematician (up until then he had been contemplating a career in philology). In all likelihood something similar must have happened to Sophus Lie after he made this discovery that seemed so full of new possibilities for studying imaginary quantities geometrically. Without question he was enthralled by the prospects this new system seemed to open up. He engaged a printer who published an eight-page paper entitled "Repräsentation der Imaginären der Plangeometrie" in which he announced these results. And in a request for financial support from the *Collegium Academicum* in Christiania, Lie submitted a copy of this paper to which he appended some additional, hand-written remarks, including the following:

> Algebra introduced the concepts of "imaginary point, imaginary line" into geometry. The word imaginary is here, as always, to be translated as unconceived rather than inconceivable. Geometry has not dismissed these ideas, but reasons with them, using

them as an intermediary tool for propositions concerning conceived geometric entities. This reasoning is algebraic. One has searched for a representation of the imaginary in geometry, that is, one has searched for higher viewpoints and more comprehensive definitions under which the imaginary entities would also exist. Were this goal attained, one would nevertheless still be able to pose geometric problems that could only be solved algebraically. However, each such problem would correspond to a more comprehensive problem that could be formulated algebraically, just like the other, but which would have a geometric solution.[21]

There can be no question that Lie thought that such an approach to imaginary elements lay within his grasp when he wrote these words. In both his first paper and the expanded version that he wrote before coming to Berlin, he even included a short section setting forth the "metaphysical" implications of his theory. There he emphasized that "every theorem in plane geometry [i.e., in $\mathbf{P}^2(\mathbb{C})$] is a special case of a stereometric double-theorem [since points and planes are dual objects] in the geometry of line-congruences."[22] In order to realize this program, Lie introduced into his new system notions analogous to cross-ratio, orthogonality, angle measure, etc. With these he was able to derive new results utilizing metrical concepts as well as purely projective theorems.

Much of what appears in these first papers on the imaginary is very difficult to understand, as Lie's presentation is at once rather loose and extremely pithy. Indeed, the commentary by Engel and Heegaard in Lie's *Gesammelte Abhandlungen* runs to 130 pages, and even they were unable to decipher everything. Little wonder that this work exerted next to no influence on contemporary mathematicians. Still, it provides a very suggestive picture of the kind of ideas Lie was grappling with during these early years. A number of these ideas were to surface again in a different guise or become crystallized in more refined form in the work immediately following. In particular, one encounters several examples of contact transformations, a notion that only became explicit in his later work where it assumed an ever more important role. Most characteristic of all is Lie's constant search for mappings that enable one to transfer familiar properties and constructs from one space over to another. This was the key idea that motivated his theory of the imaginary, and it was an approach that he would turn to again and again. Murky as these early papers may have been, they certainly testify to Lie's uncanny powers of spatial intuition.

4. LIE AND KLEIN IN BERLIN

The focal point of Lie's and Klein's activities during their brief stay in Berlin centered around Kummer's seminar on general ray systems and the differential geometry of congruences, a field that owed much to W. R. Hamilton's publications of the 1840s. They were fortunate to have come to Berlin when they had, as ten years earlier Kummer's research was exclusively devoted to number theory. By the 1860s, however, he had shifted rather abruptly over to geometry, and in the course of these studies discovered the well-known surface that bears his name. Lie and Klein both took an avid interest in Kummer's seminar and made a number of presentations to its members. Since he had no doctoral students at the time, it appears that Klein and Lie were the only ones who had a real command of the subject matter being covered.[23] Not surprisingly, Kummer was highly impressed with both of them. In a conversation with Klein he remarked that Lie appeared to be an altogether unusual talent.[24] As for Klein's performance, Kummer and Weierstrass sent this report to the Prussian Ministry of Culture:

> Dr. phil. Felix Klein from Düsseldorf... participated in the seminar in the most avid manner and held a number of lectures in it that, from the formal stand-point, were altogether excellent. As he is also under-taking scientific researches with lively ambition and untiring diligence supported by good talent, it is to be expected that he will distinguish himself as a mathe-matics teacher and will exercise a very beneficial influ-ence.[25]

Klein's foremost talent was an ability to quickly absorb new ideas and find hidden interconnections between them. Consequently he was interested in taking in as much of Berlin mathematics as he could during his stay there. He attended Kronecker's lectures on quadratic forms and number theory and met privately with Weierstrass to discuss recent work of Clebsch and Noether in algebraic geometry.[26] Although Weierstrass's lectures were by far the most popular among the advanced courses offered at Berlin, Klein never attended them regularly, a decision he came to regret later in life.[27] Part of the reason for this was his aversion to what he felt was a "cult" surrounding the "master of us all."[28] Klein felt that many of those who attended Weierstrass's lectures did so in a spirit of ritual piety in which they paid homage to the teachings of a modern-day sage rather than developing a critical understanding of the subject matter he imparted. He described Weierstrass to Max Noether as a "very imposing" personality, even in private conversation, but he also thought this aspect of his nature tended to contribute to the authoritarian air in his classes, which he found to be in sharp contrast with the milieu that surrounded Clebsch in Göttingen.[29]

Aside from his friendship with Lie, Klein's most important contact in Berlin was with the Austrian mathematician Otto Stolz, who became professor of mathematics at Innsbruck two years later.[30] Having read the works of Bolyai and Lobachevsky, Stolz was able to give Klein an excellent grounding in non-Euclidean geometry, a field that was still very unfamiliar to most mathematicians at this time. Stolz had also made a thorough study of von Staudt's difficult, yet fundamental work on projective geometry. Staudt had introduced projective coordinates without making use of an underlying metric, thereby freeing the purely projective concept of cross-ratio from its former reliance on metric geometry. This construction turned out to be of fundamental importance for Klein in his subsequent work on the foundations of geometry.

Around the same time Klein was becoming acquainted with these ideas through Stolz, he also became familiar with Arthur Cayley's "Sixth Memoir on Quantics," probably through his reading of the 1860 edition of the Salmon-Fiedler textbook on conics. In this paper Cayley introduced a generalized metric in projective space by fixing an "absolute figure"—a conic section in the plane—and making use of the invariance of the cross-ratio under projective transformations. In the case of ordinary metric geometry, this absolute figure degenerates into the circular points at infinity. Cayley summarized the significance of his findings in the following words:

> ... the theory, in effect is, that the metrical properties of a figure are not the properties of a figure considered *per se* apart from everything else, but its properties when considered in connexion with another figure, viz. the conic termed the Absolute. The original figure might comprise a conic; for instance, we might consider the properties of the figure formed by two or more conics, and we are then in the region of pure descriptive [i.e., projective] geometry: we pass out of it into metrical geometry by fixing upon a conic of the figure as a standard of reference and calling it the Absolute. Metrical geometry is thus a part of descriptive geometry, and descriptive geometry is *all* geometry...[31]

These ideas, particularly Cayley's construction of a generalized metric defined by referring to a fixed 2nd-degree curve, intrigued Klein, and he soon began to speculate that there might be some connection between certain special cases of the Cayley metric and the non-Euclidean geometries. In later years, he was fond of telling how he had been rebuffed by Weierstrass when he suggested this possibility during his final seminar lecture in Berlin.[32] This alleged incident, however, is one of those heroic tales to which one should not attach undue significance. Klein might well have thought that there *might* be some connection between the Cayley metric and non-Euclidean geometry, but he apparently never pursued this question seriously until more than a year later. Thus, even if there is no reason to believe this story is apochryphal, its significance could easily have become embellished over the years.

All that can be said with certainty is that Klein did lecture on the Cayley metric in mid-March, about two weeks after Lie had already left Berlin for Göttingen.[33] It was not until the summer of 1871, however, after he had returned to Göttingen as a *Privatdozent*, that he took up this question again in earnest. According to his later accounts, the presence of Otto Stolz in Göttingen played a decisive role in spurning him on in this quest. By late August he was able to submit a provisional version of "Über die sogenannte Nicht-Euklidische Geometrie" for publication in the *Göttinger Nachrichten*. Here, and in two more detailed articles with the same title that were to follow, Klein showed that by utilizing an appropriate quadric surface in 3-space as absolute figure the Cayley metric yields one of three types of geometry: elliptic (when the absolute figure is imaginary); hyperbolic (when it is a real, non-ruled surface); and parabolic (when it degenerates to the spherical-circle at infinity).[34]

It should be mentioned here that Lie probably knew very little, if anything, about Klein's early reflections on the Cayley metric. Having left Berlin about two weeks prior to Klein's final lecture there, it is unlikely that he heard about these speculations then. Neither does there seem to be any evidence indicating that he knew about them later, at least not until the summer of 1871. Even if he had, it is difficult to imagine that he would have given Klein's musings on this subject more than passing notice. Lie was simply too immersed in a wealth of new ideas of his own to be distracted by vague thoughts about the connections between projective and non-Euclidean geometry, a field he knew nothing about and which would only engage his attention many years later.

These circumstances are significant for the delineation of those areas that were of mutual interest to Klein and Lie, and through which they exercised considerable influence on one another. For Klein, the Cayley metric emerged as a key tool for the systematization of geometry and an important source of inspiration for the Erlangen Program. Indeed, the Erlangen Program itself was largely motivated by Klein's desire to transcend the framework implied by Cayley's dictum that "projective geometry is all geometry." He accomplished this by generalizing Cayley's principles so as to make them applicable to a much broader category of geometrical phenomena. It is therefore difficult to see how Klein could have even formulated the leading principles of the Erlangen Program without having first uncovered the connection between projective and non-Euclidean geometry. These remarks are not meant to downplay the considerable impact that Lie's mathematics exerted on Klein's thinking, including, as we shall see, the ideas set forth in the Erlangen Program. This influence, however, had virtually nothing to do with Klein's work on non-Euclidean geometry.

While in Berlin, Lie lectured several times in Kummer's seminar on results he had obtained in the theory of tetrahedral line-complexes. Aside from this seminar, however, he took almost no interest in what Berlin had to offer mathematically. No doubt this was due in part to his strong predilection for geometry, a field that was given very short shrift in Berlin, where analysis, algebra, and number theory dominated the scene. Nevertheless, this reluctance to widen his horizons was also characteristic of Lie's working habits and general temperment. Genius is often one-sided, and in this respect Lie was certainly no exception: a strong, independent personality, he had the ability to shut out everything around him in order to devote himself exclusively to his own ideas. His originality and immense powers of concentration were manifest right from the beginning. Even at the very earliest stages of his career he immersed himself in his own research problems and cared little about ideas that were not directly related to his work. More than twenty years later, Klein drew this comparison of their mathematical styles:

> Lie was a thoroughly productive researcher; he worked
> out his own ideas and only took up something foreign
> when it was of immediate interest to him. I myself
> had already developed the ideal in Bonn, to which
> I have since remained faithful, that I wished to ac-
> complish something in science by comprehending and
> comparing various standpoints. In particular, I also
> followed then, and for a long time afterward, an in-
> terest in physics. I was by no means so exclusively
> mathematical as Lie.[35]

5. LIE'S WORK ON TETRAHEDRAL COMPLEXES

Lie's mathematical training was indeed not very solid, in part because he had vacillated over his choice of a career right up until 1868. Apparently what awoke his slumbering mathematical genius at that time was the inspiration he gained from reading Poncelet and Plücker, although his earliest work was also clearly influenced by the work of Chasles as well. By the time he met Klein, however, Lie's interests had already begun to expand beyond his early work on the representation of imaginary curves in $P^2(C)$. His research now turned to the transformation properties of tetrahedral complexes, a theme he developed by using methods related to his earlier work. He had every intention or returning to his representation theory at some point, and he even announced at the close of his final article in the *Fordhandlinger* of the Scientific Society of Christiania that a sequel would follow. But he never did, undoubtedly because he felt that the ideas that remained to be worked out in the theory proper were not nearly as rich as those that it had spawned in the meantime. At any rate, Felix Klein was certainly excited about the new vistas opened by Lie's investigations, as may be seen from the following passage taken from a letter to Max Noether dated December 7, 1869:

> First I must tell you about the investigations of Dr. Lie, the same person from Christiania who last summer sent Clebsch the notice on the representation of two complex variables in 3-space. By means of this representation, which you undoubtedly remember, he can map objects in 3-space to those in the plane. Each transformation in the plane corresponds to another in 3-space whose degree, however, is in general one greater than the degree of the plane transformation. . . . By following up on such like considerations, Lie is now in possession of a great number of space transformations that he can manipulate with extreme ease. The principal object of his investigations right now is the 2nd-degree Reye complex [tetrahedral complex]. This can be mapped onto 3-space so as to induce an involutionary relationship in the image space—this is similar to the reciprocal relationship induced by a conic section in the plane, but carried over to 3-space following Lie's method. By means of this technique, one investigates not just the image space alone, but rather the image space and complex simultaneously, that is by constantly switching or combining observations of both structures one obtains with great ease a whole host of results.[36]

As this letter reveals, in their early work Lie and Klein referred to tetrahedral complexes as Reye complexes, after Theodor Reye, who investigated their properties in the 1868 edition of his *Geometrie der Lage*. It was not until a year later, however, that H. Müller first called attention to their key defining property, namely that the lines of such a complex meet the faces of a tetrahedron in a fixed cross-ratio.[37] This characteristic was of fundamental significance in Lie's subsequent investigations, which were primarily concerned with various families of curves associated with a given tetrahedral complex. These curves, which are enveloped by lines of the complex, will, like the complex itself, remain invariant under the family of linear transformations that leave the tetrahedron fixed. This 3-fold infinite family of mappings happens to form a group, although the significance of this fact only became evident to Lie and Klein somewhat later. The intimate connection between the "action" of this group and the geometry of the tetrahedral complex is discussed in detail in §1 of Hawkins' paper.

One may study such structures most easily by employing the standard tetrahedron whose faces are formed by the three coordinate planes and the plane at infinity. In this case, the transformations assume the form:

$$x' = \alpha x, \quad y' = \beta y, \quad z' = \gamma z$$

a family containing three arbitrary parameters. Lie noted that to generate a tetrahedral complex it is sufficient to apply this family of transformations $S_{\alpha,\beta,\gamma}$ to an arbitrary line L, as one then obtains a 3-parameter family of lines that will remain invariant under these mappings.[38] A necessary and sufficient condition for the existence of a transformation $T \in S_{\alpha,\beta,\gamma}$ to map a given line L_1 to another given line L_2 is that L_1 and L_2 meet the faces of the tetrahedron in the same cross-ratio. Thus the values of this cross-ratio yield a one-parameter family of tetrahedral complexes associated with a given tetrahedron.

With characteristic boldness, Lie recognized that this construction need not be restricted to the lines of the complex, but could equally well be applied to curves enveloped by these lines (assuming that these are not W–curves which are carried into themselves by a subgroup of $S_{\alpha,\beta,\gamma}$). For by applying the full group of transformations $S_{\alpha,\beta,\gamma}$ to such a curve C, one obtains a 3-fold infinite family of covariant curves which, taken as a whole, will clearly remain invariant under this group. Lie called these image curves that arise via the action of $S_{\alpha,\beta,\gamma}$ on a given complex curve C, curves of the same *species (Gattung)*, a notion of central importance for his theory.

What does the geometry of a tetrahedral complex look like? Consider an arbitrary plane E that intersects the faces of the given tetrahedron T in four lines g_1, g_2, g_3, g_4. The lines L of the complex that lie in E therefore meet these g_i in sets of four points with the same cross-ratio; thus they determine a certain conic k to which the g_i are tangent lines. The complex must therefore be of the 2nd-degree, and the cone of complex lines determined by an arbitrary point p will contain four lines that pass through the vertices of T. This cone will be non-singular provided p does not lie on T, in which case it degenerates into two planes. Dually, the class conic k degenerates into two points if and only if the plane E contains a vertex of T. Thus for any plane E that does not contain a vertex of T, by varying the cross-ratio one obtains a one-parameter family of conic sections each of which is inscribed in the the quadrilateral determined by the lines g_i.

One of Lie's standard techniques was to study the induced mapping on a line-complex determined by a point transformation of 3-space. An important case was a point transformation of the following form:

(**) $$x_1 = \lambda x^m, \quad y_1 = \mu y^m, \quad z_1 = \nu z^m.$$

Such a mapping transforms the plane E:

$$Ax_1 + By_1 + Cz_1 + D = 0$$

onto the surface F:

$$\lambda Ax^m + \mu By^m + \nu Cz^m + D = 0,$$

a so-called tetrahedral-symmetric surface studied earlier in another context by de la Gournerie.[39] The one-parameter family of complex lines that lie in the plane E and envelope the conic k are transformed by (**) into curves of the tetrahedral complex that lie on F. Since the tangents to these tetrahedral-symmetric curves all belong to the given line-complex, the curves so obtained will all be of the same species.

Furthermore, the conic k itself will be transformed into a curve k' having the property that the infinitesimal complex-cones along the points of k' are tangent to k'. And from this it follows that the tangent plane to F along such a curve k' is identical with the osculating plane, i. e., k' is an asymptotic curve. Lie realized that this was not an isolated result, and that, in fact, one can obtain *all* of F simply by varying the cross-ratio thereby ranging over the entire one-parameter family of complexes associated with a given tetrahedron. By so doing, the conic sections inscribed in the quadrilateral determined by the lines intersecting T and the given plane E will be mapped to the asymptotic curves of the tetrahedral-symmetric surface F.

In close analogy with the above investigations, Lie also considered the effect of the family of involutions

$$\{J_{abc}\} : x' = a/x, \quad y' = b/y, \quad z' = c/z$$

on the various species of curves associated with a given tetrahedral complex. These transformations are one-to-one except on the faces of the tetrahedron, i. e. the coordinate planes and the plane at infinity, which are mapped to their opposite vertex (the corresponding point at infinity or the origin, respectively). When applied to the curves of a given species, the J_{abc} determine a three-parameter family of curves of the same species. The lines of the tetrahedral complex K will be transformed into a three-parameter family of third-order curves that pass through the vertices of the tetrahedron and whose chords belong to K. For given any two points p, q on a line $L \in K$, there is always a $T \in \{J_{abc}\}$ such that $T(p) = q$ and $T(q) = p$. Thus the curve $C = T(L)$ also passes through p and q. Conversely, if the points p and q lie on the curve C, then they also lie on $T(C) = L$ and hence the line pq belongs to the complex K. Similar considerations reveal that the second-degree cone determined by the lines of K that pass through a given point p will also determine a one-parameter family of 3rd-order curves that contain p and the vertices of the tetrahedron. If one then considers the two cones associated with points p_1, p_2 on a line $L \in K$, then these will intersect in a common 3rd-degree curve C as well as the line L. Thus each $T \in \{J_{abc}\}$ induces an involution on the complex-cones of K whereby the lines and 3rd-order curves associated with a point p_1 are mapped, respectively, to the curves and lines associated with $T(p_1)$.

Lie went on to explore the effect these trnasformations have on a second-degree surface F that passes through the vertices of the tetrahedron T. Such surfaces contain two families of lines as generators and each of these has the property that their lines meet the faces of T in a fixed cross-ratio. Each family therefore belongs to the particular tetrahedral complex with this given cross-ratio. Since a second-degree surface that contains nine given points is completely determined, the surfaces constrained to pass through the vertices of the tetrahedron form a five-parameter family. Those whose generators also belong to two given tetrahedral complexes with the same tetrahedron T thus comprise a three-parameter family, F_α. The family F_α remains invariant under the $\{J_{abc}\}$. Since these transformations take the lines of one complex to 3rd-order curves of the same species passing through the vertices of T, it follows that a given surface $F \in F_\alpha$ will contain two pairs of one-parameter families of curves (i.e., the generators and their associated 3rd-order curves) each of a different species. Assuming that F contains none of the edges of T, then there exists a unique involution J_{abc} which leaves F fixed and interchanges the two families of lines with the two families of 3rd-order curves.

This was one of the principal results Lie presented in Kummer's seminar. It was not until some 25 years later, however, that he published the above argument in *Geometrie der Berührungstransformationen.*[40] Not that this was unusual for Lie; in fact, it turned out to be a lifelong pattern. He was continually seduced into exploring new ideas before he could find time to work out the older ones carefully and write them up in a readable form. Fortunately, in this case he was able to make use of Klein's ready enthusiasm for his work, as otherwise a good deal of what Lie discovered during this formative period might never have found its way into print.

In January 1870 Klein wrote up Lie's article "Über die Reciprocitäts-Verhältnisse des Reye'schen Complexes" and sent it on to Clebsch for publication in the *Nachrichten* of the Göttingen Scientific Society.[41] As with all these early papers, it was extremely pithy and difficult to follow, being little more than a compendium of results with barely a hint as to how they were obtained, much less anything remotely resembling proofs of their validity. Besides presenting the transformations J_{abc} discussed above, the article called attention to two other families of involutions (or, to follow Lie's terminology, reciprocal transformations): g-transformations, which send lines to lines, and r-transformations, which send points to lines. Each of these form a 3-parameter family, and composing members of each in alternating order yields new types, e.g. a grg-transformation or an rgr-transformation, which happens to be equivalent to one of the point transformations J_{abc}. The reader should consult §1 of Hawkins' paper for more details.

If the transformation so obtained begins and ends with the same letter, then composing it twice produces the identity, which means that these are new reciprocal transformations. Lie showed that by means of these one can study the various curves associated with a tetrahedral complex K, as each will send curves of one given species to those of another. Something similar takes place when one allows a curve $C' \in K$ to roll along another fixed curve $C \in K$ whose species differs from C': it generates a one-parameter family of curves in K whose species varies continuously from point to point. This is merely a more vivid illustration of what lies behind this whole theory, for all of the above mappings are in fact examples of what Lie later called contact transformations, that is transformations that, in general, change the underlying space element but preserve the order of contact between tangent figures.

Lie was apparently quite satisfied with Klein's performance as his amanuensis: at any rate he found almost nothing that needed to be changed in the manuscript Klein had prepared.[42] Klein noticed, however, that Lie's theory implicitly assumed that the various curves under consideration would not be mapped into themselves by means of a sub-family of the transformations that act on the complex.[43] It could not have taken long to convince himself and Lie that this need not necessarily be the case. This realization led them to the concept of W–curves, the subject of their first collaborative effort, "Deux Notes sur une certaine famille de courbes et de surfaces."[44] These two notes, written in June 1870 during their stay in Paris, were published in the *Comptes rendus* of the Academy of Sciences. In them, Klein and Lie outlined some of the fundamental ideas in the theory of W–curves and W–surfaces. But unfortunately the language they employed was so terse and vague that no one, aside from Max Noether, seems to have understood their content. In fact, Lie and Klein themselves were still groping with ideas that only became clear after Lie developed the concepts of line-element and surface-element as part of a systematic study of the theory of contact transformations. Klein was well aware of the difficulties that had to be overcome before this work could meet with a positive reception, and he hoped to remedy the situation by preparing a longer article on W–curves in the plane and 3-space also to be co-authored with Lie. In this subsequent work, which is described briefly below, Klein gave a more explicit definition of what is meant by a W–curve, the W being an abbreviation for "Wurf," the terminology employed by von Staudt for his purely projective notion of cross-ratio.[45] This "Wurf" is a projective invariant, which is not quite true of the W–curves, as these are invariant under a one-parameter closed system of commutative linear transformations.

Klein and Lie had already discussed plans for a more extensive article in November 1870, when Lie stopped in Düsseldorf to pay his friend a brief visit on his way back to Norway. Due to his involvement in the Franco-Prussian War, Klein had not found much time for mathematics since departing from Paris the previous July. But for the next several months he worked on the W–curves paper, keeping Lie abreast of developments on a regular basis. At one point he wrote him the following:

> A short time ago I spent an evening with Clebsch and read to him what I have put together on plane W–curves. Clebsch was clearly pleased with the material, but in no way denied that he was unable to understand our article in the *Comptes rendus*. Evidently this was generally the case for everyone except Noether. I recently gave a reprint to one of my auditors who had nothing better to do than study it through: he, too, could not understand it. When he explained this to me, I told him that it wasn't written in order to be understood. By the way, in Berlin they claim to

have understood it. Someone there said, in fact, that such investigations are really too simple. In order to counter such views, I have stated fairly clearly in the introduction that we consider the work as indeed worthy of "something" [für "etwas" halten].[46]

Klein's letters to Lie reveal, on the other hand, that he had other things on his mind now that his war experiences were behind him and he was free to think about mathematics full time again. In part, it may have been simply an aversion for detailed and systematic research that caused him to put off work on his joint paper with Lie, but it was also surely due to the fact that he was hot on the trail of certain other ideas that excited his interest much more than did the theory of W–curves. On December 30, 1870, he wrote Lie that he wanted to publish a short article dealing with elliptic coordinates for a line in the *Gottinger Nachrichten*. Klein then added: "After that I'll take up the W–curves again, as I have neglected them for the moment in favor of more interesting reflections."[47] In the meantime Klein had become a *Privatdozent* in Gottingen, and he wrote Lie from there about a month later:

Since my arrival here, I still have not gotten around to doing any reasonable research of my own. Not that I haven't had sufficient free time: it is just that I definitely want to finish the W–curves, a topic that has now become so commonplace [alltäglich] for me that the work is really boring.... I have now gotten far enough to see, more or less, how matters stand in the plane, but in three-space things still look quite confusing.[48]

In the end, Klein threw up his hands and settled for a paper that was confined to the treatment of W–curves in the plane. He was clearly disgruntled when he wrote Lie on the 4th of February:

I hope to be able to send you the W–curves soon, but in fact in a completely different form than I myself had conceived of it just a few days ago. It deals only with plane W–curves: not a word is said about them in three-space. I would like to recommend that we publish the article in this form.... I have always striven to work out the details of a problem fully before publishing it. That is something I will never do to the same extent in the future. For if one goes beyond certain limits, a subject is bound to become tiresome, and with the W–curves I reached that limit long ago.[49]

And four days later, when he sent the manuscript on to Lie, his feelings had not changed: "At the moment the whole matter is so horribly repulsive to me that I am no longer able to think about it, and for that reason I consider the whole paper an abortive effort."[50]

Lie requested that some material be added on his new approach to the theory of differential equations. This Klein wrote up and appended as section 7 of the published article, which bore the title *Über diejenigen ebenen Kurven, welche durch ein geschlossenes System von einfach unendlich vielen vertauschbaren linearen Transformationen in sich übergehen.*[51] After a brief introduction aimed at demonstrating the heuristic value of the theory to sceptical Berliners and other potential critics, Klein and Lie classified the plane linear transformations into five types depending on the behavior of the fundamental triangle under the mapping. For example, a transformation of the first type: $x' = ax$, $y' = by$ will leave the three vertices of the fundamental triangle (two of which lie on the line at infinity) fixed. To obtain the desired one-parameter family of mappings, they introduced a variable $t : x' = a^t x$, $y' = b^t y$ to produce a family of transformations that is clearly closed under composition. In fact, this was apparently the first occasion when Lie referred in print to the notion of an infinitesimal transformation, which later became a cornerstone for his theory of continuous groups.

The determination of the associated W–curves amounts to eliminating the parameter t from the above equations. This leads in each of the five cases to a certain differential equation which can easily be solved. For the transformations above, this differential equation has the form: $(\log by)dx = (\log ax)dy$. Its integral curves are then the W–curves associated with this type of transformation. These include a wide variety of algebraic and transcendental curves in the plane. In particular, Klein observed that if the two infinitely distant vertices of the fundamental triangle are moved to the circular points at infinity, then one obtains the logarithmic spiral as a special type of W–curve.

W–curves are closely related to the space-curves that Lie had studied in connection with tetrahedral complexes. In fact, they possess the same characteristic property with relation to the fundamental triangle that the complex-curves satisfy with respect to their associated tetrahedron, i. e. the tangents to them determine a fixed cross-ratio. To see this, suppose T is the given coordinate triangle and the tangent line L meets the W–curve C at point p. If q, r, s are the points of intersection of L with the sides of T, then p, q, r, and s determine a certain cross-ratio. But now applying the given one-parameter family of projective transformations to C will leave both it and T invariant. The point p will consequently run through the points of C while L sweeps out the tangents to C, and the orbits of q, r, and s will simply be the lines that contain the three sides of T. It then follows immediately from the invariance of the cross-ratio under projective transformations that each of the tangent lines so obtained must yield the same cross-ratio as L.

Many years later, Klein seems to have developed more appreciation for this joint work with Lie than he did at the time of its composition. In the wake of the fame that had since attached to his "Erlanger Programm" and Lie's work on the theory of continuous groups, the status of these rather elementary investigations into the geometry of plane W–curves and the earlier notes on W–surfaces has grown considerably. Their significance has generally been related to the emergence of the group concept in geometry, a connection that is clearly lurking in the background but which only reaches center stage a year or so later. It should be observed, for example, that the groups Lie and Klein utilized were of a very special type designed for the particular geometric purposes they had in mind, namely to preserve tangency relations. This meant they even had to rule out the orthogonal group, since, to use topological language, it is not connected (reflections lie in a different component than rotations). The reader should consult T. Hawkins' article for a detailed analysis of the group-theoretic aspects of this work and a somewhat different interpretation of its significance.

In the meantime, however, both Lie and Klein were hurriedly chasing after certain other ideas that had little or nothing to do with groups, but which nevertheless turned out to play a decisive role in shaping the geometrical ideas that appear in the "Erlanger Programm." Indeed, the focal point of most of Klein's research between July 1870 and December 1872 centered around a new-found analogy between German line geometry and French metric geometry as practiced by Darboux and others. This analogy, in turn, was intimately tied together with Lie's work on the line-to-sphere transformation, and this takes us back to Paris in the summer of 1870.

6. Lie's Line-to-Sphere Transformation

Up until he came to Paris, Lie's mathematics had been almost entirely motivated by geometry in the tradition of Poncelet and Plücker. He operated freely with complex coordinates, the spherical-circle at infinity, line-complexes, all manner and art of duality relationships, and a series of other concepts that had close analogues in projective geometry. From this point onward, however, a notable shift in his thinking took place unquestionably as a result of his exposure to the French tradition, especially Monge's theory of differential equations and Darboux's work on differential geometry. In all probability, Lie was led to pursue a deeper study of Monge's geometric theory of partial differential equations after realizing that such equations formed an integral part of his own work.

One of Lie's early insights into the nature of line-complexes came in recognizing that Plücker's theory could be studied in the context of the theory of differential equations of the form $F(x, y, z, dx, dy, dz) = 0$ that are homogeneous in dx, dy, dz.[52] Such equations are named after Monge who was the first to deal with them systematically. Viewed geometrically a Mongian equation assigns to each point (x, y, z) in 3-space a certain "elementary cone" of directions emanating from the given point. A special case is the Pfaffian equation, which is linear in dx, dy, dz:

$$X(x, y, z)dx + Y(x, y, z)dy + Z(x, y, z)dz = 0.$$

It was known that when a Pfaffian equation can be written in the form

$$A(ydz - zdy) + B(zdx - xdz) + C(xdy - ydx) + Ddx + Edy + Gdz = 0,$$

then its integral curves will include a 3-parameter family of straight lines, i. e. a line-complex. This line-complex will be of the first-degree—a so-called linear complex (or to use the language of Möbius, a nullsystem), where the lines through a point all lie in a plane.

Lie was able to show that an *arbitrary* line-complex satisfies a Mongian equation of the form

$$F(ydz - zdy, zdx - xdz, xdy - ydx, dx, dy, dz) = 0$$

whose integral curves contain the 3-parameter family of lines of the complex.[53] In this case the elementary cone determined by the Mongian equation at each of the points (x, y, z) will be identical with the cone of lines of the complex that pass through (x, y, z). This was the sense in which Plücker's theory may be regarded as a special part of the general theory of Mongian equations.

One way to obtain a second-degree line-complex is to consider the collection of all lines that meet a given conic section. A particular example of this that played a key role in Lie's work is the complex of "minimal lines," those that meet the spherical-circle at infinity. This complex is closely associated with the conformal transformations, since a mapping $f(x, y, z)$ is conformal if and only if it leaves invariant the form $dx^2 + dy^2 + dz^2 = 0$. The integral curves of this Mongian equation are the so-called minimal curves, i.e. imaginary curves of length zero; a conformal transformation maps these imaginary curves to themselves. But, in fact, it must also have the property of sending minimal lines to minimal lines as well. Furthermore, since spheres are the only surfaces that contain two families of minimal lines, a conformal transformation must send spheres to spheres. The same applies to the degenerate cases (so-called null-spheres) which are really cones of lines of the form $(x-a)^2+(y-b)^2+(z-c)^2 = 0$ passing through the spherical-circle at infinity. Thus the condition that a mapping of $\mathbf{P}^3(\mathbf{C})$ sends null-spheres to null-spheres (or minimal lines to minimal lines) in fact characterizes the conformal transformations.

The structure of these transformations had been known for some time, having been worked out by Liouville in an article published in 1847.[54] Liouville showed, in fact, that each such transformation could be decomposed into a similarity mapping followed by a transformation by reciprocal radii. The conformal transformations thus form a 10-parameter group: three for translations of the coordinate axes, three more for determining rotations about these, a seventh giving the size of dilation or retraction about the new origin, and finally three for fixing the pole of the transformation by reciprocal radii (whose fixed circle may be assumed to have a radius of unit length). Lie succeeded not only in giving a very natural, straightforward proof of this important result but also in extending it to determine the conformal transformations for spaces of arbitrarily large dimension.[55]

The relationship between line geometry and certain Mongian differential equations was one of the key motivating ideas behind Lie's work during the early 1870s. Another was his generalization of a construction that Plücker had already employed some forty years earlier. In his *Analytisch-geometrische Entwicklungen*, published in two volumes in 1828 and 1831, Plücker showed that the principle of duality in the plane can be understood by regarding the linear equation $ux + vy + 1 = 0$ as representing either the points on a fixed line with coordinates (u, v) or as the lines through a fixed point with coordinates (x, y). Similarly, in 3-space the equation $ux + vy + wz + 1 = 0$ exhibits the dual nature of points (x, y, z) and planes (u, v, w). In the second volume of his study, Plücker went on to show how an arbitrary polynomial equation $F(x, y, X, Y) = 0$ leads to a dual relationship between points (x, y) and their corresponding curves in the (X, Y)-plane, as well as points (X, Y) and their corresponding curves in the (x, y)-plane.[56] The special case where $F(x, y, X, Y) = 0$ is of the form

$$X(ax + by + c) + Y(dx + ey + f) + (gx + hy + j) = 0$$

is the classical one, referred to by Lie as the Poncelet-Gergonne reciprocity theory for the plane.[57]

The conceptual significance of Plücker's generalization of the classical duality structures was that it revealed how an arbitrary curve can be viewed as the fundamental space-element rather than just points, lines, or planes. This insight made a profound impression on Lie, and in his dissertation of 1871, "Over en Classe geometriske Transformationer," he studied in a similar fashion reciprocal relationships between two 3-dimensional spaces by considering the duality system induced by two equations of the form:

$$F(x,y,z,X,Y,Z) = 0, \quad G(x,y,z,X,Y,Z) = 0.$$

Although he outlined what amounted to a general theory for such systems, he was clearly less interested in pursuing its implications than he was in developing a particularly fruitful special case.[58] This arises from the following equations:

(∗)
$$X + iY + xZ + z = 0$$
$$x(X - iY) - Z - y = 0$$

which induce a mapping $T: (x,y,z) \rightarrow (X,Y,Z)$. Fixing (X,Y,Z) in the range space and differentiating these equations, one obtains:

$$Z\,dx + dz = 0$$
$$(X - iY)dx - dy = 0.$$

By utilizing (∗), one can then eliminate X,Y,Z from the above equations to obtain $x\,dy - y\,dx + dz = 0$, a Pfaffian equation of the form required to define a linear complex C_1.

A similar procedure taking (x,y,z) as fixed leads to the equations:

$$dX + idY + xdZ = 0$$
$$x(dX - idY) - dZ = 0.$$

This time eliminating x,y,z produces the equation $dX^2 + dY^2 + dZ^2 = 0$, a Mongian equation defining C_2, the 2nd-degree complex of minimal lines. Thus the transformation T induced by (∗) maps points (x,y,z) to the minimal lines of the complex C_2. Likewise, lines of the complex C_1 are mapped to points (X,Y,Z) by virtue of the fact that the image under T of the points along a line in C_1 will be a family of lines passing through a fixed point in the (X,Y,Z)-space. Similarly, the points (X,Y,Z) of the range space will be mapped by T^{-1} to lines of the complex C_1.

All this followed directly from the general theory Lie had developed. He soon went beyond it, however, to consider the essential question: "What will the image under T look like for a line that does not belong to C_1?" If g_1 is such a line, then there exists a unique line g_2, the polar reciprocal of g_1 with respect to the line complex C_1. The line g_2 may be obtained by taking any two points p and q on g_1. These two points determine two planes E_p and E_q—the planes of C_1 containing p and q, respectively—and their intersection is the line g_2. One can easily show that g_2 is independent of the choice of p and q. For if r is any point on g_2, then the lines pr and qr must belong to C_1. Thus, the plane E_r contains g_1, and every point on g_1 determines a plane through r. Since the choice of r is arbitrary, it follows that every point s on g_1 is associated with a plane E_s that contains the line g_2. Thus the intersection of any two such planes always yields this very line. Furthermore, the relationship between g_1 and g_2 is fully reciprocal: one could just as well have started these deliberations with the line g_2 as with g_1. The lines of C_1 that intersect these two polar reciprocal lines form a ray system of the first order and class.

Lie considered an arbitrary line g_1 in (x, y, z) written in the form $x = rz + \rho$, $y = sz + \sigma$, where the coordinates $(r, s, \rho, \sigma, \eta)$ may be regarded as fixed when substituted in the equations (*). Since the six variables x, y, z, X, Y, Z are subject to four independent conditions, eliminating x, y, z produces a single condition on the three remaining variables X, Y, Z, namely:

$$(X - \frac{\rho + s}{2r})^2 + (Y + i\frac{\rho - s}{2r})^2 + (Z - \frac{\eta - 1}{2r})^2 = (\frac{\eta + 1}{2r})^2.$$

This equation represents the points on an imaginary sphere centered at $(\frac{\rho + s}{2r}, -i\frac{\rho - s}{2r}, \frac{\eta - 1}{2r})$ and with radius $\frac{\eta + 1}{2r}$.

Thus the image of the line g_1 under T is a sphere in the space (X, Y, Z). In the event that g_1 belongs to C_1, the image sphere collapses to a null-sphere, i.e. a collection of minimal lines in C_2 through a fixed point, its center. In view of the reciprocal relationship between g_1 and g_2, one would expect that the induced transformation T would produce the same image in both cases. That this is indeed the case may be seen by considering the coordinates

$$r' = -r/\eta, \quad s' = -s/\eta, \quad \rho' = -\rho/\eta, \quad \sigma' = -\sigma/\eta, \quad \eta' = 1/\eta$$

of the reciprocal line g_2. For substituting these values into the equation above yields exactly the same result, so that g_1 and g_2 are mapped to the same image sphere in the space (X, Y, Z). Thus the line-to-sphere mapping is, in general, a two-to-one transformation, except for the case when the line g_1 belongs to C_1 (i. e., g_1 is identical with its polar reciprocal g_2) and the mapping is one-to-one.

In light of the dual nature of these constructions, one may view the image sphere in two distinct ways: either as comprised of ∞^2 points or as generated by a 2-parameter family of lines. The former viewpoint pertains when one considers the image under T of the ray system consisting of all lines that intersect g_1 and g_2. Each such line is mapped to a point (X, Y, Z) on the image sphere. Alternatively, one may consider a single line h that intersects g_1 and g_2 in two points p_1 and p_2, respectively. These points correspond to two minimal lines on the image sphere that pass through the point $T(h)$. Fixing the point p_1 and varying the lines h produces one system of generators for the sphere, while the opposite procedure produces the other.

If two spheres S and S' in the space (X, Y, Z) are tangent to one another at a point p, then they share the same pair of minimal lines, say L_1 and L_2 passing through p. If $T^{-1}(L_1) = p_1$ and $T^{-1}(L_2) = p_2$, and if g_1, g_2 and g'_1, g'_2 are the reciprocal polars corresponding to S and S', respectively, then it follows that the corresponding polar lines intersect each other since $g_1 \cap g'_1 = p_1$ and $g_2 \cap g'_2 = p_2$. Thus, under T, intersecting lines are mapped to tangent spheres.

Having outlined the mathematics which underlies Lie's line-to-sphere transformation, the following questions naturally arise: how was he led to study this mapping in the first place and why did it prove such an important source of inspiration both for Lie himself and for Klein? Already in 1869, Lie had constructed a transformation that maps a linear complex C onto the points of $\mathbb{P}^3(\mathbb{C})$ (a mapping that Max Noether had also discovered independently a short time earlier).[59] Lie viewed this as a transformation between two line-complexes—the one given and the other formed by the lines that meet a fixed conic section K. Thus the lines of C that pass through a given point p map to the points on a line that intersects K, and the points of C map to the lines of a 2nd-degree complex that is special. Furthermore, the congruence D determined by the lines common to C and an nth-degree complex C' has the important property that it maps to a surface of degree $2n$ that contains the conic K as a curve of multiplicity n. In particular, if C' is linear, then the congruence D contains a single line through p, and this line is mapped to a 2nd-degree surface containing K.

Sometime during his stay in Paris, if not earlier, Lie was inspired to shift the conic K to the spherical-circle at infinity. This produced a new transformation which sends lines to spheres, for, as noted earlier, these are the only quadric surfaces that contain the spherical-circle at infinity. Furthermore, this mapping had the property that intersecting lines in the first space corresponded to tangent spheres in the second. According to Felix Klein, Lie had a deep-seated belief that this transformation supplied an essential connection between the metric geometry of the French and the line geometry of the Germans. It was this faith that enabled him eventually to discover its most important structural feature, namely, that it transforms the asymptotic lines of a given surface into the lines of curvature of the image surface. Conversely, if a surface (F, θ) be given in the image space having θ as a line of curvature, then the pre-image of (F, θ) is a ray system together with a ruled surface belonging to it. This ruled surface intersects the *Brennfläche* of the ray system along one of its asymptotic curves.

Klein considered this one of the most illuminating examples of the mysterious role played by the psyche in mathematical creativity. In his notes to a lecture on the psychology of mathematical invention given ten years after Lie's death he wrote:

> What was it that led Lie to transfer the fundamental conic section that arises through the mapping of the line-complex to the spherical-circle and for weeks and months to incessantly plumb the depths of this relationship between them? "In his dark urges the good person is always conscious of the right path." Lie later said: "For a long time I 'lived' in the two spaces together."[60]

Lie's first findings regarding this transformation led to new derivations of results that had been obtained earlier by the French geometers. He also immediately recognized its power for transfering results from sphere geometry back into line geometry. For example, since a ruled surface is transformed into a surface enveloped by spheres, it follows that a quadric surface, which has two independent families of lines as asymptotic curves, must be mapped to a surface enveloped by two families of spheres—and these are the cyclides of Dupin, which are characterized by the property that their lines of curvature are circles. One can obtain the same result as follows. A hyperboloid surface is generated by a mobile straight line constrained to intersect three fixed lines. Under Lie's mapping, this configuration will be sent to three fixed spheres and a variable sphere tangent to the other three. Since the variable sphere is tangent along a curve of the transformed surface of the hyperboloid, it is therefore the envelope of this surface. But the envelope surface of a family of spheres tangent to three fixed spheres is again a Dupin cyclide.[61]

It appears likely that Lie's persistent study of the line-to-sphere transformation was at least partly inspired by the work of the French school of "anallagmaticians," whose main proponents were Laguerre, Moutard, and Darboux. These geometers had developed a new type of sphere geometry which will be discussed in more detail below. It combined concepts from classical differential geometry (lines of curvature, orthogonal systems, etc.)—a tradition going back to Monge and his pupils and later advanced by Liouville and Bonnet—with the projective methods developed by Chasles and his followers, who adroitly exploited the properties of the spherical-circle at infinity. Lie evidently hoped to develop fruitful analogues in line geometry to some of the key results that had since been obtained by the French "anallagmaticians," an approach strikingly similar in spirit to the one that guided his early work on the geometry of imaginary elements. Darboux had a masterful command of both differential and projective geometry, and Lie and Klein were both fortunate that he imparted many of his latest findings to them during their short stay in Paris. One of these was a discovery that he and Moutard had made independently of one another concerning the existence of a 3-parameter orthogonal system of confocal cyclides.

Such generalized cyclides are 4th-order surfaces with the property that they contain the spherical-circle at infinity twice over. Soon after realizing that the line-to-sphere mapping sends the asymptotic curves of one surface to the lines of curvature of another, Lie also saw that he could utilize the known lines of curvature of a generalized cyclide to determine the asymptotic curves of the Kummer surface. All he had to do was note that a 2nd-degree complex C', when intersected with the linear complex C above, determines a congruence D of the second order and class. This was precisely the type of ray system that Kummer had shown possesses a Kummer surface as its *Brennfläche*. It followed, then, that the image of this Kummer surface must be a 4th-order surface that contains the spherical-circle at infinity twice over. But these were just the generalized cyclides studied by Darboux and Moutard, who had determined that their lines of curvature are 8th-order algebraic curves that intersect the spherical-circle in eight points. From this fact, Lie could easily deduce that if (F, θ) was such a cyclide with line of curvature θ, then the inverse image of (F, θ) was a congruence of the second order and class together with an 8th-order ruled surface that intersects the associated Kummer surface of the congruence along an algebraic curve of degree sixteen.[63]

This discovery, one of the most dramatic and significant of Lie's entire career, was made early in July 1870, only a short time before the outbreak of the Franco-Prussian War forced Klein's departure from Paris. Reflecting back on this fateful day, Klein wrote:

> ...one morning I got up early and wanted to go out right away when Lie, who still lay in bed, called me into his room. He explained to me that relationship he had found during the night between the asymptotic curves of one surface and the lines of curvature of another, but in such a way that I could not understand a

word (it had to do with the line-to-sphere transformation, but instead of spheres he operated part-visually with straight-lined hyperboloids that passed through a fixed conic section). In any case, he assured me that this meant that the asymptotic curves of the Kummer surface must be algebraic curves of degree sixteen. That morning, while I was visiting the *Conservatoire des Arts et Métiers*, the thought came to me that these must be the same curves of degree sixteen that already appeared in ... my "Theorie der Linienkomplexe ersten und zweiten Grades," and I quickly succeeded in showing this independent of Lie's geometric considerations. Whe I returned around four o'clock in the afternoon, Lie had gone out, so I left him a summary of my results in a letter.[64]

The curves Klein now recognized as identical to the asymptotic curves of the Kummer surface were, of course, those enveloped by the singular lines of the surface that point in an asymptotic direction. Not long after this Klein presented a detailed analysis of the singularities and intersection properties of these asymptotic curves in a joint article with Lie entitled "Über die Haupttangentenkurven der Kummerschen Fläche vierten Grades mit 16 Knotenpunkten."[65]

Meanwhile, Lie sent a letter to the Scientific Society in Christiania in which he gave a brief summary of his own results. During the nineteenth century, this was still a commonly practiced measure for ensuring one's priority in the event of a dispute, and Lie was always on the lookout for those who might be inclined to "steal" his ideas. In this case, no such squabble arose, and the letter was soon forgotten. It was published shortly after Lie's death in 1899. As an historical document, it not only captures the freshness of an important discovery but it also offers a fascinating glimpse into the plethora of ideas that Lie was generating at this triumphant period in his career. Lie wrote:

To the Scientific Society of Christiania:

I take the liberty of communicating the following scientific results with the purpose of possibly securing my priority.

1. Through my theory of the imaginary I have found a geometric transformation that converts a descriptive proposition concerning straight lines into one concerning spheres. Thereby two straight lines that intersect one another correspond to two spheres that are tangent to one another.

2. From here I have deduced that it is always possible through algebraic operations to determine the asymptotic curves of one surface by means of the lines of curvature of another surface, and vice-versa.

3. The Kummer surface of the fourth order and fourth class has algebraic asymptotic curves of the sixteenth order and sixteenth class. It follows, of course, from this that these curves are also algebraic for wave surfaces, the Plücker complex-surface, and so on [these being special cases of the Kummer surface].

4. On a ruled surface whose generator belongs to a linear complex the asymptotic curves can be determined by quadrature.

5. Every minimal surface can be described in two ways by translation motions of an imaginary line of length zero.

6. If a minimal surface A is rolled along a minimal surface B, then each point that is fixed on A describes a third minimal surface. It is assumed that the above motion is a translation motion.

7. Let $F(x, y, z) = 0$ be the general Cartesian equation for a minimal surface. The family of surfaces: $F(\log x, \log y, \log z) = 0$ is of special interest for the study of the Reye line-complexes. Here belong the surfaces that I treated in my work "Über die Reziprokitätsverhältnisse des Reyeschen Complexes. Göttingen." All the observations made therein have analogues in the study of the above family of surfaces.

Most respectfully yours,
Paris, 5 July, 1870   M. Sophus Lie, Cand. real.[66]

One gathers from this note and from Klein's account quoted above that Lie followed a rather indirect and convoluted path in arriving at his principal discoveries (numbers 2 and 3). His results on minimal surfaces, on the other hand, served as the starting point for the important contributions to this field that he published in the late 1870s. In particular, the significance of the logarithmic transformation (mentioned in point 7) is discussed at length in §1 of T. Hawkins' article. Klein shed further light on the background to these discoveries in an unpublished manuscript written in 1892:

Lie's investigations took a sudden turn in the beginning of July when he simultaneously obtained two fundamental results. In his studies of the tetrahedral complex Lie already had busied himself for a long time with surfaces in which the complex curves (which cover every surface twice over) are conjugate to the asymptotic curves. He had found a simple kinematic method for generating these surfaces, which he studied all the more thoroughly, as it turned out that a large number of remarkably interesting special cases were included among them. Lie utilized the logarithmic transformation [$x' = \log x$, $y' = \log y$, $z' = \log z$] to transfer these surfaces to the complex of lines that meet a conic section, respectively, the spherical-circle, and found to his great surprise that by so doing he obtained by way of *geometric generation* the most familiar *minimal surfaces* that Monge had given earlier by means of analytic formulas.

This was the first result, which Lie, however, only followed up somewhat later. Afterward, in fact, it appeared rather obvious once one is at all accustomed to following imaginary elements on the spherical-circle. Indeed, the accomplishment is that he even undertook to interpret the Monge formulas at all. But this accomplishment arose in an indirect manner; it did not take place because Lie intended to interpret Monge as Lie's interpretation was already completed before he became familiar with Monge's work in the original. The prevailing view was that [Monge's formulas] were not useful geometrically.

But I must report further. The lines of curvature on minimal surfaces had been the subject of various prior investigations. Lie now attempted to carry over the theory of minimal surfaces and, in particular, their lines of curvature to the space of a line-complex. And this revealed that the lines of curvature turn into asymptotic curves of the transformed surfaces; moreover, this relation was in no way bound to the property that the original surfaces were minimal—it was a general theorem in surface theory that had been discovered.

I have described the course of this discovery, which took place over a few days time, in detail, as I think it is in itself interesting. *It was not the grasp of definite questions known in advance that propelled it, but rather the spontaneous development of subjectively given starting points.*[67]

Model of the Kummer Surface with 16 Real Nodal Points.
Courtesy of Springer-Verlag Archive.

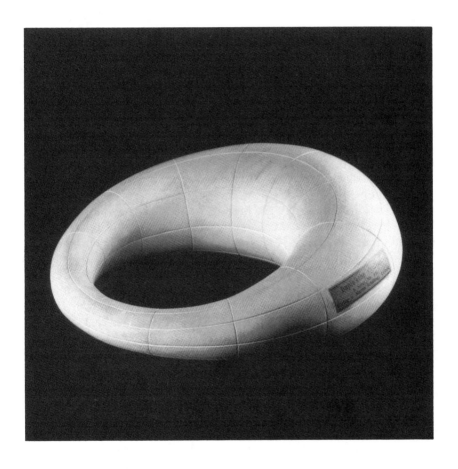

Model of a Dupin Cyclide showing Lines of Curvature.
Courtesy of Gerd Fischer, *Mathematische Modelle*,
Akademie-Verlag, Berlin, 1986.

7. The Role of "Übertragungsprinzipien" in Klein's Thought

The determination of the lines of curvature on surfaces had long pre-occupied French geometers, because only in a few specific cases could they be determined by direct integration. Early on, Darboux had made the surprising discovery that an imaginary line of curvature can be produced on *every* surface by making use of the developable surface that circumscribes both the given surface and the spherical-circle at infinity. One can obtain this developable kinematically by utilizing planes that simultaneously roll along the surface and the spherical-circle, thus giving rise to a family of surface normals that intersect the surface along an imaginary line of curvature.[68] Lie's result, like Darboux's, provided a purely geometric method for finding lines of curvature. But more important still, it opened the way for studying the interrelationship between the metric properties of surfaces and the analogous structures in line geometry.

Beyond this, Lie's discovery also had a deep and far-reaching impact on Klein's philosophical views regarding the foundations of geometry. As Klein later related, his predilection for projective thinking had led him to the conviction that the study of asymptotic curves deserved precedence over lines of curvature, since the former seemed to depend on the surface alone, whereas the latter, as Darboux had shown, were related to the spherical-circle at infinity. Lie's transformation revealed, however, that these two structures were, in fact, on an altogether equal footing. For Klein, this meant that the projective standpoint, an outlook which he had adopted from Clebsch and which his reading of Cayley had reinforced, was not the only possible foundation for geometry. Another system, one in which spheres rather than lines were regarded as the basic elements, could be studied with equal justification alongside projective geometry. When he gave his habilitation lecture in Göttingen the following January, Klein defended this thesis against Clebsch and won him over completely to this position.[69]

As was pointed out earlier, Klein had interrupted his work on W–curves in December 1870 in order to pursue various ideas that he found more interesting. The most significant of these was inspired not only by Lie's key discovery but also by certain information that had been conveyed to him earlier by Darboux. In order to appreciate the importance Klein attached to this information, however, it is necessary to present a few details concerning French sphere geometry.[70]

Suppose
$$(x - \alpha)^2 + (y - \beta)^2 + (z - \gamma)^2 - r = 0$$

is the equation for a sphere S in $\mathbf{P}^3(\mathbf{C})$ and (x, y, z) is any point outside it. The "power" (*potenz*) of S is given by

$$P = x^2 + y^2 + z^2 - 2\alpha x - 2\beta y - 2\gamma z + C,$$

where $C = \alpha^2 + \beta^2 + \gamma^2 - r^2$. Given any line L tangent to S through the point (x, y, z), P measures the length of the segment of L that runs from (x, y, z) to the point of tangency.

Now let S_i, $i = 1, \ldots, 5$ be any five spheres in $\mathbf{P}^3(\mathbf{C})$. To introduce pentaspherical coordinates x_i, one chooses fixed $k_i \in \mathbf{C}$ and lets

$$\sigma x_i = k_i P_i, \qquad i = 1, \ldots, 5,$$

i. e.

$$\sigma x_i = a_i(x^2 + y^2 + z^2) + b_i x + c_i y + d_i z + e_i.$$

As with line coordinates, the x_i are homogeneous and dependent on a homogeneous 2nd-degree expression $\Omega = 0$. If it assumed that the S_i are real (i. e. given by an equation with coefficients in $\mathbf{R}$) and orthogonal to one another, then it can easily be shown that exactly one of them has an imaginary radius. Furthermore, by choosing $k_i = 1/\rho_i$, where ρ_i is the radius of S_i, one obtains $\sigma x_i = S_i/\rho_i$ and $\Omega = 0$ takes the form

$$\Omega = \sum_{i=1}^{5} x_i^2 = 0.$$

It is worth noting that this result is formally the same as the one utilized by Klein in his work on 2nd-degree line-complexes except that Klein's sum had one less variable in it.

But what advantages do these pentaspherical coordinates offer over ordinary Cartesian ones? Assuming that the determinant $|abcde|$ associated with the five equations

$$\sigma x_i = a_i(x^2 + y^2 + z^2) + b_i x + c_i y + d_i z + e_i$$

is nonzero, one can solve for $\sigma, x^2 + y^2 + z^2, x, y, z$ in this linear system. Suppose the solutions are

$$Z_1/N = x^2 + y^2 + z^2, \; Z_2/N = x, \; Z_3/N = y, \; Z_4/N = z,$$

and consider an arbitrary sphere S written in the form $A(x^2 + y^2 + z^2) + Bx + Cy + Dz + E = 0$. Then substituting in the above solutions produces a linear equation for $S : AZ_1 + BZ_2 + CZ_3 + DZ_4 + EN = 0$, where N and the Z_i are homogeneous linear functions of the x_i. It therefore follows that every sphere can be represented as an equation in pentaspherical co-ordinates of the form $\sum_{k=1}^{5} \alpha_k x_k = 0$, and conversely every such equation represents a sphere.

If $\sum_{k=1}^{5} \alpha_k^2 = 0$, then the sphere has degenerated to a point for which the α_k are its pentaspheric coordinates. On the other hand, a sphere will have infinite radius, i.e. determine a plane, if $\sum_{i=1}^{5} \alpha_k/\rho_k = 0$. These two special cases follow immediately from the formula for the radius ρ of a sphere:

$$\rho^2 = \sum \frac{\alpha_k^2}{\left(\sum \alpha_k/\rho_k\right)^2}.$$

Now suppose $S : \sum_{k=1}^{5} \alpha_k x_k = 0$ and $S' : \sum_{k=1}^{5} \alpha'_k x_k = 0$ are any two spheres, then the condition for them to intersect one another orthogonally is simply $\sum_{k=1}^{5} \alpha_k \alpha'_k = 0$. These and other elegant results formed the basis of French sphere geometry, which differed considerably from the sphere geometry developed by Lie. In particular, Lie's approach made essential use of the radius ρ (although he neglected to differentiate between the two cases $\rho > 0$ and $\rho < 0$, the so-called "oriented sphere" introduced by Laguerre in 1880). In the French system, on the other hand, only the quantity ρ^2 enters into the equations.

The generalized cyclides of Darboux and Moutard can now be obtained by simply considering 2nd-order equations in $x^2 + y^2 + z^2, x, y, z, 1$ with constant coefficients. Such an equation $F(x_1, x_2, x_3, x_4, x_5) = 0$ will in general be a 4th-degree surface, but it is important to note that this form also contains arbitrary quadric surfaces as special cases. The equation $F = 0$ is, of course, also subject to the condition $\Omega = 0$, and just as Klein showed how to simultaneously diagonalize two similar forms to obtain a canonical representation for a 2nd-degree line-complex, so could Darboux show that by means of an appropriate linear transformation one could write F in the form

$$F = \sum_{k=1}^{5} y_k^2/\alpha_k = 0, \quad \Omega = \sum_{k=1}^{5} y_k^2 = 0.$$

This led him to consider the one-parameter family of cyclide surfaces

$$\sum_{k=1}^{5} \frac{y_k^2}{\alpha_k - \lambda} = 0.$$

These have the same form as the confocal quadric surfaces given by $\sum_{i=1}^{3} \frac{x_i^2}{a_i - \lambda} = 1$. The latter were long known to comprise an orthogonal system and thus served as the classical illustration of Dupin's Theorem which states that a system of three mutually orthogonal families of surfaces intersect one another along lines of curvature. Darboux and Moutard were able to show that precisely the same property holds true for the system of confocal cyclides given by $F = 0$. This meant that they had discovered a broad class of surfaces, containing the quadric surfaces and Dupin cyclides as special cases, whose lines of curvature were readily determined.

But what about the connection with line geometry? Klein first became aware of this possibility through a conversation with Darboux who pointed out the striking similarity between the equations $F = 0$, $\Omega = 0$ above and the ones that Klein had developed for the one-parameter family of 2nd-degree line-complexes that share the same Kummer surface as their singularity surface.[71] Since the confocal cyclides were intimately connected with Dupin's Theorem, it was only natural that Klein's attention should have been drawn to the possibility of establishing an analogous theorem for line geometry, an intriguing prospect in view of the fact that there were scant few links between Plücker's new subject and classical differential geometry. Klein wrote about this surprising new avenue for research in the following terms:

> These investigations [of the French geometers] were communicated to us in numerous meetings with Darboux. They excited our interest all the more, as they revealed a surprising analogy between the theory of cyclides and that of the second-degree line-complex as presented shortly before this in my article... ["Zur Theorie der Linienkomplexe ersten und zweiten Grades"]. I remember how one day Darboux showed me a manuscript giving a detailed treatment of the theory of cyclides, and how he added the remark that he had obtained the same formula as I had earlier only with five variables instead of six. There was no doubt that here there must exist a transfer principle ["Übertragungsprincip"] connecting line geometry with metric geometry, and from here, by further developing this line of thought, precisely those ideas would emerge which I set forth in 1871 in... ["Über Liniengeometrie und metrische Geometrie"]. In the meantime this question took a new turn and assumed a great deal more gravity through Lie's discovery, made in those very

days, of his wonderful mapping which transforms line geometry into a sphere geometry, and by so doing sends the
asymptotic curves of one surface to the lines of curvature of
another. This last fact was so remarkable that our immediate
interest necessarily concentrated upon it.[72]

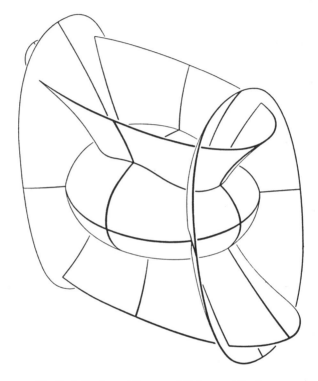

Fig. 2. A Conformal System of Orthogonal Surfaces

Klein's involvement in the Franco-Prussian War also prevented him
from returning to this question immediately. But he clearly had been
mulling over the matter, and in a letter to Lie dated December 17, 1870
he conjectured the following result: "The ruled surface belonging to three
complexes with the same Kummer surface intersects each of the *Brenn-
flächen* associated with the congruences determined by any two of the
three complexes along an asymptotic curve."[73] Two weeks later he wrote
to say that this conjecture was indeed correct, and that his proof showed
"...that the theorem is valid not just for second-degree complexes but for
general complexes which lie in involution, and thus represents a generalization of Dupin's Theorem with one additional variable."[74] Klein published his generalization of this classical result, "Über einen Satz aus der

Theorie der Linienkomplexe, welcher dem Dupinschen Theorem analog ist," in a March 1871 issue of the *Göttinger Nachrichten*. There he pointed out that his theorem not only represented a direct analogue of Dupin's Theorem but that his proof was a direct translation into line-geometry of the proof given in Salmon-Fiedler's *Raumgeometrie*.[75]

Klein's original formulation of the line-geometric analogue to Dupin's Theorem, as expressed in his letter to Lie of December 17 quoted above, suggests that at this time he was still in the process of working out what the appropriate statement of the theorem should be. Clearly, the assumption that three 2nd-degree complexes share the same Kummer surface as their singularity surface was far too particular to serve as an analogue to the notion of orthogonal surfaces in 3-space. As it turned out, Klein's intimate knowledge of line geometry quickly led him to the "correct" notion. The concept he needed, that of involutionary line complexes, was, in fact, an idea he himself had introduced in another guise in his article of 1870 entitled "Zur Theorie der Linienkomplexe des ersten und zweiten Grades."[76] There, however, he had developed this notion only for the case of linear complexes, where it has the following geometric significance.

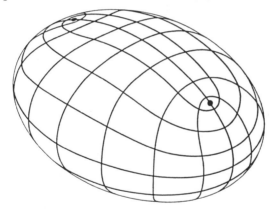

Fig. 3. Dupin's Theorem: Intersection is along Lines of Curvature

The congruence determined by two linear complexes C and C' has two directrices, say g and h. If an arbitrary plane E is taken, then it intersects g and h in two points p and q which lie on the line L determined by the null points r and s associated with C and C'. These two complexes are then said to lie in involution if r and s are harmonic with respect to p and q, i. e., if their cross-ratio equals -1. Alternatively, consider the bundle of planes determined by the line L, which clearly belongs to both C and C'. As these planes rotate about L, they determine pairs of null points which are in involution provided C and C' are. This is essentially the definition Klein employs in considering two arbitrary line-complexes that lie in involution with respect to one another. If one allows a plane E to rotate about a line L common to the two complexes, each position assumed by E determines two points on L where the curves in E associated with the two complexes are tangent to L. Then the point pairs so obtained lie in involution if and only if the complexes themselves are in involution.

Klein and Lie utilized the notion of involutionary line-complexes in a wide variety of settings. One of the most striking, however, was Klein's reformulation of Dupin's Theorem for line geometry:

> If four complexes lie pairwise in involution with respect to a common line L, and if furthermore any three of the four also lie pairwise in involution with respect to the next [infinitesimally close] nearby line common to all three, then their common ruled surface intersects the *Brennfläche* belonging to any two of the three complexes near the line L in the direction of an asymptotic curve.[77]

Although this "higher-dimensional" line-geometric version of Dupin's Theorem is more complicated than other possible formulations, it has the advantage of being easiest to prove. It also led Klein to another key insight that put him well on the way to the *"Erlanger Programm."* After acknowledging his debt to Lie, whose line-to-sphere transformation exhibited a "complete parallelism" between the geometry of a linear complex and the metric geometry of 3-space, Klein noted that his search for a suitable line-geometric formulation of Dupin's Theorem led him to the insight that "line geometry is equivalent to metric geometry in four variables," a point he promised to elaborate on another occasion.[78]

Klein spelled out the import of this idea soon afterward in an article entitled *"Über Liniengeometrie und metrische Geometrie,"* which he wrote in October 1871 and which appeared in *Mathematische Annalen* the following year.[79] This paper already contained allusions to many of the fundamental leitmotifs and examples that Klein took up in more detail a year later in the *"Erlanger Programm."*

He started with the six homogeneous line coordinates (p_{ij}) introduced by Plücker for describing the 4-parameter set of lines in $\mathbf{P}^3(\mathbf{C})$. These are subject to the relation:

$$P = p_{12}p_{34} + p_{13}p_{42} + p_{14}p_{23} = 0.$$

Klein regarded the (p_{ij}) as point coordinates in $\mathbf{P}^5(\mathbf{C})$, and he viewed those satisfying $P = 0$ as a 4-dimensional submanifold $M_4^{(2)}$ (the superscript (2) designates the degree of P). He then considered all the mappings $T: \mathbf{P}^5(\mathbf{C}) \to \mathbf{P}^5(\mathbf{C})$ induced by linear transformations of the coordinates (p_{ij}) subject to the single condition that $P = 0$ remains invariant. Klein showed that these mappings T send lines to lines and points to either points or planes. When restricted to $M_4^{(2)}$ they were thus identical with the totality of all collineations and dual transformations on the lines in space. He concluded that: "All of line geometry can thereby be reduced to the following problem: one investigates the projective properties of the $M_4^{(2)}$ in $\mathbf{P}^5(\mathbf{C})$; then one translates the results into the language of line geometry."[80]

This mode of argumentation bears a striking resemblance to Lie's treatment of the line-to-sphere transformation. In the version of his work that he published in *Mathematische Annalen*, Lie showed that: (1) the linear and dualistic transformations of the domain space correspond to the transformations of the range space that preserve tangency along lines of curvature; (2) the 10-parameter group of linear transformations that map a linear complex onto itself correspond to the group of conformal transformations of the range space; and (3) the 7-parameter group of linear transformations that map a special linear complex onto itself give rise to the conformal homographic transformations that leave the sphericalcircle at infinity invariant.[81] Of course it was not the groups themselves that were of interest here but rather the way in which a familiar group of transformations acting on one space induces another familiar group action on another space. This theme was later highlighted by Klein in section 4 of the "Erlanger Programm," which was entitled "*Übertragung durch Abbildung.*"

Even if he did not stress it significance immediately, Klein was certainly aware of another fundamental principle later set forth explicitly in the "Erlanger Programm." He already recognized that studying the geometry of a space like $M_4^{(2)}$ directly was no different than viewing it as embedded in a higher-dimensional space X and considering the appropriate group of transformations on X that leave it invariant.[82] In this particular context, a linear equation in $\mathbf{P}^5(\mathbf{C})$, i. e., a 4-dimensional hyperplane, is analogous to a first-degree line-complex (which is 3-dimensional). This complex will be special if the hyperplane is tangent to the hypersurface $M_4^{(2)}$. If two such hyperplanes are reciprocal with respect to $M_4^{(2)}$, then they lie in involution with respect to one another. And if their intersection is tangent to $M_4^{(2)}$, then the congruence they determine has two identical directrices. This approach was later developed systematically by Corrado Segre in the 1880s.[83]

The fact that the dimensions of the corresponding objects—hyperspace and linear complex—do not agree brings to mind the following passage from the "Erlanger Programm":

> Instead of the points of a line, plane, space, or any manifold under investigation, we may use instead any figure contained within the manifold: a group of points, curve, surface, etc. As there is nothing at all determined at the outset about the number of arbitrary parameters upon which these figures should depend, the number of dimensions of the line, plane, space, etc. is likewise arbitrary and depends only on the choice of space element. *But so long as we base our geometrical investigation on the same group of transformations, the geometrical content remains unchanged.* That is, every theorem resulting from *one* choice of space element will also be a theorem under any other choice; only the arrangement and correlation of the theorems will be changed. The essential thing is thus the group of transformations; the number of dimensions to be assigned to a manifold is only of secondary importance.[84]

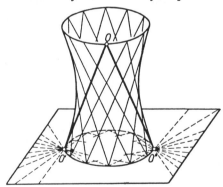

Fig. 4. Stereographic projection of a Quadric Surface onto $\mathbf{P}^2(\mathbf{C})$

In section 1 of "*Über Liniengeometrie und metrische Geometrie*," Klein also mentioned other matters that later occupied an important place in the "Erlanger Programm." He pointed out, for example, that the projective geometry of $\mathbf{P}^1(\mathbf{C})$ (or equivalently, the theory of binary forms) amounted to studying the linear transformations of $\mathbf{P}^2(\mathbf{C})$ that leave a given conic section fixed, or the linear transformations of $\mathbf{P}^3(\mathbf{C})$ that leave a certain 3rd-degree curve invariant. In this connection, Klein also called attention to the role played by Hesse's "*Übertragungsprinzip*," which enables one to transfer plane geometry over to the geometry of point pairs on a line.[85] These examples followed Klein's assertion that "in the final analysis, line geometry is nothing but the projective geometry of space... [whereby] the linear and dualistic trnasformations of $\mathbf{P}^3(\mathbf{C})$ are replaced by the transformations of a higher-dimensional space which leave a certain structure in this space unchanged."[86]

If these statements have a familiar ring, it is because Klein made a number of similar pronouncements in the "*Erlanger Programm*." Indeed, anyone who reads the first two sections of "Über Liniengeometrie und metrische Geometrie" will find it difficult to escape the conclusion that Klein was already in possession of most of the really critical ideas and examples he needed in order to write his much more famous work.[87]

One of the key examples cited in the "Erlanger Programm" is the following. Consider the stereographic projection $f: S \rightarrow \mathbf{P}^2(\mathbf{C})$ from a fixed point P on a 2nd-degree surface S in $\mathbf{P}^3(\mathbf{C})$. This mapping is one-to-one except for two points p and q in the range which are the image of the two generators of S that pass through P. Now the group of linear transformations of $\mathbf{P}^2(\mathbf{C})$ that leaves p and q fixed, when pulled back by f^{-1}, yields the group of linear transformations of S that fix P, i. e., the restrictions to S of those linear transformations of $\mathbf{P}^3(\mathbf{C})$ that leave both S and P invariant. Furthermore, since any two points of $\mathbf{P}^2(\mathbf{C})$ are projectively equivalent, one can take p and q to be the circular points at infinity. But, by Cayley's principle, the linear transformations of $\mathbf{P}^2(\mathbf{C})$ that leave these points fixed are precisely the transformations of Euclidean plane geometry. Klein, therefore, concludes that the study of projective invariants on a 2nd-degree surface with a single point held fixed—or treated as a designated figure ("*ausgezeichnetes Gebilde*")—is equivalent to studying ordinary Euclidean geometry in the plane.[88]

It is interesting to observe that this argument is merely the simplest case of a general theorem proved by Klein in section 2 of "*Über Linien-geometrie und metrische Geometrie*." There he observed that if $M_{n-1}^{(2)}$ is a hypersurface in $\mathbf{P}^n(\mathbf{C})$ given by a 2nd-degree equation, then one can exhibit a mapping $f: M_{n-1}^{(2)} \rightarrow \mathbf{P}^{n-1}(\mathbf{C})$ which is analogous to the one given above. Thus, in the domain space it has a singular point P, the projection point, and elsewhere it is one-to-one. In the range space $\mathbf{P}^{n-1}(\mathbf{C})$, on the other hand, the singular set will be a hypersurface $M_{n-3}^{(1,2)}$ determined by one linear and one quadratic equation. This is the same type of hypersurface that arises in the ordinary metric geometry of $\mathbf{P}^{n-1}(\mathbf{C})$, and since any two such hypersurfaces of the type $M_{n-3}^{(1,2)}$ are projectively equivalent, it follows that one can replace this $M_{n-3}^{(1,2)}$ by S^{n-3}, the $(n-3)$-dimensional

imaginary sphere at infinity, which is simply the higher-dimensional analogue of the spherical-circle at infinity. Klein then showed that the group
of linear transformations of $\mathbf{P}^n(\mathbf{C})$ that leave $M_{n-1}^{(2)}$ invariant precisely corresponds to the group of all similarity mappings and transformations by
reciprocal radii acting on $\mathbf{P}^{n-1}(\mathbf{C})$. When one further restricts the group
of linear transformations acting on $\mathbf{P}^n(\mathbf{C})$ to those which leave P as well
as $M_{n-1}^{(2)}$ fixed, then one obtains exactly those linear transformations of
$\mathbf{P}^{n-1}(\mathbf{C})$ that leave S^{n-3} invariant, i. e. the motions of $(n-1)$-dimensional
Euclidean geometry.[89]

Klein gave another simple but suggestive illustration of the application of *Übertragungsprinzipien* in a short note that was written about one
year after the composition of the "Erlanger Programm." There he showed
how one could use Hesse's transfer principle to obtain a 3-dimensional
analogue of Pascal's Theorem, the classic result of projective geometry
which states that the three points of intersection obtained by extending
the opposite sides of a hexagon inscribed in a conic section will lie on a
common line. One way to prove this is by showing that any such figure
can be mapped projectively to a regular hexagon inscribed in a circle,
since in this case the opposite sides meet on the line at infinity. Plücker
gave the following elegant proof:[90] Let the six lines be given by linear
equations $x_i = 0, i = 1, 2 \ldots, 6$ where $x_i \cap x_{i+1} = p_i$, the six points lying
on the conic C (the "hexagon" of Pascal's Theorem need not be a convex
figure). Each member of the one-parameter family of cubic curves

$$C_\mu : x_1 x_3 x_5 - \mu x_2 x_4 x_6 = 0$$

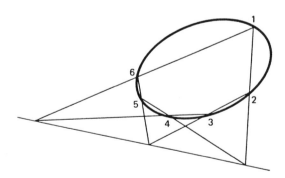

Fig. 5. Pascal's "hexagrammum mysticum." (See Dirk J. Struik, ed., *A Source Book in
Mathematics, 1200–1800* (Princeton: Princeton Univ. Press, 1986), pp. 163–167).

must contain all six of the p_i plus the three points $q_{i,i+3} = x_i \cap x_{i+3}$ where
the opposite sides of the hexagon meet. Thus if $\hat{p}$ is any seventh point of

C, there must exist some μ' for which $C_{\mu'}$ also passes through $\hat{p}$. But by Bezout's Theorem an irreducible cubic intersects a quadratic curve in at most six distinct points, from which it follows that $C_{\mu'}$ decomposes into C and a linear factor that contains the three points $q_{i,i+3}$.

In "Eine Übertragung des Pascalschen Satzes auf Raumgeometrie," Klein employed Hesse's transfer principle to obtain a representation of the lines in $\mathbf{P}^2(\mathbf{C})$ as pairs of points on a line L ($= \mathbf{P}^1(\mathbf{C})$).[91] Each line $l \in \mathbf{P}^2(\mathbf{C})$ will meet a fixed conic C in two points. If one then takes a point $p \in C$ and uses stereographic projection to map C onto L, the image of l will be two points λ_1, λ_2 on L. In applying this mapping to the conic C of Pascal's Theorem, Klein viewed C as a class curve enveloped by a one-parameter family of real lines in $\mathbf{P}^2(\mathbf{C})$. An inscribed "hexagon" is then determined by any six of these enveloping lines, $h_i, i = 1, 2 \ldots, 6$. He then treated L as a real plane and employed stereographic projection to map it one-to-one and onto a punctured quadric surface F in 3-space. If now Hesse's map is followed by this mapping, then the lines $l_i \in \mathbf{P}^2(\mathbf{C})$ go over to pairs of points on the surface F since the points λ_1^i, λ_2^i will correspond to lines in $\mathbf{P}^3(\mathbf{C})$ that meet F in two points. The h_i, however, being tangents to C will correspond to lines that intersect F in a single point. Thus, Klein succeeded in establishing a correspondence between the lines of $\mathbf{P}^2(\mathbf{C})$ and the *real* lines of $\mathbf{P}^3(\mathbf{C})$ that intersect F. Moreover, this correspondence is projective so that cross-ratio will be preserved. In particular, lines that are conjugate with respect to the conic C will correspond to lines that are conjugate with respect to F, or, to employ Cayley's language, lines which are "perpendicular" in the plane will be mapped to "perpendicular" lines in 3-space.

By utilizing this transference apparatus, Klein merely needed to translate Pascal's Theorem into the language appropriate for exhibiting its counterpart in 3-space. Thus if $x_i = 0, x_{i+3} = 0$ are the opposite sides of the hexagon inscribed in C and $y = 0$ is the line containing their three points of intersection $q_{i,i+3}$, then the polars of the $q_{i,i+3}$ must contain the pole of $y = 0$. In other words, the line $y = 0$ is perpendicular to each of the polars of the $q_{i,i+3}$. But since each of the three points $q_{i,i+3} = x_i \cap x_{i+3}$, the polar to $q_{i,i+3}$ is, in turn, perpendicular to both x_i and x_{i+3}. Thus Pascal's Theorem can be reworded to say that to three lines "perpendicular" to the three pairs of opposite sides of a hexagon inscribed in a conic section there always exists a fourth line "perpendicular" to the other three. Klein could now carry this theorem over word for word to the case of a general "hexagon" inscribed in a closed quadric surface F. Moreover, since the six points on F are completely arbitrary, the "hexagon" need not even be a plane figure.

8. Concluding Remarks

The foregoing analysis of the early geometrical works of Klein and Lie is far from exhaustive. Nevertheless, it should suffice to make clear that during this relatively brief period they developed a wealth of interesting ideas, techniques, and results that are all but forgotten today. As Thomas Hawkins points out in the essay that follows, this work was of decisive importance for the emergence of Lie's theory of transformation groups in 1873-74. Moreover, as indicated above, this research also represented a critical step toward the composition of Klein's "Erlanger Programm," which is largely unintelligible apart from these background developments. The vast majority of writers who refer to the "Erlanger Programm" appear to derive their knowledge of it from secondary sources. Unfortunately, most of these accounts tend to trivialize its content, in particular by restricting their discussion to examples in which familiar transformations groups act on real spaces.[92] This interpretation necessarily overlooks all the deeper geometrical results—which were the only ones Klein bothered to present—since these all require that the base space be a complex manifold.

But if this work is so unfamiliar today, what can be said about its reception during the period immediately following its publication? Here we are confronted with an interesting, but very complex historical problem to which I can offer little more than some provisional observations. Curiously enough, with a few important exceptions, these early investigations of Lie and Klein appear to have been ignored for the most part until the early 1890s. As Hawkins has shown, this was even the case with Klein's "Erlanger Programm," which appears to have been largely unknown until after 1890 when it was republished and translated into a number of languages.[93] Like several other classics in the history of mathematics, its immediate reception was practically nil. What circumstances can be cited that would account for this?

The first, and most obvious, is that Klein wrote it to fulfill a formal requirement incumbant upon every new *Ordinarius* who entered the Erlangen faculty. Its name, in fact, had nothing to do with a new program for mathematical research, but rather derives from having been a "Programm zum Eintritt in die philosophische Fakultät und den Senat" at Erlangen. One notes here certain parallels with Riemann's *Habilitationsvortrag*. For although Klein chose a different topic for his inaugural address to the Erlangen faculty, the written "Programm" that accompanied it was also intended for a broad audience.[94] Klein's style of presentation was novel, something of a cross between a research article and a purely expository account. Written in bold, general terms, it contained few details and a good deal of geometrical "folklore," but most of all a series of brilliant insights about the nature of the geometrical enterprise, richly illustrated by a panoply of recent discoveries. Clearly, this was not the sort of thing one published in a journal, at least not if you are a 23-year-old mathematician at the outset of your career. Klein disseminated his "Programm," which was published locally, to the members of the Erlangen faculty and to a few close friends and colleagues, like Lie and Noether. Most of all he was eager to get Clebsch's reaction to his ideas, but his mentor's death came only weeks after Klein composed it.[95] Certainly the immediate circumstances were not very propitious, but the puzzling question remains: what was it that prevented Klein from following up on the ideas in the "Erlanger Programm" over the next 20 years in order to bring its underlying philosophy to the attention of his contemporaries? And what accounts for his sudden resurgence of interest in this work during the early 1890s?

Klein's lectures of 1892-93 on "höhere Geometrie" were in fact largely intended as a vehicle for calling attention to these largely overlooked ideas that he and Lie had set forth 20 years earlier. He even hoped at this time to persuade Lie to republish many of his earlier papers in *Mathematische Annalen,* but eventually he had to abandon this plan and along with it his friendship with Lie who by this time felt that he had to break with Klein or spend the rest of his career in his shadow.[96] Despite this estrangement however, Klein continued to promote Lie's "forgotten work" at nearly every opportunity.[97] And in his 1892–93 lectures he bemoaned the fact that, although the French and Italian geometers took up Lie's sphere geometry, the subject was largely ignored by the Germans who, following the lead of Theodor Reye, confined their investigations to the more elementary sphere geometry that had been developed by Darboux, et al.[98]

What makes these matters so problematic is that they appear to be part of a larger phenomenon, namely the demise of geometry in Germany and the gradual emergence of a flourishing new geometrical school in Italy, a transformation clearly reflected in J. J. Gray's survey of key developments in algebraic geometry. Throughout the 1870s and 80s the Berlin school largely dictated what was mathematically fashionable in German mathematics, and this had an influence not only on the fields mathematicians chose to research but also the style in which they presented their work. In both regards, Klein and Lie, different as they were, came from a tradition that was poles apart from the one practiced by Weierstrass, Kronecker, and the leading spokesman for the Berlin school after their departure, Frobenius. Still, it is very difficult to assess just how inhibiting these circumstances may have been for two young men who, after all, had emerged as the stars of Kummer's seminar during their single semester in Berlin. What we do know is that neither made much effort to circulate the results of their early collaboration until some twenty years later, and that Klein in particular made a strong push in this direction beginning around 1892. It seems to have been more than just a coincidence that this effort came precisely at the time when the old guard in Berlin—Weierstrass and Kronecker—were no longer on the scene.

NOTES*

1. F. Klein to M. Noether, 10 Oct. 1869, *Klein Nachlass XII, Niedersächsische Staats- und Universitätsbibliothek Göttingen (NSUB)*. This remark suggests that the influence of the geometer Jakob Steiner (1796-1863) and his school had by now run its course.

2. This terminology was popularized through Klein's lecture, "Ueber die Arithmetisierung der Mathematik," held in Göttingen on 2 Nov. 1895, see *Gesammelte Mathematische Abhandlungen (GMA)*, 3 vols. (Berlin/Heidelberg/New York: Springer, 1921–23, reprinted 1973), vol. 2, pp. 232–240.

3. F. Klein to his mother, 31 Oct. 1869, quoted in S. Lie, *Gesammelte Abhandlungen (GA)*, 7 vols. (Leipzig: Teubner and Oslo: Haschehoug, 1924–60), Anmerkungen zum ersten Bande, p. 636. Lie had sent a copy of his eight-page article "Repräsentation der Imaginären der Plangeometrie"—originally printed privately shortly before it was published in *Crelle's Journal*—to Clebsch in Göttingen earlier that year.

4. No trace remains of Lie's letters to Klein from before 1877 as Klein had burned all his correspondence from these early years. Friedrich Engel had access to Klein's letters to Lie and published excerpts from them in Lie's *GA*, but these too seem to have disappeared afterward.

5. See H. Wussing, *The Genesis of the Abstract Group Concept*, trans., by Abe Schneitzer (Cambridge, Mass./London: MIT Press, 1984), pp. 167–229.

6. Engel admitted, however, that Sylow's course had no real influence on Lie (F. Engel, "Zur Erinnerung an Sophus Lie," *Bericht über die Verhandlungen der königlich sächsischen Gesellschaft der Wissenschaften zu Leipzig*, 51(1899), 11–61, p. 14).

7. F. Klein, *GMA*, vol. 1, p. 51.

8. See F. Klein, *Vorlesungen über die Entwicklung der Mathematik im 19. Jahrhundert*, vol. 1 (Berlin: Springer, 1926), p. 130.

* The author wants to thank Thomas Hawkins and Karen Parshall for their suggestions for improving an earlier version of this paper.

9. The following exposition is based on F. Klein, *Elementarmathematik vom hoheren Standpunkte aus*, 3rd ed., vol. 2 (Berlin: Springer), 1925, pp. 126–130.

10. F. Klein, *Entwicklung der Mathematik*, p. 144.

11. Plücker introduced the fifth coordinate in his paper "On a new Geometry of Space," *Philosophical Transactions of the Royal Society of London*, 155(1865), 725–791.

12. An excellent treatment of the connection between line geometry and mechanics can be found in Renatus Ziegler, *Die Geschichte der geometrischen Mechanik im 19. Jahrhundert* (Stuttgart: Steiner, 1985); Möbius's theory of null-systems can also be found in S. Lie and G. Scheffers, *Geometrie der Berührungstransformationen (GdBT)*, (Leipzig: Teubner, 1896), pp. 181–247.

13. Malus gave a detailed presentation of these structures in his "Optique," *Journal de l'Ecole Polytechnique*, 7(14), 1808.

14. E. E. Kummer, "Über die Flächen vierten Grades mit sechszehn singulären Punkten," *Monatsberichte der Akademie der Wissenschaften zu Berlin*, (1864), 246–260.

15. Klein's dissertation, "Ueber die Transformation der allgemeinen Gleichung des zweiten Grades zwischen Linien-Coordinaten auf eine canonische Form," was republished along with background information in F. Klein, *GMA*, vol. 1, pp. 3-49.

16. K. Weierstrass, "Zur Theorie der quadratischen und bilinearen Formen," *Monatsberichte der Akademie der Wissenschaften zu Berlin*, 1868, pp. 310–338; *Mathematische Werke*, vol. 2 (Hildesheim: Georg Olms; New York: Johnson Reprint, 1967), pp. 19–44.

17. J. Plücker, *Neue Geometrie des Raumes gegründet auf die Betrachtung der geraden Linie als Raumelement*, 2 vols. (Leipzig: Teubner, 1868–69), Nr. 170.

18. F. Klein, "Zur Theorie der Linienkomplexe des ersten und zweiten Grades," *GMA*, vol. 1, pp. 53–80.

19. J. Plücker, *Neue Geometrie*, Nr. 311, 320.

20. "Repräsentation der Imaginären der Plangeometrie," in S. Lie, *GA*, vol. 1, pp. 1–11.

21. Lie's handwritten notes are reproduced in the "Anmerkungen zum ersten Bande," *GA*, vol.1, 537–539; those quoted are on p. 539.

22. Lie's "metaphysical observations" were as follows:

A geometric theorem usually can be viewed algebraically, that is, by expressing the equality of certain quantities as the consequence of given equalities. The theorems of plane geometry that cannot be derived through algebra are easy to verify in our theory. Since algebra contains imaginary quantities, it is clear that every theorem of plane geometry (that one can express algebraically) is valid for our imaginary objects. *Every theorem of plane geometry is a special case of a stereometric double theorem in the geometry of line congruences.*

S. Lie, *GA*, vol. 1, "Repräsentation der Imaginären der Plangeometrie," pp. 14–66, on p. 20. His dissertation appeared in two installments of the *Fordhandlinger i Videnskabs-Selskabet i Christiania* in 1870, *GA*, vol. 1, pp. 105–210.

23. This was the opinion expressed by F. Klein in a letter to M. Noether, 17 Dec. 1869, *Klein Nachlass XII, NSUB*.

24. F. Engel, "Zur Erinnerung an Sophus Lie," p. 16.

25. Quoted in R. Tobies, *Felix Klein*, Biographien hervorragender Naturwissenschaftler, Techniker und Mediziner, vol. 50. Leipzig: BSB B. G. Teubner, 1981, p. 22.

26. F. Klein to M. Noether, 17 Dec. 1869, *Klein Nachlass XII, NSUB*.

27. Klein expressed this regret on numerous occasions, one being in his *Entwicklung der Mathematik*, p. 284.

28. These words of praise for Weierstrass were reportedly spoken by Hermite; see G. Mittag-Leffler, "Une page de la vie de Weierstrass," *Compte Rendu du Deuxième Congrès International des Mathématiciens* (Paris: Gauthier-Villars) 1902, pp. 131–159, on p. 131.

29. F. Klein to M. Noether, 17 Dec. 1869. Klein's views on Clebsch can be found in *Entwicklung der Mathematik*, p. 297.

30. On Stolz's role in helping Klein see *GMA*, vol. 1, pp. 50–52.

31. A. Cayley, "Sixth Memoir on Quantics," *The Collected Mathematical Papers*, vol. 2 (Cambridge, 1889), p. 592.

32. F. Klein, *Entwicklung der Mathematik*, p. 152.

33. Lie left Berlin for Göttingen on 28 Feb. 1870; cf. *GA*, "Anmerkungen zum ersten Bande," p. 636.

34. F. Klein, "Über die sogenannte Nicht-Euklidische Geometrie" (Vorl. Mitt.), *GMA*, vol. 1, pp. 244-254: (*Erster Aufsatz*), pp. 254–305; (*Zweiter Aufsatz*), pp. 311–343.

35. F. Klein, "Über Lie's und meine ältere geometrische Arbeiten," unpublished manuscript, *Klein Nachlass XXII G, NSUB*.

36. F. Klein to M. Noether, 17 Dec. 1869, *Klein Nachlass XII, NSUB*.

37. H. Müller, "Zur Geometrie auf den Flächen zweiter Ordnung," *Mathematische Annalen*, 1(1869), pp. 407–423.

38. The above and what follows is based on S. Lie and G. Scheffers, *GdBT*, pp. 311–356.

39. De la Gournerie, *Recherches sur les surfaces réglées tétraèdrales symmétriques* (Paris: Gauthier-Villiars), 1867.

40. This theorem and its proof first appeared in *GdBT*, pp. 348–351.

41. S. Lie, "Ueber die Reziprozitätsverhältnisse des Reyeschen Komplexes," *GA*, vol. 1, pp. 68–77.

42. F. Engel, "Zur Erinnerung an Sophus Lie," p. 18.

43. Klein recalled that it was he who first noted the existence of the *W*–curves in *GMA*, vol. 1, p. 415. In a letter from Klein to Lie written on 8 Feb. 1871. Klein wrote: "Es ist seltsamerweise morgen genau ein Jahr, dass wir damals bei Voelker auf die *W*–Kurven gerieten." (*GA*, "Anmerkungen zum ersten Bande," p. 744).

44. F. Klein, *GMA*, vol. 1, pp. 415–423; S. Lie, *GA*, vol. 1, pp. 78–87.

45. See M. Noether, "Sophus Lie," *Mathematische Annalen*, 53(1901), 1–41, pp. 6–7.

46. F. Klein to S. Lie, 4 Feb. 1871, *GA*, "Anmerkungen zum ersten Bande," p. 744.

47. F. Klein to S. Lie, 30 Dec. 1870, *Ibid.*, p. 743. Klein's paper utilizing elliptic coordinates appeared in mid-January of the following year under the title "Zur Theorie der Kummerschen Fläche und der zugehörigen Linienkomplexe zweiten Grades," *Göttinger Nachrichten*, Jan. 1871. For an explanation of elliptic coordinates and their role in geometry, see F. Klein, *Vorlesungen über höhere Geometrie*, 3rd. ed. revised and edited by W. Blaschke. New York: Chelsea, 1949, pp. 19–25.

48. F. Klein to S. Lie, 28 Jan. 1871, *Ibid.*, pp. 743–744.

49. F. Klein to S. Lie, 4 Feb. 1871, *Ibid.*, p. 744.

50. F. Klein to S. Lie, 8 Feb. 1871, *Ibid.*

51. F. Klein and S. Lie, "Über diejenigen ebenen Kurven, welche durch ein geschlossenes System von einfach unendlich vielen vertauschbaren linearen Transformationen in sich übergehen," *GMA*, vol. 1, pp. 424–459.

52. The reader should note that the following presentation, which is based on *GdBT*, pp. 249–255, 414–429, does not adhere precisely to the chronological order in which Lie made his discoveries. My principal concern here is to give a cogent presentation of some of the main ideas Lie developed in his dissertation and the lengthy article based upon it that appeared in vol. 5 of *Mathematische Annalen*. Afterward, I will consider more carefully the path of discovery that Lie actually followed.

53. This result was first presented in the sequel to Lie's dissertation of 1871, "Über eine Klasse geometrischer Transformationen," *GA*, vol. I, pp. 153–210, on pp. 155–157.

54. J. Liouville, "Note au sujet de l'article précédent," *Journal de Mathématiques pures et appliquées*, 12(1847), p. 265.

55. See S. Lie, "Über diejenige Theorie eines Raumes mit beliebig vielen Dimensionen, die der Krümmungstheorie des gewöhnlichen Raumes entspricht," *Göttinger Nachrichten*, May 1871 (*GA*, vol. 1, pp. 215–228). In this paper Lie also presented a generalization of Dupin's Theorem to spaces of arbitrary dimension.

56. J. Plücker, *Analytisch-geometrische Entwicklungen*, 2 vols. Essen, 1828, 1831.

57. See S. Lie, "Over en Classe geometriske Transformationer," *GA*, vol. 1, pp. 105–152 on pp. 150–151, on p. 106.

58. The following discussion is based on *GdBT*, pp. 444–475.

59. M. Noether, "Zur Theorie der algebraischen Funktionen mehrerer komplexer Veränderlichen," *Göttinger Nachrichten*, 1869.

60. From the protocol book to Klein's seminar of 1909/1910 on mathematics and psychology, *Bibliothek des Mathematischen Instituts der Universität Göttingen*.

61. See S. Lie, "Over en Classe geometriske Transformationer," *GA*, vol. 1, pp. 139–141.

62. Klein discussed this incident in his unpublished manuscript, "Ueber Lie's und meine ältere geometrische Arbeiten," *Klein Nachlass, XXII G, NSUB*.

63. See Lie's dissertation, *GA*, vol. 1, pp. 144-145, and Klein's discussion in S. Lie and F. Klein, "Über die Haupttangentenkurven der Kümmerschen Fläche vierten Grades mit 16 Knotenpunkten," *GMA*, vol. 1, pp. 95–96.

64. F. Klein, *GMA*, vol. 1, p. 97.

65. Republished in F. Klein, *GMA*, vol. 1, pp. 90–97, and S. Lie, *GA*, vol. 1, pp. 97–104. Friedrich Engel criticized Klein for not sufficiently emphasizing Lie's accomplishments in this article ("Anmerkungen zum ersten Bande," p. 674).

66. The original Norwegian version was translated into German by L. Sylow in 1899. My translation is from Sylow's German text which appears in S. Lie, *GA*, vol. 1, pp. 86–87.

67. F. Klein, "Über Lie's und meine Arbeiten aus den Jahren 1870–72," unpublished manuscript, *Lie Nachlass, Ms. fol. 3839, LXVII: 11, Universitetsbiblioteket, Oslo.*

68. S. Lie, "Über Komplexe, insbesondere Linien- und Kugelkomplexe, mit Anwendung auf die Theorie partieller Differentialgleichungen," *GA*, vol. II, pp. 1–121, on p. 37. See also F. Klein, *Entwicklung der Mathematik*, p. 146.

69. F. Klein, "Über Lie's und meine Arbeiten aus den Jahren 1870–72," *op cit.*

70. The following discussion is based on F. Klein, *Vorlesungen über höhere Geometrie*, 3rd. ed., pp. 49–58.

71. See the discussion in §2 above.

72. F. Klein, "Über Lie's und meine Arbeiten aus den Jahren 1870–72," *Lie Nachlass, Universitetsbiblioteket, Oslo.*

73. F. Klein to S. Lie, 17 Dec. 1870, *GA*, "Anmerkungen zum ersten Bande," p. 734.

74. F. Klein to S. Lie, 30 Dec. 1870, *Ibid.*, p. 735.

75. F. Klein, *GMA*, vol. 1, p. 99.

76. F. Klein, *GMA*, vol. 1, p. 56.

77. F. Klein, "Über einen Satz aus der Theorie der Linienkomplexe, welcher dem Dupinschen Theorem analog ist," *GMA*, vol. 1, p. 99.

78. *Ibid.*

79. F. Klein, *GMA*, vol. 1, pp. 106–126.

80. *Ibid.*, p. 112.

81. Lie summarized these results in the paper "Über Komplexe, insbesondere Linien- und Kugelkomplexe, mit Anwendung auf die Theorie partieller Differentialgleichungen," *Mathematische Annalen*, 5(1)(1872), 145-208, 5(2)(1872), 209-256; *GA*, vol. 2, pp. 1–121, on p. 45.

82. F. Klein, "Vergleichende Betrachtungen über neuere geometrische Forschungen," *GMA*, vol. 1, pp. 460–497, on p. 465.

83. See C. Segre, "Sulle geometrie metriche dei complessi lineari e delle sfere," *Atti della Reale Accademia dei Scienze di Torino*, vol. XIX(1883), 159–186; C. Segre, "Studio sulle quadriche in uno spazio lineare ad un numero qualunque di dimensioni," *Opere*, vol. III. Rome: Edizioni Cremonese, 1961.

84. *Ibid.*, pp. 470–471.

85. For a detailed study, see T. Hawkins, "Hesse's Principle of Transfer and the Representation of Lie Algebras," *Archive for History of Exact Sciences*, 39(1)(1988), 41–73.

86. F. Klein, "Über Liniengeometrie und metrische Geometrie," *GMA*, vol. 1, p. 112.

87. This applies particularly to the first six sections of the "Erlangen Program." As Klein noted in preparing his collected works, his paper "Über Liniengeometrie und metrische Geometrie" rightfully belonged in the section entitled "Zum Erlanger Programm." Owing to its fundamental importance for line geometry, however, it was placed in the section "Zur Liniengeometrie" (*GMA*, vol. 1, p. 411).

88. See F. Klein, "Erlanger Programm," *GMA*, vol. 1, pp. 469–470.

89. F. Klein, "Über Liniengeometrie und metrische Geometrie," pp. 113–114.

90. See F. Klein, *Entwicklung der Mathematik*, p. 122.

91. F. Klein, *GMA*, vol. 1, pp. 406–408.

92. This criticism applies, for example, to I. M. Yaglom, *Felix Klein and Sophus Lie. Evolution of the Idea of Symmetry in the Nineteenth Century*, trans. from the Russian by S. Sossinsky. Boston/Basel: Birkhäuser, 1988.

93. See T. Hawkins, "The *Erlanger Programm* of Felix Klein: Reflections on its Place in the History of Mathematics," *Historia Mathematica*, 11(1984), 442–470. The contrary opinion has recently been forwarded in Garrett Birkhoff and M. K. Bennett, "Felix Klein and his 'Erlanger Programm'" in W. Asray and P. Kitcher, eds. *History and Philosophy of Modern Mathematics*. Minneapolis: Univ. of Minnesota Press, 1987, pp. 145–176. The authors, however, fail to advance any new information that would refute Hawkins' position, and merely assert that the "Erlanger Programm" met with a delayed reception because "[it] was 20 years ahead of its time"(p. 152).

94. This speech appears in D. E. Rowe, "Felix Klein's *Erlanger Antrittsrede*: A Transcription with English Translation and Commentary," *Historia Mathematica*, 12(1984), 123–141.

95. Klein composed an early draft of the "Erlangen Program," entitled "Methoden der Geometrie," in December 1871. Lie was apparently the only one who read this manuscript, which is no longer extant. This early version is discussed in F. Klein, "Über Lie's und meine ältere geometrische Arbeiten," *Klein Nachlass XXII G, NSUB*.

96. For more on the reasons behind their estrangement, see D. E. Rowe, "Der Briefwechsel Sophus Lie – Felix Klein, eine Einsicht in ihre persönlichen und wissenschaftlichen Beziehungen," *NTM. Schriften-reihe für Geschichte der Naturwissenschaften, Technik und Medizin*, 25(1) (1988), 37–47.

97. Besides the detailed presentation of Lie's work given in his 1892–93 lectures, Klein devoted two of his twelve Evanston Colloquium lectures to Lie's early geometrical researches and wrote a glowing *Gutachten* that helped secure for him the first awarding of the Loba-chevsky prize (see F. Klein, "Gutachten, betreffend den dritten Band der Theorie der Transformationsgruppen von S. Lie anlässlich der ersten Verteilung des Lobatschewsky-Preises," *GMA*, vol. 1, pp. 384–401).

98. F. Klein, *Einleitung in die höhere Geometrie*, 1st ed. Leipzig: Teubner, 1893, p. 228. The same remarks also appear in F. Klein, *Vorlesungen über höhere Geometrie*, 3rd. ed. pp. 115–116.

Felix Klein (1849–1925)
Courtesy of Springer-Verlag Archives

Line Geometry, Differential Equations
And the Birth of Lie's Theory of Groups

Thomas Hawkins

The purpose of this essay is to present some of the conclusions and implications of a research project which in scope is too broad and mathematically complex to permit a detailed discussion within the framework of these proceedings.[1] By Lie's own account, the birth of his theory of groups occurred during the winter of 1873-74. At that time he posed the problem of determining all continuous groups of transformations acting on n-dimensional space, that is all groups of transformations generated by a finite number of infinitesimal transformations of n-dimensional space. Although he never resolved this problem for $n > 3$—it is much too difficult—it was in the course of attempting to resolve it that he introduced the mathematical apparatus that constitutes his theory of groups. Thus, mathematically speaking, Lie's theory originated in that fateful winter. From a historical point of view, however, this is not the case. What happened during that winter represented the culmination of certain mathematical experiences Lie had during the first years of his career as a mathematician—from 1869 to 1873. During this formative and fertile period, his work on line geometry and the work on differential equations which evolved from it set the stage for the events of the winter of 1873-74: by bringing to Lie's attention the notion of a continuous group (a novel notion at the time) and by convincing him of the mathematical importance of this notion in applications to the study of geometry and differential equations. Furthermore, Lie acquired along the way the mathematical tools that convinced him of the feasibility of the above group classification problem. The purpose of this essay is to convey something of the flavor of this early work and its significance for the birth of Lie's theory.

Lie's prodigious mathematical research activity between 1869 and the winter of 1873-74 was not dominated by group-related considerations but involved a diverse spectrum of mathematical ideas. Nonetheless, with the wisdom provided by historical hindsight, a path can be traced through Lie's mathematics along which he unwittingly moved towards his destiny as creator of the theory of continuous transformation groups. Although Lie was influenced by many ideas and by several mathematicians along

THE HISTORY OF
MODERN MATHEMATICS
275

the way, the central driving force leading towards the events of the winter of 1873-74 was his own remarkable creative energy. For this reason, the stages along the way are best characterized in terms of the changing focus of his research interests: (1) the geometry of tetrahedral complexes (1869-70); (2) the sphere mapping (1870-71); and (3) the theory of partial differential equations in any number of variables and the related "invariant theory" of contact transformations (1872-73). Here the emphasis will be upon (1) and (2) which are discussed in §1 and §2, respectively. Some of the conclusions from the analysis of (3) are utilized in §3, where the events of the winter of 1873-74 are discussed.[2] Much of Lie's work on (1) and (2) was done in close contact with Felix Klein. Lie saw Klein on a daily basis in Berlin (where they first met) from the end of October 1869 to end of February 1870, in Paris from the end of April to mid-July 1870, and in Göttingen and Erlangen during September and October 1872. During these periods they freely discussed and shared their mathematical ideas. Klein's interaction with Lie turns out to be historically important in relation to the birth of Lie's theory and is considered carefully in what follows.

The historical material analyzed here, the viewpoint taken in considering the material and the conclusions presented are related to two other papers in these proceedings: David Rowe's "The Early Geometrical Works of Felix Klein and Sophus Lie," and Erhard Scholz's "Crystallographic Symmetry Concepts and Group Theory (1850-1880)." Rowe's paper is also concerned with the work of Klein and Lie during 1869-72, although his approach to the material and the one adopted here are considerably different. Here the purpose in considering the early work of Klein and Lie is to examine how group related notions and problems emerged from their investigations of problems involving geometry and differential equations and to assess their relative significance for the events of the winter of 1873-74, when the creation of the theory of continuous groups commenced in earnest. Much of the work done by Klein and Lie during 1869-72 is of no relevance to our objective and has consequently been ignored in the following presentation. This could leave the mistaken impression with the reader that all or most of Klein's and Lie's interests in this period were related to groups. Rowe's paper is in a sense written to correct, or prevent, such a misunderstanding. Rather than viewing the material with an eye towards future events (as is done here), he "stresses the paths [Klein and Lie] came from rather than those that emerged in the wake of their achievements." As a consequence, his paper and what follows here are largely complementary presentations of the same material. His exposition is referred to throughout §§1-2 as a source for further background and another perspective. Scholz's paper is concerned with Camille Jordan's memoir [1869a] classifying groups of rigid motions in ordinary space, a memoir which marked an important advance in the development of crystallography. In the final section of his paper, Scholz suggests that Jordan's memoir was an appreciable factor in the developments leading to Klein's Erlangen Program and to Lie's theory of groups. A differing assessment of the significance of Jordan's memoir in this respect is implicit in the following presentation.

In making bibliographic references, such as "Lie [1871e]" or "[Klein 1872a]", the customary practice of historians of mathematics has been followed, albeit with one difference. Since Klein's and Lie's publications during the period under consideration are numerous and since the dates when they wrote them are important, the year in the bibliographic references for Klein and Lie refers not to the year of publication but the year in which the paper was submitted. Thus Lie's paper [1871e] was submitted in 1871 , although published in 1872. Within a given year the papers are ordered by the dates of submission. Thus Lie's paper [1871c] was submitted before his [1871e]. The exact dates of submission and the years of publication are given in the list of references. Several manuscripts are also referred to in the paper. They are cited in the text by a notation similar to that for publications. For example, "[Klein MS1892]" refers to an unpublished manuscript by Klein written in 1892. A full description of these documents, including present location, is given in the list of manuscript references. Except for [Lie MS1879], all the manuscripts listed there were brought to my attention by David Rowe, who also graciously facilitated my study of them.

1. The Geometry of Tetrahedral Complexes

Sophus Lie, a Norwegian, first became interested in mathematical research in 1868 when, at the age of 26, he discovered a way to set up a correspondence between complex projective geometry in the plane and line geometry in real projective 3-space (Rowe, §3). As a result of this display of mathematical talent, he was awarded a stipend enabling him to visit leading centers of mathematics. Lie began his travels in the fall of 1869, and his first stop was the University of Berlin. There he met Felix Klein. Klein, who was only 20, had received his doctorate from the University of Bonn in 1868 for a dissertation on the line geometry of his mentor, Julius Plücker. (Lie had no doctorate yet; his doctoral dissertation [1871a] was based on his work on the sphere map (§2).) Klein and Lie were drawn together by their common interest in geometry in general and line geometry in particular. Line geometry and its connections with the development of projective geometry are discussed in detail by Rowe (§1). Here it will suffice to explain that line geometry is an approach to the study of geometrical objects based upon considering straight lines (rather than the customary points) as the basic elements and studying configurations comprised of lines, such as what Plücker called "line complexes." One way to assign coordinates to a line ℓ is as follows. Let $x_1, \ldots, x_4$ and $y_1, \ldots, y_4$ denote the homogeneous coordinates of two points lying on ℓ and hence determining it. If for $i \neq k$, $p_{ik} = x_i y_k - x_k y_i$, then $p_{ik} = -p_{ki}$. The 6 quantities $p_{12}, p_{13}, p_{14}, p_{23}, p_{42}, p_{34}$ may be chosen as (homogeneous) line coordinates of ℓ. They are related by the equation $\Omega = p_{12}p_{34} + p_{13}p_{42} + p_{14}p_{23} = 0$. Line complexes are sets of lines whose coordinates p_{ik} satisfy a homogeneous algebraic equation (in addition to $\Omega = 0$). The degree of the equation is the degree of the complex.

Although Klein and Lie had common interests, they represented diametrically opposite mathematical personalities.[3] Klein had a solid mathematical background and was interested in all types of mathematics. He enjoyed learning about the work of others because he wanted to understand it from his own point of view, to see it as a part of a broader, conceptually unified mathematical picture. By contrast, Lie, whose mathematical background was limited, was not disposed to be interested in the work of others. He tended to focus rather exclusively upon developing his own ideas and became interested in the work of others only when it was clear that it was relevant to his research program. This difference in personalities, reinforced by the difference in their ages, dictated the way they related to each other as mathematicians: Lie developed his ideas, explained them to Klein (who was interested in hearing about them), Klein reacted to Lie's ideas, Lie (sometimes) responded to Klein's reactions, and so on. The first such interaction involved the research that Lie had underway when he arrived in Berlin. He was studying the geometry of tetrahedral complexes.

A tetrahedral line complex may be defined as follows. Let Δ denote a fixed tetrahedron, with faces determined by planes $\pi_1, \ldots, \pi_4$. Each line ℓ in complex projective space ($\mathbb{P}^3(\mathbb{C})$) meets these planes in four points $p_1, \ldots, p_4$. A tetrahedral complex T consists of all lines ℓ for which the cross ratio of these points is the same. Lie was not the first to consider tetrahedral complexes (which are second degree complexes). His approach, however, was profoundly original and is exemplified by the way he conceived of them. He introduced the totality all projective transformations of space which leave the vertices of Δ fixed. These transformations take each plane π_i into itself and thus take Δ into itself. The totality of these transformations will be denoted here by $\mathfrak{G}$. For Lie, a tetrahedral complex may be characterized as follows: Fix any line ℓ_0. Then a tetrahedral complex T is obtained by applying all transformations of $\mathfrak{G}$ to ℓ_0. That is, $T = \{T[\ell_0] : T \in \mathfrak{G}\}$.

Sophus Lie (1842–1899).
Courtesy of Springer-Verlag Archives.

In his study of the geometry related to T, Lie made essential use of many properties of $\mathfrak{G}$ which can now be seen to relate to its structure as a continuous group. For example, $\mathfrak{G}$ contains a threefold infinity of transformations. This fact becomes evident if Δ is taken to be the tetrahedron with faces determined by the three coordinate planes and the plane at infinity. Then the transformations of $\mathfrak{G}$ are given by

(1.1) $$x' = \lambda x, \quad y' = \mu y, \quad z' = \nu z.$$

Each transformation corresponds to a choice of the parameters λ, μ, ν ($\lambda\mu\nu \neq 0$) and so there are "∞^3" of them. Further properties of $\mathfrak{G}$ utilized by Lie follow readily from (1.1). For example, for all $T_1, T_2 \in \mathfrak{G}$, $T_1(T_2(p)) = T_2(T_1(p))$ for all points p. Lie and Klein subsequently described this property by using the language of permutation groups and saying the transformations of $\mathfrak{G}$ are commutative (*vertauschbar*). It is also evident from (1.1), that if $T \in \mathfrak{G}$ then $T^{-1} \in \mathfrak{G}$. Furthermore, if p and q are points in "general position" – that is, points not lying in any of the four planes determining the faces of Δ – then there is precisely one $T \in \mathfrak{G}$ such that $T(p) = q$. Finally, it is obvious from the definition of $\mathfrak{G}$ that it is closed under composition: if T_1 and T_2 are transformations of $\mathfrak{G}$, then so is their composite $T_1 \circ T_2$. In sum, and in the permutation-theoretic terminology introduced by Lie after he created his theory of groups, $\mathfrak{G}$ has properties which make it a simply transitive, 3-parameter, commutative transformation group.

In his work on the geometry of tetrahedral complexes, Lie made fundamental use of these properties of $\mathfrak{G}$. To illustrate how this occurred something must be said about his notion of a complex curve. According to Lie, any "continuous succession" of lines from T in which each intersects its predecessor is a complex curve [1870:69]. The simplest example of a complex curve is a space curve with tangent lines belonging to T; as a point traces out the curve, the corresponding tangent lines succeed each other continuously. But his conception of a complex curve was more general. For example, the cone of all lines from T passing through a "stationary" point p (called the *complex cone with vertex p*) was also regarded as a complex curve.[4] Lie divided complex curves into *species* (*Gattungen*). Two complex curves, c_1 and c_2 are of the same species if $T[c_1] = c_2$ for some $T \in \mathfrak{G}$. The species containing a given complex curve c_0 consists of all curves of the form $T[c_0]$ as T runs through $\mathfrak{G}$. This species will be denoted by $\mathcal{S}(c_0)$. In effect, $\mathcal{S}(c_0)$ is the orbit of c_0 under $\mathfrak{G}$. If $c_0 = \ell_0 \in T$, then $\mathcal{S}(c_0) = T$. Since there are ∞^3 transformations in $\mathfrak{G}$, Lie assumed that each species contains ∞^3 distinct curves. As we shall see below, this assumption turns out to admit certain mathematically interesting exceptions (W-curves). Based on this assumption, Lie concluded that the correspondence $T[c_0] \leftrightarrow T[\ell_0]$ between T and the species $S(c_0)$ is therefore one to one onto. Thus whenever convenient he would identify T with a species $S(c_0)$; under this identification, T was conceived in the spirit of Plücker as comprised of complex curves from $S(c_0)$ as "elements."

Lie's study of complex curves was by means of certain geometrical transformations which take T, suitably conceived, into itself. These transformations, which he called p-, g- and r-transformations, were constructed by means of a uniform procedure with the aid of the transformations of $\mathfrak{G}$.[5] Let p_0 and q_0 be two points "of general position." Then any point of general position, p, has a unique representation of the form $p = T_p(p_0)$ for some $T_p \in \mathfrak{G}$, and letting $P : p \to T_p^{-1} q_0$ defines a point transformation of space. P is a "reciprocal" transformation in the sense that a twofold application of P leaves everything unchanged ($P^2 = I$). The transformation P is a p-transformation. If the point q_0 is replaced by a straight line ℓ_0, then $R : p \to T_p^{-1}[\ell_0]$ defines an r-transformation which generates an example of what Lie later called a contact transformation (§2). If q_0 is replaced by a plane π_0, then $G : p \to T_p^{-1}[\pi_0]$ defines a g-transformation which also generates a contact transformation. As Lie explained [1870:69], all these transformations take T, suitably conceived, into itself, and he used them to study properties of complex curves. Although he was more concerned with the g- and r-transformations, both of which are also reciprocal, his results and methods are most simply illustrated for the p-transformation P.

That the transformation P takes T into itself is a consequence of the fact that P takes one species of complex curves into another. Thus P takes $T = S(\ell_0)$ into some species $S(c_0)$ which from Lie's perspective can also be identified with T, as explained above. Hence P takes T into itself. As an illustration of the type of results he had obtained, Lie gave the following theorem [1870:71].

THEOREM 1.1.. *Let S be a fixed species of complex curves. Fix a point p_1 and consider all the curves in S which pass through p_1. These curves describe a surface, $\Sigma(S, p_1)$. Let $S' = P[S]$, the "p-reciprocal" species. Then the curves of S' which pass through p_1 describe the same surface; that is, $\Sigma(S, p_1) = \Sigma(S', p_1)$.*

As an application of Theorem 1.1, Lie took $S = S(\ell_0) = T$. Then the surface $\Sigma(S, p_1)$ is the above mentioned complex cone with vertex p_1 consisting of all lines of T passing through p_1. When $S = T$, the transformed species S' turns out upon calculation to consist of third degree curves which pass through the vertices of the tetrahedron [Lie 1896:332,340]. The fact that $\Sigma(S, p_1) = \Sigma(S', p_1)$ implies that at each point p_1 the complex cone of T is covered by the third order curves of the species S' which pass through its vertex. Lie provided no proof of Theorem 1.1, or of any other theorems in [1870] – a brief announcement of some of his results written up by Klein – but it is not difficult to imagine how he might have arrived at it since it follows readily from the construction of P by means of $\mathfrak{G}$.[6] The commutativity of the transformations, which is essential to the proof, was used in most of the reasoning involving groups associated to tetrahedral complexes. Thus the system of transformations $\mathfrak{G}$, and specifically the above mentioned properties, were fundamental to Lie's exploration of the geometry of T. There is, however, no evidence to indicate that, before meeting Klein in Berlin, Lie had imagined that the reasoning he based on the properties of the particular system $\mathfrak{G}$ might extend with profit to other families of transformations with the same properties and that these properties defined a type of mathematical object of general significance to geometry and to mathematics in general. In particular, there is no evidence that he saw a significant analogy between his work and the theory and application of permutation groups.

In August, just before coming to Berlin, Klein submitted a paper [1869] on line complexes which are related to a type of surface first studied by Kummer in 1864. If new homogeneous line coordinates $p_{ik} \rightarrow x_1, \ldots, x_6$ are introduced so that $\Omega = \sum_{i=1}^{6} x_i^2 = 0$ is the condition for $x_1, \ldots, x_6$ to represent a line, then these complexes are given by an equation of the form $\Phi = \sum_{i=1}^{6} k_i x_i^2 = 0$. The k_i are assumed to be distinct and nonzero, so that the complex $\mathcal{K}$ defined by $\Omega = 0$, $\Phi = 0$ will be referred to as a *generic complex*. The locus of all points p such that the cone of $\mathcal{K}$ with vertex p degenerates into a pair of planes is a Kummer surface, and Klein showed that the line geometric study of $\mathcal{K}$ is tantamount to the study of Kummer surfaces. (See Rowe, §2)

Klein's paper also contained a division of lines and planes into "closed systems" that is reminiscent of Lie's division of complex curves into species. Consider for example lines. They are divided into closed systems of 32 lines each, the line $(a_1, \ldots, a_6)$ being associated with the 32 lines of the system $(\pm a_1, \ldots, \pm a_6)$ [1869:64]. The significance of such a grouping of lines – as well as groupings of other geometrical elements naturally related by the choice of coordinates x_i – was emphasized by Klein in his introductory remarks: objects so grouped together have identical geometrical relationships to the complex and the associated Kummer surface. Thus, in particular, the lines in a closed system have the same geometrical properties *vis à vis* $\mathcal{K}$. A simple illustration of this fact is that a line $(a_1, \ldots, a_6)$ belongs to $\mathcal{K}$ if and only if all the lines $(\pm a_1, \ldots, \pm a_6)$ of the closed system belong to $\mathcal{K}$. Klein's study also tacitly involved a collection $\mathfrak{G}$ of distinguished projective transformations, namely the $2^6 \div 2 = 32$ transformations of the form

$$(1.2) \qquad x_1' = \pm x_1, \ x_2' = \pm x_2, \cdots, \ x_6' = \pm x_6.$$

At least superficially, these transformations have properties similar to those of the system $\mathfrak{G}$ in Lie's study of T. In addition to being closed under composition and commutative, the transformations take $\mathcal{K}$ into itself; and each closed system is the orbit of any one of its members. As with lines of the same closed system, complex curves of the same species possess identical geometrical properties relative to T and the associated tetrahedron. For example, if one curve of a species passes through the vertices of the tetrahedron, then they all do by virtue of the fact that the transformations of $\mathfrak{G}$ fix the vertices.

Eventually, Klein and Lie recognized the above mentioned similarities, but it is unclear when this occurred. Soon after meeting in Berlin, they apparently formulated the common element in their work as deriving from a simple proposition which may be stated as follows:

THEOREM 1.2.. *Suppose that a geometrical configuration $\mathcal{G}_0$ is taken into itself by a transformation T. If another configuration $\mathcal{G}$ has a certain geometrical relation to $\mathcal{G}_0$, denoted by $\mathcal{G} \sim \mathcal{G}_0$, and if $\sim$ is preserved under T, so that $T[\mathcal{G}] \sim T[\mathcal{G}_0]$, then one can conclude that $T[\mathcal{G}] \sim \mathcal{G}_0$* [Klein & Lie 1871:50].

The conclusion of Theorem 1.2 follows immediately since $T[\mathcal{G}_0] = \mathcal{G}_0$. As Klein explained to Max Noether, by virtue of their above described work on line complexes the principle implicit in Theorem 1.2 lay quite close at hand for both him and Lie when they met in Berlin.[7] In Lie's work, however, the principle was applied systematically to transformations T with additional properties which in retrospect can be summed up by saying that the totality $\mathfrak{G}$ of T's form a continuous commutative simply transitive group. It is unclear exactly when Klein and Lie realized the fundamental importance in Lie's work of the totality $\mathfrak{G}$ and its properties as a group. When Klein arrived in Berlin, he may have had in the back of his mind the thought that "groups" (meaning permutation groups) are geometrically relevant. In Göttingen, where Klein spent over seven months before going to Berlin, Clebsch was encouraging a young French mathematician in Paris by the name of Camille Jordan. Jordan was in the process of writing a treatise on the theory of permutation groups and its applications to algebra and geometry, his *Traité des substitutions* [1870]. In connection with this enterprise, he was applying Galois' theory to various algebraic equations which determine geometrical objects of interest. For example, in a paper in Crelle's journal [1869b] he considered the polynomial equation of degree 16 which yields the singular points of a Kummer surface. The geometry of the situation enabled him to determine its Galois group, and he then used its structure to conclude that the resolution of the equation of degree 16 reduces to resolving equations of degrees 6 and 2. Clebsch actively encouraged Jordan's application of group theory to geometry by supplying him with many further geometrical equations to analyze in this manner.[8] The inaugural volume of Clebsch's new journal, *Mathematische Annalen*, also contained Jordan's "Commentaire sur Galois" [1869c]. Thus during Klein's stay in Göttingen when he composed [1869], he may have overheard some general talk about Galois' theory and its application to geometry through Clebsch and his students. In any case, Klein referred to Jordan's paper [1869b] in his own paper on the generic line complex (written while in Göttingen) [1869:n.16].

Thus when Klein arrived in Berlin, he may have been more disposed than Lie to perceive groups arising in a geometrical context. As he learned the details of Lie's ongoing investigation of the geometry of tetrahedral complexes, his interest in it increased to the point of active participation. Two episodes in that participation (while in Berlin) served to focus attention on the fact that Lie's work involved systems of transformations (later called continuous groups) which were recognized as analogs of groups of permutations: (i) Klein's discovery of W-curves and (ii) Klein's observations about Lie's study of differential equations related to T.

Episode (i) appears to have occurred first. According to Lie's recollections, the discovery of W-curves happened as follows.[9] Lie had posed to Klein the question: Do curves c exist with the property that whenever a point $p \in c$ is taken by a $T \in \mathfrak{G}$ into another point $T(p)$ which also lies on c, then the tangent line to c at p is taken by T into the tangent line to c at $T(p)$? Lie realized that if such curves existed, they would possess many interesting geometrical properties. Of course if a curve c existed which is taken into itself by all $T \in \mathfrak{G}$, then the tangent property would follow and c would be a curve of the type Lie sought. However, no such curves exist. In general as T runs through the ∞^3 transformations in $\mathfrak{G}$, $T[c]$ represents ∞^3 different curves. Lie, however, overlooked the possibility that in some cases there might be fewer than ∞^3 curves. By means of the analytical set up of (1.1), Klein showed that curves exist which are taken into themselves by precisely ∞^1 transformations of $\mathfrak{G}$. It turns out that these curves actually have the property stipulated by Lie in his question. The curves Klein discovered are special examples of what he and Lie called W-curves in German and V-curves in French. (See Rowe, §5, for an explanation of this terminology.)

Some of the discoveries Lie and Klein made about these curves, and about the associated W-surfaces, were briefly described in two notes presented in the *Comptes Rendus* during their stay in Paris [Klein & Lie 1870]. Already in these notes and even more so in their detailed study of W-curves in the plane [Klein & Lie 1871], we witness a shift in emphasis from Klein's [1869] and Lie's [1870], a shift which marked a significant step towards the emergence, as an independent mathematical entity, of the notion of a continuous group of transformations and a concomitant recognition of an analogy with the theory of permutation groups. The shift begins to be evident from the definition in their *Comptes Rendus* note of a W-curve in space. Whereas in Klein's [1869] and Lie's [1870] groups of transformations had arisen naturally from a given geometrical object, now the starting point is any family $\mathfrak{H}$ of ∞^1 projective transformations of space T_λ which are continuously related in the sense that they are generated by an infinitesimal projective transformation [1870:1222]. In homogeneous coordinates $x = (x_1, \ldots, x_4)$, such an infinitesimal transformation $x \rightarrow x + dx$ is defined by the differential equation $dx/d\lambda = Ax$, where A is a fixed 4×4 matrix. Hence $T_\lambda = e^{\lambda A}$. Of course Klein and Lie did not use such notation; it is used here for brevity.

From the nature of the T_λ it thus follows that

$$(1.3) \qquad\qquad\qquad T_\lambda T_\mu = T_{\lambda + \mu}.$$

From (1.3) it follows that the family $\mathfrak{H} = \{T_\lambda\}$ has the following further properties:

$$(1.4) \qquad \text{For all } T, U \in \mathfrak{H}: \text{ (a) } TU \in \mathfrak{H}; \text{ (b) } T^{-1} \in \mathfrak{H}; \text{ (c) } TU = UT.$$

Klein and Lie realized that $\mathfrak{H}$ possesses these properties, although without the uniform representation afforded by matrix exponentiation, these properties are not immediately apparent. Later, in their joint paper [1871], they were to say that families with property (1.4a) form *closed systems*. Property (1.4a) together with (1.4b) states that $\mathfrak{H}$ is a group in the modern sense. (For a long time Klein and Lie took it for granted that in general for infinite systems of transformations, property (1.4a) implies (1.4b), as it does for finite permutation groups.[10])

Given such a family $\mathfrak{H}$, then the W-curve through point p_0 in space is the locus of points $c_0(\lambda) = T_\lambda(p_0)$ obtained by varying λ. Since $T_\mu(c_0(\lambda)) = T_{\mu+\lambda}(p_0) = c_0(\mu + \lambda)$, every $T_\mu \in \mathfrak{H}$ takes c_0 into itself. Every such family $\mathfrak{H}$ defines a family of W-curves. Furthermore, associated to every such $\mathfrak{H}$ is a family of ∞^3 transformations, $\mathfrak{G}$, containing $\mathfrak{H}$. The transformations in $\mathfrak{G}$ are the projective transformations which fix the vertices of a (possibly degenerate) tetrahedron (determined in effect by a canonical form of matrix A). Even in the degenerate cases, these families $\mathfrak{G}$ share the properties of the transformations (1.1) in Lie's study of tetrahedral complexes, namely the properties in (1.4) and simple transitivity. The starting point of the theory is thus the family $\mathfrak{H}$ and the associated family $\mathfrak{G}$, which they later (in [1871]) called closed systems of commuting transformations and which are commutative groups in the modern sense.

Prior to 1872, Lie and Klein used the words "closed system", "cycle" and "group" interchangeably with regard to systems of transformations closed under composition – and assumed to be also closed under inversion (as was always the case in the examples they considered).[11] In the following discussion of the pre-1872 work, the word "group" will be employed, but the reader should be mindful that such terminological consistency is to be found with Lie and Klein only after 1871.

Further subgroups of the "tetrahedral" groups $\mathfrak{G}$ arise from the consideration of W-surfaces, which were introduced by Lie. They were defined as follows. Consider distinct systems of W-curves determined, respectively, by two families $\mathfrak{H} = \{T_\lambda\}$ and $\mathfrak{K} = \{S_\mu\}$, associated with the same "tetrahedron" and therefore included in the same collection $\mathfrak{G}$ of ∞^3 transformations leaving its vertices invariant. Fix a W-curve c_0 associated with $\{S_\mu\}$. Then through each point of this curve passes exactly one W-curve determined by $\{T_\lambda\}$. The totality of these curves as p varies along c_0 generates a surface called a W-surface. (If c_0 is given by $c_0(\mu) = S_\mu(p_0)$, then the W-curve through $c_0(\mu)$ is the locus of points $\lambda \to T_\lambda(c_0(\mu))$ so that the W-surface is the locus of points $\sigma(\mu, \lambda) = T_\lambda c_0(\mu) = T_\lambda S_\mu(p_0)$.) It follows immediately from the commutativity of all the transformations involved that a W-surface is taken into itself by the ∞^2 transformations $\mathfrak{L} = \{T_\lambda S_\mu\}$ (which form a subgroup of $\mathfrak{G}$).

The study of W-curves and surfaces thus involved the consideration of many continuous families of commuting projective transformations closed under composition, and these families formed the starting point for the theory of W-curves and surfaces. As Klein explained to Max Noether in a letter of 17 December 1870:

> At the moment I conceive of [the work on W-curves and surfaces] ... like this: .... In connection with the above stated general proposition [Theorem 1.2 above], now we do not think of the geometrical configuration as given and ask about the transformations; rather we consider the system of transformations as given and and ask about the geometrical configurations which are left invariant by it. Then among the most varied conceivable types of transformation cycles, the systems of commutative linear transformations which depend upon one ... or more continuously varying parameters lie closest at hand. Our theory of W-curves and surfaces, as well as that of the [tetrahedral] ... and linear complexes, and that of the complex whose lines intersect a fixed conic, are now nothing but an implementation of the general mode of reasoning for the special case of commutative linear transformations.[12]

Thus Klein envisioned the work on tetrahedral complexes and W-configurations as a small aspect of a far reaching geometrical method, according to which the starting point was a group of transformations. Indeed, the above quotation suggests Klein conceived the possibility of applying their mode of reasoning to other "cycles of transformations" besides the systems of commuting projective transformations which arose readily from the study of tetrahedral complexes. Since Theorem 1.2 does not assume the commutativity or linearity of transformations, it seems likely that in the above passage Klein did not envision his and Lie's "general mode of reasoning" as necessarily limited by either of these assumptions. The choice of linear, commuting transformation cycles was just the simplest cycle type with which to begin developing the more general principles. This seems to be what Klein is saying. Nonetheless, in their joint paper on W-curves in the plane [1871] both assumptions play a fundamental role in the reasoning. Indeed, in all the work on the tetrahedral complex, the commutativity of the groups involved is critical to most of the reasoning utilizing them (as e.g. in Theorem 1.1). It is not at all clear that Klein saw how to obtain interesting geometrical results without these assumptions.

During the first few months of 1871, while separated from Lie, Klein took on the responsibility for composing their joint paper [1871]. Initially he hoped to treat both W-curves in the plane and W-curves and surfaces in space, but the latter subject proved to be so extensive and complex that he did not have the patience to work through the details. Dealing with the situation in the plane he found tedious enough. (See the quotations from Klein's letters to Lie presented by Rowe (§7).) In March of 1871, after Lie had studied Klein's draft and made suggestions for changes, the paper was finally submitted to *Mathematische Annalen*. In [1871] a system of transformations is defined to form a closed system if it is closed under composition [1871:235]. Furthermore, they realized that their definition is the exact analog of the customary definition of a group of substitutions (i.e., permutations) [1871:235n.2]. They were fully aware that what they were doing was analogous to the theory of substitutions and algebraic equations (Galois' theory), for which by now the standard reference was Jordan's *Traité des substitutions* [1870], a work to which they referred [1871:233n.1], and which seems to have impressed Klein.[13] Thus they wrote: "The following considerations of closed systems of commutative transformations have an intimate relation to investigations which occur in the theory of substitutions and the attendant theory of algebraic equations." Since the group concept was central to the latter theory, in making the analogy Klein and Lie certainly regarded closed systems of commutative transformations as likewise central to their own theory. But why then did they decide to use the term "closed system" rather than "group"? Although they did not provide a reason, the continuation of the above quotation suggests one: "There is however a far reaching difference between the form in which these problems occur in the above mentioned discipline and here, namely in that discipline the variables are always discrete, while here it is a matter of continuously variable quantities" [1871:233]. This fundamental difference may have prompted the choice of a different terminology. Klein and Lie boldly proceeded to suggest that since the continuous theory was more tractable than the discrete, it might be helpful to first treat and resolve problems in the continuous case and then apply the results to the discrete case by restricting the values of the variables.

Although Klein gave up the intention of doing for space what he had done for the plane in their joint paper [1871], probably shortly after it was completed he sent off to Lie a draft of an essay describing the highlights of the three dimensional case [Klein MS1871b]. The introductory remarks to this essay are of interest. In their joint paper [1871], Klein and Lie had begun their investigation of W-curves by first determining all continuous groups of commuting projective transformations in the plane generated by infinitesimal transformations.[14] Then all W-curves were determined. The analogous problem for space, Klein observed, was to investigate all configurations taken into themselves by a continuous family of commuting projective transformations. "If we wished to provide an exhaustive overview of the totality of these configurations, then first of all we would have to carry out for space the investigation that at the time we placed in the forefront in the plane: *What closed systems of commuting linear [=projective] transformations exist in space?*"[15] Here we have an example of what may be referred to as a (vaguely articulated) group classification problem. Klein felt that the solution of this classification problem, posed for the plane, for space or for a domain of any fixed dimension presented no essential difficulties; all the finite number of possibilities could be checked out one after the other. Undoubtedly Klein could have done this for dimension $n = 3$ had he the patience, but it is doubtful he thought he could check out in this manner all the possibilities for a large value of n. In any case, he did not see how to deal with a space of arbitrary dimension n. The group classification problem for an indeterminate n, Klein admitted, was a problem of a different sort and very difficult.[16]

The joint paper [1871] was the last publication of Klein and Lie on W-configurations. To carry the work further in the direction taken in [1871] meant dealing with the associated group classification problem. Even for a relatively small value of n it was clear that the number of commutative closed systems involved is considerable. Each such system of ∞^k transformations ($k < n$) defines a class of k-dimensional "W-configurations"

(*W-Gebilde*) with potentially interesting geometrical properties by virtue of the same sort of reasoning that had been used to deal exhaustively with $n = 2$ and partially with $n = 3$. Evidently Klein was not inclined to get bogged down in such considerations, and certainly Lie was similarly inclined. They felt their ideas in this direction were important, but importance is a relative notion; they were also attracted by other important mathematical ideas. Another factor that probably encouraged them to suspend their work on W-configurations in favor of other research projects was the fact that their work had not been well received by other mathematicians. Neither the French nor Clebsch understood the *Comptes Rendus* notes, while the word Klein received from Berlin was that the mathematics of the notes was "really too simple."[17]

The problem was that the work on tetrahedral complexes and W-configurations involved many radically new geometrical ideas of Lie's. This was also true, and more so, of the material that they finally decided not to write up for publication. For example, regarding Lie's results on integral surfaces of partial differential equations associated to tetrahedral complexes (a bit of which is discussed below), Klein declared: "No one understands these matters" [Klein MS1871a:1]. Thus, although Klein and Lie felt their work on W-configurations was important, they apparently did not wish to give priority to the laborious task of educating mathematicians about it. The joint paper on W-curves in the plane [1871] was perhaps partly intended as related to that task, but, despite his pedagogical talents, Klein had clearly had enough of such work. After all, most of the results were Lie's anyway. Being preoccupied with other problems (the sphere mapping and its implications – §2) Lie himself had neither the time nor the temperament and talent for systematic exposition. It was not until the end of his life that Lie, with the assistance of his student, G. Scheffers, presented some of his results in [1896]. Thus the development of ideas related to the group concept might have ended with the publication of [1871]. That it did not was due to two reasons.

The first reason was that Klein, so to speak, put a bee in Lie's bonnet, that remained there and eventually was to play an important role in the events of the winter of 1873-74. This occurred already in Berlin, and involved another aspect of Lie's work on tetrahedral complexes. The study of W-surfaces had led Lie to introduce a logarithmic mapping which played a significant role in encouraging his hopes of a group-theoretic study of partial differential equations. When the tetrahedron is the one corresponding to (1.1), the mapping is given by

$$(1.5) \qquad\qquad X = \log x, \quad Y = \log y, \quad Z = \log z.$$

By means of (1.5) a correspondence is established between a space $\mathbf{r}$ in coordinates x, y, z and a space $\mathbf{R}$ in coordinates X, Y, Z.[18] By virtue of (1.5)
the transformations (1.1) of $\mathbf{r}$ comprising $\mathfrak{G}$ correspond in $\mathbf{R}$ to translations,

(1.6) $X' = X + \alpha, \quad Y' = Y + \beta, \quad Z' = Z + \gamma,$

where $\alpha = \log \lambda$, $\beta = \log \mu$, $\gamma = \log \nu$. (Thus $\mathfrak{G}$ and the group $\mathfrak{G}^*$ of
translations (1.6) are *similar*, as Klein and Lie would later say.[19]) The W-
curves in $\mathbf{r}$ associated with $\mathfrak{G}$, being curves of the form $c(t) = e^{tA} p_0$ with
A diagonal, thus correspond to straight lines in $\mathbf{R}$ left invariant by the ∞^1
translations along this line, and the W-surfaces in $\mathbf{r}$ therefore correspond
in $\mathbf{R}$ to planes. In this way the equations of the W-surfaces associated
with the above tetrahedron are obtained. That is, each such surface in
$\mathbf{r}$ corresponds to some plane $AX + BX + CZ = D$ in $\mathbf{R}$ so by (1.5) its
equation is of the form $x^A y^B z^C = D'$. As Klein and Lie pointed out
in their joint note [1870], surfaces with such equations had been studied
by Serret, who determined their lines of curvature. The equations of W-
curves for any proper tetrahedron then follow readily since it may be
transformed projectively into the above one.

  Lie's approach to the geometry of the tetrahedral complex through
the transformations of $\mathfrak{G}$ afforded him many avenues of investigation besides the ones so far related. For example, associated with each point p
in space is the complex cone with vertex p, $C(p)$, which consists of all
the lines of T which pass through p. The complex cone $C(p)$ was a basic
element in Plücker's theory of quadratic line complexes. Among other
things, he used this notion to define surfaces, called complex surfaces,
which are associated to a given quadratic line complex [1868:159;1869:335].
In a similar spirit, Lie also studied surfaces associated to the particular
quadratic line complex T. Unlike Plücker's approach, Lie's involved infinitesimal geometry. For example, he considered the problem of determining all surfaces σ with the property that at each point $p \in \sigma$, the cone
$C(p)$ "touches" σ at p. That is, the tangent plane to σ at p meets the cone
$C(p)$ in exactly one straight line (as in Figure I). This condition on σ translates readily into a first order differential equation (see [Lie 1896:260,169]),
that is, an equation of the form

$$f(x, y, z, p, q) = 0, \quad p = \frac{\partial z}{\partial x}, \quad q = \frac{\partial z}{\partial y}.$$

The solutions to this equation are the surfaces σ satisfying the condition of
the problem.[20] It is geometrically evident that any $T \in \mathfrak{G}$ takes a solution
surface σ into another solution surface $\sigma' = T[\sigma]$. For if $p' = T(p)$, T takes
the tangent plane to σ at p onto the tangent plane of σ' at p' and the cone
$C(p)$ onto the cone $C(p')$. Thus $C(p')$ touches σ' at p' and σ' also satisfies

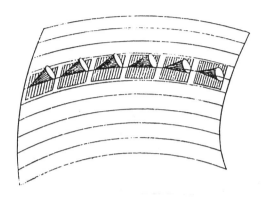

Figure I

partial the conditions of the problem. The transformations of $\mathfrak{G}$ thus take solutions of the partial differential equation into solutions, or, as Lie would later say, *the differential equation admits the transformations of* $\mathfrak{G}$. For the sake of brevity, Lie's subsequent terminology will be used in what follows.

Lie solved this problem by using the logarithmic mapping (1.5).[21] That is, assume without loss of generality that the tetrahedron defining T is the one corresponding to (1.1). The mapping (1.5) then transforms the partial differential equation $f(x, y, z, p, q) = 0$ into another such equation, $F(X, Y, Z, P, Q) = 0$, where P, Q denote partial differentiation of Z with respect to X and Y, respectively. Solving this new equation is tantamount to solving the original, since the solution can be pulled back by (1.5) to yield the solution to the original equation. Since $F(X, Y, Z, P, Q) = 0$ admits $\mathfrak{G}^*$ and hence all translations, Lie concluded that F is a function of P and Q alone (cf. Lie [1896:584ff]). The transformed differential equation is therefore of the special form $F(P, Q) = 0$ and can be solved directly and the solutions described geometrically. (Cf. Lie [1896:265].) Lie also solved other partial differential equations arising from the geometry of T by similar considerations.

When Klein learned (in Berlin) of this aspect of Lie's research on the tetrahedral complex, he made an observation which was to have a great and lasting impression upon Lie. He pointed out that there seemed to be an analogy between Lie's method of integrating differential equations admitting ∞^3 commuting transformations (illustrated here by the above example) and Abel's work on what became known as Abelian polynomial equations.[22] Abel had shown that Gauss' method for proving that cyclotomic equations can be solved by radicals could be generalized to any polynomial $f(x)$ with distinct roots $a_1, \ldots, a_n$ which possess the following two properties: (i) every root is a rational function of one of them, say $a_i = \theta_i(a_1)$ for all i; (ii) $\theta_i(\theta_j(a_1)) = \theta_j(\theta_i(a_1))$ for all i and j. That is, Abel showed that if $f(x)$ satisfies (i) and (ii) then it is solvable by radicals [1829]. The proof involved showing the resolution of $f(x) = 0$ reduces to that of an equation with roots a_i satisfying the following condition: (iii) $a_i = \theta^{i-1}(a_1)$, where $\theta(x)$ is rational and θ^{i-1} denotes its $(i-1)$-fold composite. Polynomials satisfying these conditions came to be known as Abelian ([Kronecker 1853:6; 1877:65-6; Jordan 1870:286-92]).

Property (ii) is a sort of commutativity property, as is (iii). It is possible this is all Klein had in mind when he called Lie's attention to the analogy, although anyone familiar with the rudiments of Galois' theory would have been in a position to see that the Galois group associated to an Abelian $f(x)$ consists of commuting permutations. Little is known about Klein's knowledge of Galois' theory in 1869. As was noted earlier in this section, during his stay in Göttingen prior to coming to Berlin he may have overheard some general talk about Jordan's work on Galois' theory and its application to geometry through Clebsch and his students. In any case he was familiar with one example of such work: Jordan's paper [1869b] on the singularity equation of a Kummer surface. Jordan's exposition of Galois' memoir [1846], his "Commentaire sur Galois," which appeared in the inaugural volume of Clebsch's *Mathematische Annalen*, makes no reference to Abel's theorem. But anyone who understood Theorem I in Jordan's "Commentaire" (which corresponds to Proposition I of Galois' memoir [1846]), and who asked whether the Galois group of an Abelian $f(x)$ consists of commuting permutations, would see readily that the answer is affirmative. Thus Klein may have picked up this piece of information in Göttingen. In any case, in Berlin he saw some analogy between Abel's theorem and Lie's method of "resolving" differential equations; and eventually he and Lie perceived this analogy to be that in each case knowledge of a group admitted by an equation yields information about its resolution. (Since the permutations in the Galois group permute the roots of the associated polynomial equation, which are the solutions of the equation, they take solutions into solutions.)

That Klein and, following him, Lie should see an analogy with the Galois theory of commutative permutation groups on the basis of the above sort of example may at first seem surprising since the success of Lie's method would seem to depend on the special form (1.1) of the transformations comprising $\mathfrak{G}$ rather than on their commutativity *per se*.

However, they showed that Lie's method had a broader scope. For example both Lie and Klein confirmed that all of the groups of ∞^3 commuting projective transformations associated to some (possibly degenerate) "tetrahedron" can be transformed into groups of translations.[23] Thus, for example, any partial differential equation $f(x, y, z, p, q) = 0$ which admits any such group of transformations could be transformed into the form $F(P, Q) = 0$ and therefore integrated.

One reason Klein's observed analogy was received with such enthusiasm by Lie was that his own deliberations had prepared him to believe in such an analogy. Independently of Klein's observation, Lie had noted that special types of differential equations with known methods of solution admit known infinitesimal transformations; and he suspected that the existence of these transformations was the "real" reason why they could be integrated. A simple example is afforded by the first order ordinary differential equation $dy/dx = f(y/x)$. This equation can be integrated by making the variable change $v = y/x$ which transforms it into a separable equation in x, v. This equation admits the ∞^1 commuting transformations $x' = ax$, $y' = ay$ which Lie saw as the underlying reason why the above technique is successful [Klein & Lie 1871:§7]. Klein's observation thus supported Lie's opinion that knowledge of transformations, "finite" or infinitesimal, admitted by a differential equation was the true basis for its solution or at least for its reduction to a simpler equation. To Lie it thus seemed plausible that a bone fide theory of differential equations admitting known, commuting infinitesimal transformations could be developed which would serve a function analogous to that served by Galois' theory for polynomial equations. In this connection, he extended his method to yield further results about the integration of first order partial differential equations admitting a group of ∞^2 or ∞^1 commuting projective transformations. For easy reference all these results are summarized as

THEOREM 1.3.. (i) If $f(x, y, z, p, q) = 0$ admits three commuting infinitesimal projective transformations, then it can be transformed into $F(P, Q) = 0$. (ii) If $f(x, y, z, p, q) = 0$ admits two commuting infinitesimal projective transformations, then it can be transformed into $F(Z, P, Q) = 0$. (iii) If $f(x, y, z, p, q) = 0$ admits one infinitesimal projective transformation, then it can be transformed into $F(X, Y, P, Q) = 0$.

In each of cases (i)-(iii), information about the solution was provided by knowledge of infinitesimal transformations the equation admitted. For example, from the conclusion of (ii) Lie concluded from known results that the integration of the transformed equation reduces to a quadrature.[24]

Although the prospect of a meaningful analogy between Galois' theory and the type of results on differential equations suggested by Theorem 1.3 was to play an important role in Lie's eventual creation of his theory of transformation groups, his understanding of Galois' theory was minimal. The mathematical apparatus of Galois' theory was far removed from his interests which, until 1872, were focused on geometrical matters. In 1862 Lie had attended Sylow's lectures on permutation groups while a student at the University of Christiania, but he understood very little of these lectures. It therefore seems likely that he realized what a group of permutations was and had some vague idea of what Galois' theory accomplished but that his understanding went no further. Probably Klein had a better, though limited, grasp of the theory and refreshed Lie's memory on these matters.[25] What Klein might have known about Galois' theory in 1869 was discussed above. Whether Klein told Lie anything about Jordan's paper [1869b] is not known, but Lie's subsequent remarks about the gist of Galois' theory suggest that Jordan's application of groups in [1869b] was the sort of service Lie saw Galois' theory as performing: from knowledge of a group to obtain *a priori* information about the resolution of an equation, such as for example that the resolution of a certain equation of degree 16 reduces to the resolution of equations of degrees 6 and 2. Thus in a letter received by Adolph Mayer on 3 February 1874 Lie wrote: "Before Galois, in the theory of algebraic equations, one only posed the question: Is the equation solvable by radicals and how is it solved? Since Galois, one also poses ... the question: How is the equation solved *most simply* by radicals? It can be proved e.g. that certain equations of degree six are solvable by means of equations of the second and third degrees and not, say, by equations of the second degree" [Lie 1873-4:586]. Likewise, Theorem 1.3 above gave specific *a priori* information about partial differential equations based upon knowledge of a group of commuting projective transformations admitted by the equation (the continuous group generated by the infinitesimal transformations). As will be seen in §3, Lie aspired to more general results of this type; they would constitute for him an analog of Galois' theory. In particular, starting in 1872, he began to consider the case of differential equations admitting groups of noncommuting transformations (possibly stimulated in this regard by the developments of §2). The expectation that some sort of an analog with Galois' theory of algebraic equations could be developed on a scale far greater than that implied by Theorem 1.3 became an *idée fixe* with Lie and played a role in the events of the winter of 1873-74. In this way the group element in Lie's work was kept alive, albeit in the background, while he pursued his geometrical ideas.

The other immediately more compelling reason why an interest in group related ideas continued and indeed expanded in a conceptual sense, was due to the rash of new research directions that were generated by

Lie's discovery of his sphere mapping. Lie's research on the sphere mapping led Klein and Lie, through an interesting web of mutual influences, to appreciate the importance of noncommutative groups of transformations. Without the work motivated by the discovery of this mapping it is difficult to imagine how the events of the winter of 1873-74 could have occurred.

2. THE SPHERE MAPPING

Lie discovered the sphere mapping during his stay in Paris. (See Rowe, §6, for further background.) By the time he returned to Norway in December, 1870, the discovery and some of its consequences had been announced to the academies of science in Christiania, Paris and (jointly with Klein) Berlin. The first detailed presentation of Lie's results occurred in his doctoral dissertation [1871a], submitted on 25 March 1871, and in a lengthy sequel [1871c], submitted to the scientific society of Christiania in the summer of 1871. At Clebsch's request, in October and November 1871 Lie submitted (through Klein) slightly revised versions of [1871a] and [1871c] to *Mathematische Annalen* [1871e]. Unless otherwise noted, the *Annalen* version will serve as the basis for the presentation here.

The mathematical starting point of Lie's theory is the system of bilinear equations

$$(2.1) \qquad (X + iY) - zZ - x = 0, \quad z(X - iY) + Z - y = 0$$

which establishes a "reciprocity" between a space r with point coordinates (x, y, z) and a space **R** with point coordinates (X, Y, Z) [1871e:25]. Lie allowed all these coordinates to assume, whenever convenient, complex values and regarded the spaces **r** and **R** from the viewpoint of projective geometry. The reciprocity defined by (2.1) was used by Lie to establish various correspondences between different types of geometrical objects in **r** and **R**. For example, for each fixed point $P = (X, Y, Z) \in \mathbf{R}$, (2.1) represents a system of linear equations in x, y, z and the solutions $p = (x, y, z)$ form a line $\ell \subset \mathbf{r}$. The equations (2.1) establish a one to one correspondence between points of **R** and some of the lines of **r**. For notational convenience the correspondence will be denoted by Ψ. The lines $\ell = \Psi(P)$ obtained by varying P form a line complex $\mathcal{L} = \{\ell : \ell = \Psi(P), \ P \in \mathbf{R}\}$. Lie expressed lines ℓ in terms of the nonhomogeneous line coordinates r, s, ρ, σ, introduced by Plücker, where ℓ is defined by the equations

$$(2.2) \qquad x = rz + \rho, \quad y = sz + \sigma.$$

In these line coordinates the equation of $\mathcal{L}$ is $r + \sigma = 0$. Lines parallel to the xy plane and lines in the plane at infinity cannot be expressed by means of (2.1) so that $r + \sigma = 0$ does not completely describe $\mathcal{L}$, as Lie realized. In homogeneous Plücker coordinates p_{ik} (which Lie did not employ in this connection), the complete equation of $\mathcal{L}$ is $p_{14} + p_{32} = 0$. Thus $\mathcal{L}$ is a linear line complex. In a similar fashion, (2.1) defines a correspondence Φ between points $p = (x, y, z) \in \mathbf{r}$ and lines $L = \Phi(p) \subset \mathbf{R}$, and $\mathcal{Q} = \{L : L = \Phi(p), \ p \in \mathbf{r}\}$ is a quadratic line complex with line coordinate equation $R^2 + S^2 + 1 = 0$ (or $P_{14}^2 + P_{24}^2 + P_{34}^2 = 0$). Since $R = dZ/dX$, $S = dZ/dY$, $dX^2 + dY^2 + dZ^2 = 0$ which shows that the lines of $\mathcal{Q}$ are imaginary.

The correspondences Φ and Ψ can be used to set up correspondences between curves in the two spaces and between surfaces as well, but the most important discovery that Lie made was that the reciprocity (2.1) also establishes a correspondence between all the lines ℓ in $\mathbf{r}$ and spheres $S(\ell)$ in $\mathbf{R}$ [1871e:§9,28ff]. Given any line ℓ, let $\mathcal{C}(\ell) \subset \mathcal{L}$ consist of all lines $\ell' \in \mathcal{L}$ which intersect ℓ. $\mathcal{C}(\ell)$ is an example of a line congruence, a basic notion in Plücker's line geometry. If $\ell \in \mathcal{L}$ the congruence $\mathcal{C}(\ell)$ is called special. If $\ell \notin \mathcal{L}$, a second line $\ell^* \notin \mathcal{L}$ exists, called the polar reciprocal of ℓ, such that every line $\ell' \in \mathcal{L}$ which meets ℓ also meets ℓ^* and conversely. Hence $\mathcal{C}(\ell^*) = \mathcal{C}(\ell)$. Lie discovered that all the points $P \in \mathbf{R}$ which correspond through (2.1) with lines of the congruence $\mathcal{C}(\ell)$ form a "sphere" $S(\ell)$. That is, if $S(\ell) = \{P' : \Psi(P') \in \mathcal{C}(\ell)\}$, and if r, s, ρ, σ are the line coordinates of ℓ in accordance with (2.2), then $P' = (X', Y', Z') \in S(\ell)$ if and only if

$$(X' - X)^2 + (Y' - Y)^2 + (Z' - Z)^2 = H^2$$

where:

$$(2.3) \qquad X = \frac{\rho + s}{2}, \quad Y = \frac{\rho - s}{2i}, \quad Z = \frac{\sigma - r}{2}, \quad \pm H = \frac{\sigma + r}{2}.$$

Since $S(\ell) = S(\ell^*)$, Lie spoke of the mapping: $\ell \to S(\ell)$ as two-to-one, although later it was realized that the mapping may be considered one-to-one by distinguishing between spheres with positive and negative orientation.

By means of the correspondence $\ell \to S(\ell)$, Lie sought to transform the line geometry of $\mathbf{r}$ into a sphere geometry in $\mathbf{R}$ [1871e:21ff]. He showed, for example, that line intersection in $\mathbf{r}$ corresponds to spheres touching in $\mathbf{R}$: if $\ell_1 \cap \ell_2 \neq \emptyset$, then $S(\ell_1)$ and $S(\ell_2)$ touch. And he showed that a "planar ray bundle" in $\mathbf{r}$—that is, a set of all coplanar lines through a fixed point—corresponds to the family of all spheres which touch in a common point [1871e:30-31]. The above mentioned properties of the correspondence $\ell \to S(\ell)$ illustrate (as Lie realized) the fact that the reciprocity (2.1) determines "a space transformation for which contact is an invariant relation" [1871e:§5], a type of transformation that, beginning with his paper [1872a], Lie called a contact transformation. Since the general notion of a contact transformation and the fact that the sphere mapping defines one are important to the viewpoint of this essay, it is necessary to go further into these matters.

Although Lie's earlier work had involved examples of what could be interpreted as contact transformations (such as the g- and r-transformations related to tetrahedral complexes), he first introduced the general concept in his publications on the sphere mapping. The fact that he envisioned these transformations as potentially important for the study of partial differential equations seems to have encouraged him to introduce it. Consider a first order partial differential equation $F(X, Y, Z, P, Q) = 0$, where $P = \partial Z / \partial X$, $Q = \partial Z / \partial Y$. It was commonplace in the study of such equations to consider a variable change

$$(2.4) \qquad X = g_1(x, y, z), \quad Y = g_2(x, y, z), \quad Z = g_3(x, y, z).$$

From these equations one obtains expressions for $P = \partial Z / \partial X$ and $Q = \partial Z / \partial Y$ of the form

$$(2.5) \qquad P = g_4(x, y, z, p, q), \quad Q = g_5(x, y, z, p, q),$$

where $p = \partial z / \partial x$ and $q = \partial z / \partial y$, which allow the original partial differential equation to be transformed into a new, equivalent one, $f(x, y, z, p, q) = 0$. Occasionally mathematicians such as Euler, Lagrange, Ampère and Legendre had utilized more general transformations of the form

$$(2.6) \qquad X = g_1(x, y, z, p, q), \quad Y = g_2(x, y, z, p, q), \quad Z = g_3(x, y, z, p, q),$$

such as the Legendre transformation $X = -q$, $Y = p$, $Z = z - px - qy$. Since the Legendre equations imply that $Z = z - Yx + Xy$, and hence that $P = \partial Z / \partial X = y$ and $Q = -x$, they also allow the original equation to be transformed into an equivalent equation, $f(x, y, z, p, q) = 0$.

Legendre type transformations had been given special attention in Paul du Bois-Reymond's geometrically oriented book on partial differential equations [1864:163-173] which was familiar to Lie. Du Bois-Reymond had sought to determine further transformations similar in form to the Legendre transformation which could also be used to transform partial differential equations but he went no further. Lie, however, considered the class of all transformations (2.6) with the property that these equations imply the further equations (2.5), transformations he called contact transformations [1871e:15]. (Lie tacitly limited his attention to invertible transformations.) At the time Lie did not realize that Jacobi, in a posthumously published paper on first order partial differential equations [1862:§57], had sought to determine analytically all such transformations. Eventually Jacobi's paper was to exert a considerable influence on him, but not before he had begun to develop his own theory of contact transformations.

As the name "contact transformation" suggests, Lie's approach, by contrast to Jacobi's, was geometrical. He conceived of (x, y, z, p, q) as coordinatizing (in the spirit of Plücker) a "surface element," that is, an infinitesimal surface ds in space r surrounding a point $a = (x, y, z)$ and with tangent plane π at a given by $z^* - z - p(x^* - x) - q(y^* - y) = 0$. (Later Lie formally defined a surface element as any point-plane pair (a, π) such that $a \in \pi$.) A contact transformation (2.6)-(2.5) was regarded as a transformation of surface elements $T:(x, y, z, p, q) \rightarrow (X, Y, Z, P, Q)$. If a surface s is conceived as composed of elements ds, then a contact transformation T generally takes the surface s so conceived into a surface S, which can regarded as composed of the elements $dS = T(ds)$. That is, the locus of all points (X, Y, Z) given by (2.6) with $ds = (x, y, z, p, q)$ an element of s, is generally a surface S with surface elements $dS = (X, Y, Z, P, Q)$, where P and Q are given by (2.5).[26] If two surfaces s and s' touch at a point, this means they have a common surface element at that point. A contact transformation, being a transformation of surface elements will therefore take s, s' into surfaces S, S' which share the transformed surface element and therefore also touch. Contact transformations thus preserve the contact of surfaces and hence Lie's name for them. It should be noted that "point transformations" (2.4) are contact transformations since (2.4) implies (2.5).

Lie saw that contact transformations were implicit in many geometrical theories. This insight is evident from the fact that all contact transformations are determined from reciprocities defined by one, two or three equations of the form $G(x, y, z, X, Y, Z) = 0$ [1871e:16].[27] Thus, for example the equation $xX + yY + zZ + 1 = 0$ which establishes the duality between points and planes in space, determines a contact transformation. Likewise, the two equations (2.1) determine a contact transformation $\Sigma:(x, y, z, p, q) \rightarrow (X, Y, Z, P, Q)$. This can be seen geometrically as follows (cf. Lie [1871e:31]). Given $ds = (x, y, z, p, q)$, let a and π denote the point and tangent plane associated with this surface element. Then, as noted above, the totality of lines ℓ lying in π and passing through a correspond, by the mapping $\ell \rightarrow S(\ell)$, to all spheres which touch in a common point. If $A = (X, Y, Z)$ denotes this common point and

$Z^* - Z - P(X^* - X) - Q(Y^* - Y) = 0$ defines the common tangent plane at A, then Σ sends ds into $dS = (X, Y, Z, P, Q)$.

Using his powerful geometrical intuition, Lie derived many interesting properties of the mapping Σ. He showed for example that if Σ takes a surface s into S, then it takes the asymptotic curves (principal tangent curves) on s to lines of curvature on S [1871e:§12]. While still in Paris, he discovered an impressive application of this result. If s is a Kummer surface then S is a type of surface (called a generalized cyclide) that had been studied by Darboux and Moutard in 1864. (See Rowe, §6) In particular, Darboux and Moutard had studied the lines of curvature of these cyclides. By means of the correspondence with Kummer surfaces established by Σ their results implied a new result about the asymptotic curves on a Kummer surface, namely that they are algebraic of degree 16. This result made a great impression on Klein, who had studied Kummer surfaces line geometrically (as noted in §1) and had already encountered the curves which turned out to be the asymptotic curves. Lie himself was interested in the application of the above mentioned property of Σ to the study of differential equations. This fact is illustrated by the following result [1871e:48ff]. Consider the following two classes of first order partial differential equations, each of which is characterized by the nature of the characteristic curves—curves lying on the solution surfaces which figured prominently in the Lagrange-Monge theory of the integration of first order equations [Lie 1896:Ch.11]. The first class (D_{11} in Lie's notation) consists of all first order equations such that the characteristics are asymptotic curves. The second class (D_{12}) consists of first order equations such that the characteristics are lines of curvature. By means of the map Σ, Lie showed that these two classes are equivalent in the sense that any equation of one class could be transformed by Σ into an equation of the other class. Since Σ is a contact transformation this means that integral surfaces of the first equation are taken into the integral surfaces of the second. Thus information about the integration of equations of one class could be transferred to the other class.

Although, as the above result illustrates, Lie's study of partial differential equations focused on equations that can be related to the sphere map, he emphasized that his approach was more generally valid: "To a certain extent, the following developments certainly possess a particular character, in that I am only concerned with special classes of differential equations. Nonetheless I would stress that the path taken here, namely that of linking the treatment of partial differential equations to broader geometrical concepts, seems to be a method from which general progress in the direction taken by Monge may be expected" [1871e:48]. Among the "broader geometrical concepts" that Lie had in mind when he wrote these words was certainly the concept of a contact transformation as well as ancillary concepts such as that of surface element. Indeed, the above result is illustrative of the following general problem linked to the general concept of a contact transformation: Given two first order partial differential equations or, more generally, two systems of such, determine when the one can be transformed into the other by means of some contact transformation. Ultimately, as we shall see, this problem was to inspire Lie's invariant theory of contact transformations.

The reciprocity defined by (2.1) sets up a correspondence between the two spaces **r** and **R** which has many aspects, characterized by various mappings such as Ψ, $\ell \rightarrow S(\ell)$, and Σ. Following Lie, all of these mappings will be referred to collectively as the sphere map. In Paris Klein had been struck by the analogy between results of French geometers such as Darboux in the realm of differential geometry and results of his own on line geometry [Klein MS1892:7]. The sphere map revealed the basis of such a connection. For example, Kummer surfaces "live" in **r** since, as Klein had shown in [1869], they could be studied line geometrically as the singularity surface of a quadratic line complex; but they correspond to surfaces in **R** which had been studied by the French. It was by virtue of this correspondence that the algebraic nature of the asymptotic curves on Kummer surfaces was revealed. In Lie's work, the space **r** was regarded as the space of projective line geometry in the spirit of Plücker, and **R** the space of the metrical, infinitesimal French geometry in the spirit of Monge, Chasles and Darboux. A guiding idea behind his geometrical study of the sphere map was to utilize it to bring out connections between Plückerian line geometry and French geometry and to use results known in one space to obtain results in the other. That is, he sought to investigate more generally how the line geometry of **r** corresponds to the "metrical geometry" of **R**. In this connection, Lie observed that the contact transformation Σ sets up a correspondence between contact transformations $t: (x, y, z, p, q) \rightarrow (x', y', z', p', q')$ of **r** and contact transformations $T: (X, Y, Z, P, Q) \rightarrow (X', Y', Z', P', Q')$ of **R** [1871e:§13]. For example, given T, the corresponding t is obtained as follows. Map the surface element (x, y, z, p, q) by Σ to (X, Y, Z, P, Q), then apply T and then pull back $T(X, Y, Z, P, Q)$ by Σ^{-1} to a surface element of **r**. In symbols (which Lie did not employ) $t = \Sigma^{-1} T \Sigma$.

In studying the correspondence $t \leftrightarrow T$, Lie restricted his attention to t's which are projective transformations of lines. In homogeneous Plücker coordinates p_{ik} such a t can be identified with a $t \in \mathbf{PGL}(6, \mathbf{C})$ which leaves the defining equation of a line,

(2.7) $\Omega = p_{12}p_{34} + p_{13}p_{42} + p_{14}p_{23} = 0,$

invariant—thereby taking lines into lines. Lie showed that the totality $\mathfrak{g}$ of such t corresponds to the totality $\mathfrak{G}$ of all T which take surfaces which touch along lines of curvature into surfaces with the same property. Among such T are transformations of surfaces considered by French differential geometers [1871e:42-43]. He also showed that the totality (which we denote by $\mathfrak{h}$) of ∞^{10} $t \in \mathfrak{g}$ which take $\mathcal{L}$ into itself corresponds to the totality ($\mathfrak{H}$) of ∞^{10} conformal transformations of $\mathbf{R}$, and that the totality ($\mathfrak{k}$) of ∞^7 $t \in \mathfrak{h}$ which fix a certain line $\ell_\infty \in \mathcal{L}$ (the intersection of the xy plane and the plane at infinity) corresponds to the totality ($\mathfrak{K}$) of ∞^7 projective transformations of $\mathbf{R}$ which take the imaginary circle at infinity (the intersection of any sphere in $\mathbf{R}$ with the plane at infinity) into itself. As Lie realized, the transformations of $\mathfrak{K}$ are precisely those generated by composition of translations, rotations and similarity transformations. Lie pointed out that the correspondence $\mathfrak{h} \leftrightarrow \mathfrak{H}$ can be used to give a line geometric proof of Liouville's theorem of 1846 that all conformal transformations of space are obtained by composition from rigid motions and transformations by reciprocal radii [1871e:43]. Thus by means of the sphere map, line geometry can be used to prove "metrical" theorems in $\mathbf{R}$.

For an application of the sphere mapping which proceeds in the opposite direction, Lie turned to Jordan's memoir [1869a] classifying rigid motions of ordinary space. Remarking that "one can formulate the problem of determining all groups of projective transformations," he pointed out that the results of Jordan's memoir [1869a] yields, by means of the sphere mapping, a partial solution to this problem.[28] Jordan's results, like the results of Darboux and Moutard on lines of curvature of generalized cyclides, live in the space $\mathbf{R}$. By the sphere mapping the group $\mathfrak{M}$ of ∞^6 rigid motions in $\mathbf{R}$, the group generated by composition of rotations and translations, corresponds to a group $\mathfrak{m}$ of projective transformations in $\mathbf{r}$. The group $\mathfrak{m}$ is a subgroup of the group $\mathfrak{k}$ and like it can be characterized geometrically as Lie noted. Thus each of the groups of rigid motions, continuous or discrete, enumerated by Jordan corresponds to a group of projective transformations from $\mathfrak{m}$. Jordan's enumeration of groups of rigid motions therefore transfers to an enumeration of subgroups, continuous or discrete, of $\mathfrak{m}$, just as the results of Darboux and Moutard transfer to a result about Kummer surfaces. The main significance of Jordan's memoir here was that its results could be used to illustrate the virtues of the sphere map. It is doubtful the results themselves were of great interest to Lie, for he never displayed any interest in discrete groups. He was never interested in the problem of finding all groups of projective transformations. (See in this connection Scholz's differing opinion of the importance of Jordan's memoir.)

Lie only used the word "group" at the end of his paper, in the above reference to Jordan's memoir. Nonetheless, he certainly realized that the types of transformations discussed earlier in his paper and designated there by $\mathfrak{g}$, $\mathfrak{h}$, $\mathfrak{t}$, $\mathfrak{G}$, $\mathfrak{H}$, $\mathfrak{K}$, as well as the totality $\mathfrak{B}$ of all contact transformations, have the group property. This fact, however, seems to have first gained in significance for Lie and Klein as a result of Klein's mathematical reaction to the sphere map work. Impressed by Lie's discovery concerning the asymptotic curves on Kummer surfaces, Klein considered Dupin's Theorem, which asserts that three one-parameter families of surfaces which are mutually orthogonal in the sense that surfaces from two distinct families always intersect orthogonally have the further property that the intersections are along lines of curvature. Dupin's Theorem lived in **R** and was closely related to the generalized cyclides studied by Dupin and Moutard (which define families to which Dupin's Theorem applies). Just as Lie had transferred the results of Darboux and Moutard on lines of curvature of these surfaces to Kummer surfaces in **r**, Klein sought the line-geometric analog of Dupin's Theorem. (See Rowe, §7) The theorem he ended up with, after dropping unnecessary restrictions, could, he observed, be proved by an argument that paralleled Salmon's proof of Dupin's Theorem, except that his proof involved four variables rather than the three in the proof of Dupin's theorem [1871a:99]. This circumstance led Klein (in [1871c]) to pose and answer a question of considerable historical importance. That is, the formulation of the question was important because of the new underlying conceptual viewpoint involved.

The viewpoint consisted in implicitly regarding all the geometries or geometrical modes of treatment relevant to his question as completely determined by specifying a "manifold" of elements and a distinguished set of transformations acting on the elements of the manifold. Thus Klein interpreted Lie's above described correspondence $\mathfrak{h} \leftrightarrow \mathfrak{H}$ as establishing an "analogy" between the "geometry of a linear complex" $\mathcal{L}$ and "ordinary metrical geometry." The former geometry is determined by the manifold of lines $\mathcal{L}$ and the transformations of $\mathfrak{h}$; the latter is determined by the manifold of points **R** ($\mathbf{P}^3(\mathbb{C})$) and the transformations of $\mathfrak{H}$. The analogy is established by the mapping $\Psi: R \to \mathcal{L}$, by means of which the transformations of $\mathfrak{H}$ go over into those of $\mathfrak{h}$. Motivated by this analogy, on the one hand, and by the analogy between his line geometric proof and Salmon's proof of Dupin's Theorem on the other, Klein asked: does a similar analogy exist between all of line geometry and the metrical geometry of four dimensional space? From Klein's viewpoint line geometry is determined by the four dimensional quadratic hypersurface—$M_4^{(2)}$ in Klein's notation—of elements $(p_{12}, \ldots, p_{42}) \in \mathbf{P}^5(\mathbb{C})$ satisfying $\Omega = 0$, where Ω is defined in (2.7), and by the totality $\mathfrak{g}$ of all projective line transformations. Lie's correspondence between line geometry in r and sphere geometry in **R** had been based on the mapping $\ell \to S(\ell)$ with $\mathfrak{g} \leftrightarrow \mathfrak{G}$, but Klein sought a correspondence with "the metrical geometry of a space of four dimensions" which he defined, by analogy with 3-dimensional metrical geometry, by specifying the manifold of four dimensional complex projective space ($\mathbf{P}^4(\mathbf{C})$) and the totality $\mathfrak{G}_1$ of all projective transformations of $\mathbf{P}^4(\mathbf{C})$ which take the imaginary sphere at infinity into itself. (In homogeneous coordinates $y_1, \ldots, y_5$ this sphere is defined by the equations $y_1^2 + \cdots + y_4^2 = 0$, $y_5 = 0$; thus the transformations of $\mathfrak{G}_1$ can be identified with all $T \in \mathbf{PGL}(5,\mathbf{C})$ which leave these equations invariant.)

To establish such a correspondence Klein constructed a map $\Phi: M_4^{(2)} \to \mathbf{P}^4(\mathbf{C})$. (See Rowe, §7) He showed that under Φ the totality $\mathfrak{g}_1$ of all $T \in \mathfrak{g}$, such that $T(\ell_0) = \ell_0$ for a fixed line $\ell_0 \in M_4^{(2)}$, corresponds to $\mathfrak{G}_1$, whereas all of $\mathfrak{g}$ corresponds to the "transformation cycle" ($\mathfrak{G}^*$) obtained by composing the $T \in \mathfrak{G}_1$ with the 4-dimensional analog of transformations by reciprocal radii. Klein's correspondences $\mathfrak{g} \leftrightarrow \mathfrak{G}^*$ and $\mathfrak{g}_1 \leftrightarrow \mathfrak{G}_1$ are analogous to Lie's $\mathfrak{k} \leftrightarrow \mathfrak{K}$ and $\mathfrak{h} \leftrightarrow \mathfrak{H}$, respectively. However, Klein's conception of a "geometry" is not to be found as such in Lie's sphere map papers. Not only did Klein characterize known geometries (such as line geometry) in this conceptually new way, he also defined new geometries by specifying a manifold and its distinguished transformations. That is, he generalized his characterization of line geometry to obtain the "geometry of a $M_{n-1}^{(2)}$" in $\mathbf{P}^n(\mathbf{C})$ defined by the manifold of all $(x_1, \ldots, x_{n+1}) \in \mathbf{P}^n(\mathbf{C})$ such that $\Omega = x_1^2 + \cdots + x_{n+1}^2 = 0$ and by all $T \in \mathbf{PGL}(n+1,\mathbf{C})$ leaving $\Omega = 0$ invariant. He also showed that there is an analogous correspondence between this geometry and the "metrical geometry" of $\mathbf{P}^{n-1}(\mathbf{C})$.

Thus by the time he submitted his paper, "Über Liniengeometrie und metrische Geometrie" [1871c], in October of 1871, Klein had arrived at something close to the viewpoint he was later to present in his Erlangen Program [1872b]. All that was missing was an explicit statement that the transformations defining a geometry always form a group and the declaration that any specification of a (continuous) group of transformations acting on a manifold defines a geometry. According to Klein's recollections, these points were made in an essay (no longer extant) entitled "Methoden der Geometrie" which he composed in December 1871 [MS1892:20]. Although the work on the sphere mapping was, as indicated, a major source of inspiration for this essay, Klein's work on non-Euclidean geometries [1871b] was certainly another significant source, since his "projective derivation" of these geometries by means of metrics inspired by the work of Cayley fitted nicely into the above described conceptual framework [Klein MS1892:19]. It was in his second major paper on non-Euclidean geometries, submitted in June 1872, that Klein first wrote up his ideas on geometrical methods for public presentation [1872a:316-325].[29] As we shall see, the ideas of Klein's "Methoden der Geometrie" made a strong impression on Lie because he perceived a connection with the theory of contact transformations and partial differential equations. To fully appreciate why this was so, it is first necessary to return to March of 1871, when Klein published his line geometric analog of Dupin's Theorem [1871a]. As will be shown, Lie's reaction to this paper and the research project it inspired are relevant to his reaction, nine months later, to Klein's essay "Methoden der Geometrie."

What impressed Lie about Klein's line geometric analog of Dupin's Theorem was not the underlying analogy between line geometry and 4-dimensional metrical geometry which it suggested, and which Klein explicitly developed in his seminal paper [1871c] some seven months later. Lie was impressed by the link Klein's work implied between Dupin's Theorem in dimensions 3 and 4. This led him to pose to himself the problem of extending to n-dimensional space those aspects of differential geometry related to Dupin's Theorem, which Lie referred to as "curvature theory," and to establishing a connection between the curvature theory of dimensions n and $n + 1$. Lie's results are contained in presentations written between April and December of 1871 [1871b,1871d,1871f,1871g]. Although they cannot be described here, it is noteworthy that they involve the n-dimensional analog of a contact transformation and the theory of first order partial differential equations in n independent variables. At the time, the theory of such equations was dominated, especially in Germany, by the ideas of Jacobi's posthumously published essay, "Nova methodus , aequationes differentiales partiales primi ordinis inter numerum variabilium quemcumque propositas integrandi" [1862]. Even while he was in Paris with Klein working on W-curves and the sphere mapping, Lie had shown a serious interest in Jacobi's theory. According to Klein's recollections, at that time Lie was excitedly studying "to the point of satiety" the exposition of Jacobi's "Nova methodus" and related material published by Imschenetsky [1869].[30]

Jacobi's theory was quintessentially analytical, whereas Lie's approach to partial differential equations, in his work on tetrahedral complexes and on the sphere map, was geometrical. Jacobi's analytical innovations were eventually to benefit Lie considerably, but at this time (1871), Lie's approach to the study of partial differential equations continued to be thoroughly geometrical. Now, however, the geometry was conceived on a level of abstraction commensurate with Jacobi's theory. That is, in his work on n-dimensional curvature theory, Lie made the transition from what he later called higher dimensional geometry in the sense of

Plücker to higher dimensional geometry in the sense of Grassmann's *Ausdehnungslehre*. Following Plücker's lead, Lie had previously considered n-dimensional "manifolds" (*Mannigfaltigkeiten*) for $n > 3$, but they always had a concrete intuitive interpretation. For example, the manifold of straight lines in space is a four dimensional manifold, but it corresponds to objects (lines) in ordinary space. Now he saw the value of considering n-dimensional manifolds abstractly, in the sense that its elements, $(x_1, \ldots, x_n)$, did not necessarily correspond to the coordinatization of some configuration in ordinary space. For the purposes of the following discussion this sort of n-dimensional geometry will be referred to as "Grassmannian geometry." Lie regarded the formulation of the theory of partial differential equations within the framework of Grassmannian geometry as "a step forming the natural continuation and development of Monge's L'application de l'analyse à la géométrie."[31]

It was during the same period (April-December 1871) in which Lie's work on curvature theory in n dimensions (done in Norway) was leading to this conclusion, that Klein's work (in Germany) was leading him towards the conclusions expressed in his essay, "Methoden der Geometrie." According to Klein's recollections [MS1892:20], he sent a copy of this essay to Lie, who at first did not understand what was meant by "geometrical method" but soon wholeheartedly agreed with its contents and provided Klein with "new rich material" drawn from his work on partial differential equations and supporting Klein's viewpoint. When Klein and Lie were together again in the fall of 1872, Lie, with Klein's editorial assistance, presented some of this material in the pages of *Göttinger Nachrichten* [1872b:§II], and Klein, with Lie's encouragement, composed his Erlangen Program, with an entire section devoted to Lie's work [1872b:§9]. Let us consider what these two publications had in common.

Within the context of Grassmannian geometry, Lie reformulated and generalized his notions of surface element and contact transformation. Thus corresponding to first order partial differential equations,

$$f(z, x_1, x_2, \ldots, x_n, p_1, p_2, \ldots, p_n) = 0, \quad p_i = \frac{\partial z}{\partial x_i},$$

is the $(2n + 1)$-dimensional space R_{2n+1} of (hypersurface) elements

$$(z, x, p) = (z, x_1, \ldots, x_n, p_1, \ldots, p_n),$$

where the p_i are now regarded abstractly as variables rather than partial derivatives of some function z of the x_i. Within this geometrical context, Lie discovered that contact transformations could be conceived as those

invertible transformations $T: (z, x, p) \rightarrow (Z, X, P)$ of R_{2n+1} which leave
the Pfaffian equation

$$dz - (p_1 dx_1 + \cdots + p_n dx_n) = 0$$

invariant. If $\mathfrak{B}$ denotes the totality of all contact transformations, it is
evident that the composition of two contact transformations is a contact
transformation (as in the case of the contact transformations of ordinary
space introduced in Lie's sphere map papers). Thus $\mathfrak{B}$ has the group
property. Lie also realized that the inverse of a contact transformation
is a contact transformation so that $T \in \mathfrak{B}$ implies that $T^{-1} \in \mathfrak{B}$.[32] The
manifold R_{2n+1} together with the group $\mathfrak{B}$ thus determines a geometrical
method in the sense of Klein. In this connection, it should be pointed out
that Klein's ideas on geometrical method were also formulated in the context of Grassmannian geometry, so that the new geometrical conceptual
apparatus that Lie was creating fit in perfectly with Klein's ideas. This instance of intellectual harmony must have been mutually encouraging.[33]
To appreciate the full extent of the harmony, it is necessary to keep in
mind why Lie was interested in general contact transformations. As can
be seen from his introduction of contact transformations in his sphere
map work, the reason such transformations were of interest to him was
that they can be used to transform one first order partial differential equation (or system of such equations) into another which is equivalent to the
first. Thus the question arises: given two equations, or, more generally,
two systems of equations, when are they equivalent in this sense? If one
system has a property (P) which is preserved under all contact transformations, it can only be equivalent to a system possessing (P). Thus
a related problem is to determine the invariant properties of systems of
one or more first order partial differential equations with respect to all
contact transformations. Ideally one would like to determine enough invariants to establish necessary and sufficient conditions that two systems
are equivalent.

By the time of Lie's paper [1872b], he had discovered that any two
first order partial differential equations could be transformed into one
another by contact transformations so that a single equation has no such
invariants. He also realized that systems of more than one equation do
possess invariants such as the property of being in involution. Systems
with this property had occurred in Lie's study of n-dimensional curvature theory, and they were fundamental to the theory of Jacobi's "Nova
methodus." Consider, for simplicity, a system of equations not involving
the variable z, $f_i(x_1, \ldots, x_n, p_1, \ldots, p_n) = 0$, $i = 1, \ldots, m$. They are in
involution if, for all $i \neq j$, $(f_i, f_j) = 0$, where (f_i, f_j) denotes the bracket

operation of Jacobi's "Nova methodus." It is defined for any two functions $\varphi(x,p)$, and $\psi(x,p)$ of $2n$ variables $x_1, \ldots, x_n$ and $p_1, \ldots, p_n$ by:

$$(2.8) \qquad (\varphi, \psi) = \sum_{i=1}^{n} \left(\frac{\partial \varphi}{\partial p_i} \frac{\partial \psi}{\partial x_i} - \frac{\partial \varphi}{\partial x_i} \frac{\partial \psi}{\partial p_i} \right).$$

Similar expressions had occurred much earlier in work on mechanics by Poisson and Lagrange, who used bracket notation to simplify equations.[34] Since the equation $(f_i, f_j) = 0$ is left invariant by contact transformations, a system of equations in involution is transformed by a contact transformation into a system with the same property. Thus the property of being in involution is an invariant relation, since not every system of m equations $f_i(x,p) = 0$ is in involution. Systems of $m > 1$ equations do possess invariants. The problem then is to develop the theory of such invariants. Lie perceived that this was an exemplification of Klein's views on geometrical methods. For example, in the Erlangen Program, Klein had characterized the fundamental problem of a geometrical method in the following words: "Given a manifold and a group of transformations of it, determine the theory of invariants relative to the group" [1872b:464]. For the manifold R_{2n+1} of hypersurface elements (z, x, p) and the group $\mathfrak{B}$ of contact transformations this was precisely what Lie hoped to do, and for this reason, he referred to the theory he eventually created as his invariant theory of contact transformations. Of course, when Klein spoke of "invariant theory" in the Program, he had in mind primarily, although not exclusively, the paradigm of the Clebsch-Gordan theory of algebraic invariants [1872b:464n.13(1893)], which generalizes readily to the notion of algebraic invariants associated to any linear group. Lie's invariant theory of contact transformations exemplified the ideas of the Program in a manner not anticipated by Klein and indicated that their scope was greater than originally imagined.

Lie's work on the sphere map had thus led, through the above described interplay of ideas between between him and Klein, to a more general connection between groups and geometry, namely the one articulated in the Erlangen Program. Whereas the theory of W-configurations could now be seen as an example of a geometrical method, determined by a commutative group of projective transformations, the Erlangen Program was not at all restricted to commutative groups. On the contrary, the groups that motivated the Program, such as the groups $\mathfrak{g}$, $\mathfrak{g}_1$, $\mathfrak{h}$, $\mathfrak{k}$, $\mathfrak{G}$, $\mathfrak{G}_1$, $\mathfrak{G}^*$, $\mathfrak{H}$, and $\mathfrak{K}$ discussed above, the groups defining Klein's non-Euclidean geometries, and the group $\mathfrak{B}$ of contact transformations, were all noncommutative. Now it was clear that all continuous groups, both commutative and noncommutative were relevant to the geometrical study of manifolds.

Perhaps it was this fact, made all the more impressive by the convergence of Lie's ideas on contact transformations and partial differential equations with Klein's ideas on geometrical method, that prompted Klein to call for the development of an autonomous theory of continuous transformation groups. This occurred at the end of the Erlangen Program. There Klein turned once again to the analogy with permutation groups that had been noted already in their paper on W-curves in the plane. Now, however, the analogy was more complete since noncommutative groups were definitely perceived to be of geometrical interest. He pointed out that in the treatises on the subject such as Jordan's *Traité des substitutions,* the theory of permutation groups formed a field of investigation in its own right; Galois' theory followed as an application of that theory. He therefore called for the development of a theory of transformation groups which would have as an application the theory of geometrical methods (understood as the study of invariant relations within a manifold determined by a given transformation group). Undoubtedly Lie was in full sympathy with these sentiments and encouraged Klein to include them.

In §1 it was pointed out that the work of Lie and Klein on W-configurations, despite their realization of the analogy with the theory of permutation groups and Galois' theory, did not produce any activity on their part to create an analogous theory of the continuous groups that had formed the basis of their study of these configurations. In particular, the associated group classification problem was not tackled. The situation seems to have been the same after the Erlangen Program was composed. Once again, neither Klein nor Lie turned to the development of the theory of continuous transformation groups. Klein began to work on matters related to complex function theory, and did not further develop the ideas of his Program until twenty years later (after Lie had begun to publish his monumentral treatise on the theory of continuous transformation groups [1888,1890,1893]).[35] Lie himself proceeded to develop his invariant theory of contact transformations. Thus although the work of Klein and Lie in 1869-72 led to a clear recognition of the notion of a continuous group of transformations, to an appreciation of its actual and potential importance in applications to geometry and differential equations, and even to an explicit acknowledgement of the value of a theory of these groups, developing the theory was not a project to which either of them immediately turned. The nature of such a project undoubtedly seemed just as difficult and unappealing as it had in 1871 when their joint paper on W-curves was published. This interpretation is supported by Lie's attitude towards one aspect of such a project: the group classification problem suggested by the Erlangen Program.

Since any specification of a continuous group determines, in effect, a geometrical method, it is of interest to know just how many groups there are which act on an n-dimensional manifold and what they, and the geometry they define, are like. Since similar groups[19] were regarded as determining equivalent geometries, the above considerations suggest the problem of classifying, up to similarity, all continuous groups acting on an n-dimensional manifold, for any n. Although this problem is not explicitly formulated in the Erlangen Program there is evidence that Lie had perceived this implication. The evidence is in the form of a letter Lie wrote to Klein in 1879, five years after he had begun to create his theory of transformation groups. At that time Lie was writing up an exposition of his theory for the mathematical world at large, to be published in *Mathematische Annalen* [Lie 1879].[36] He had decided to conclude the work with a section indicating all publications before 1874 which were related to his new theory [1879:§20]. Although the Erlangen Program was mentioned, it was not emphasized. Klein, to whom Lie sent the draft of [1879] for editing, apparently objected in a letter to this slighting of the Program, for in a letter to Klein, Lie responded by explaining that of course he esteemed Klein's Program and certainly it had proved stimulating to him. "But on the other hand, it must be pointed out that in your essay the problem of determining *all* groups is not posited, probably on the grounds that at the time such a problem seemed to you absurd or impossible, as it did to me. Also ... your essay gave no means for resolving my problem, at least nothing beyond what was known earlier."[37] Thus it would seem that Lie had perceived the above mentioned group classification problem suggested by the Program but had dismissed it as a problem that was not feasible since the means to resolve it were not at hand. To resolve it would have seemed to require a kind of mathematics outside Lie's (and Klein's) mathematical interests in 1872. This classification problem, like the one associated with the theory of W-configurations, therefore would have seemed difficult and unappealing to Lie—even unapproacable. Slightly over a year after the Erlangen Program was written, however, Lie completely changed his mind and decided such a classification problem was not "absurd and impossible" at all. Indeed, he had become convinced that he could solve it for transformations in any number of variables. Let us consider what caused him to change his mind.

3. THE WINTER OF 1873-74

Although the experiences relating to the sphere map and culminating in the composition of the Erlangen Program understandably did not inspire Lie to tackle the above classification problem, they encouraged him to work out his invariant theory of contact transformations. The Erlangen Program itself was of no help in this regard because "invariant" was left unspecified. Lie's study of the contents of Jacobi's "Nova methodus" provided him with the key to his theory (the notion of a function group, discussed below).[38] The emphasis upon n-dimensional geometry and noncommuting groups that emerged from the sphere map work also seems to have affected Lie's formulation of his *idée fixe* (described at the end of §1 above), which continued to occupy his attention during the period of his work on the sphere map.[39] Starting in 1872, he sought to generalize the results of Theorem 1.3 by considering infinitesimal transformations in n variables and by including the case of equations admitting noncommuting infinitesimal transformations [1872a:2;1872c:27]. In [1872c] Lie specified that, in the noncommutative case, the q infinitesimal transformations "form a group," although in a preliminary draft of the paper he had written more tentatively that they "form in a certain sense a group."[40] Precisely what he meant by either statement is uncertain. His manuscripts from 1872-73 show him dealing with infinitesimal transformations on a intuitive, geometrical level devoid of analytical computations. For example, in one such manuscript, Lie considered two arbitrary commuting infinitesimal contact transformations do_1 and do_2 "as well as all two fold infinitely many transformations $\overline{o_1 o_2}$, which arise from their composition" and which Lie referred to as a transformation group.[41] Probably Lie also conceived of noncommuting infinitesimal transformations as "forming a group" in this vague intuitive, nonanalytical way. Such an interpretation accords with the events of the winter of 1873-74. At that time, Lie had occasion to ask himself what it meant, analytically, for infinitesimal transformations to "form a group," and he apparently first thought seriously about this analytical question then, guided in this connection by the problem on which he was working, a problem generated by his *idée fixe*.

As Lie immersed himself in Jacobi's theory, he also sought to work out his *idée fixe* within that context. Clebsch had shown [1866] that the most important type of equation for Jacobi's theory and its extensions was a complete system of homogeneous partial differential equations. The integration of first order partial differential equations and Pfaffian equations reduces, by the Jacobi theory, to the integration of such systems. Expressed in the notation introduced by Jacobi, a complete system has the form $X_i(f) = 0$, $i = 1, \ldots, r$, $r \leq n - 1$, where

$$X_i(f) = \sum_{i=1}^{n} a_{ij}(x) \frac{\partial f}{\partial x_j}.$$

To be complete, the differential operators must be linearly independent and must satisfy

$$X_i(X_j(f)) - X_j(X_i(f)) = \sum_{k=1}^{r} b_{ijk}(x) X_k(f) \ \forall \ i, j = 1, \ldots, r.$$

(Fundamental to Jacobi's theory had been his observation that for any differential operators of the form of the X_i, namely for linear, homogeneous operators, the operators $X_i(X_j(f)) - X_j(X_i(f))$ are also of this form.) Motivated by his *idée fixe*, Lie formulated something like the following problem: Suppose a complete system admits q infinitesimal transformations which perhaps (in some sense) form a group. What does this fact tell us about the integration of the system; that is, what simplifications of the system (and therefore its integration) follow from knowledge that it admits the given infinitesimal transformations?[42]

Lie showed that his problem could be reduced to a special case in which the infinitesimal transformations had an additional property. Let $dT_i : x \rightarrow x + dx$ denote the ith infinitesimal transformation, $i = 1, \ldots, q$, where $x = (x_1, \ldots, x_n)$, $dx = (dx_1, \ldots, dx_n)$, and $dx_j = \xi_{ij}(x)dt$, $j = 1, \ldots, n$. Corresponding to dT_i is the differential operator

$$Y_i(f) = \sum_{j=1}^{n} \xi_{ij}(x) \frac{\partial f}{\partial x_j}.$$

Lie's additional property was that

$$(3.1) \qquad Y_i(Y_j(f)) - Y_j(Y_i(f)) = \sum_{k=1}^{q} c_{ijk} Y_k(f), \quad i, j = 1, \ldots, q,$$

where the c_{ijk} are constants. The reasoning leading to this conclusion does not involve assuming anything that relates to an assumption about the infinitesimal transformations dT_i forming a group; it simply uses the fact that the dT_i are admitted by the systems under consideration. Perhaps Lie was therefore prompted to ask whether the sort of closure of the Y_i implied by (3.1) might have something to do with the assumption that the transformations dT_i form a group. In any case, assuming that when the dT_i form a group they are closed under composition in the sense that $dT_i \circ dT_j$ is a linear combination of the dT_k, he obtained (3.1) as a consequence. He further convinced himself that (3.1) is not only necessary but also sufficient for the dT_i to form a group. Applying this fundamental theorem to the case $n = 1$ he found that there were only three nonsimilar[19] groups – and all could be represented by projective transformations. He had completely solved the group classification problem for n-variable transformation groups generated by a finite number of infinitesimal transformations in the case $n = 1$. Undoubtedly this discovery greatly encouraged him to tackle the group classification problem for $n > 1$. For $n > 1$, however, the problem becomes much more difficult. The number of nonsimilar groups turns out to be infinite as soon as $n = 2$. For $n > 1$ contact transformations also exist, and Lie wished to classify them as well. For $n > 1$ equation (3.1) does not readily yield the classification as it does for $n = 1$. Lie nonetheless believed he could deal with the case of an arbitrary n. The primary reason for his optimism derived from his discovery that there was a connection between (3.1) and his invariant theory of contact transformations, a connection which enabled him to apply the results of the latter to the group classification problem.

Briefly stated, the connection is as follows. First of all, since point transformations $T: (x_1, \ldots, x_n) \rightarrow (x'_1, \ldots, x'_n)$ may be regarded as contact transformations (as noted in §2), Lie focused on classifying contact transformations. He had also discovered that it suffices to consider without loss of generality what he called homogeneous contact transformations. These are transformations of the form $T: (x, p) \rightarrow (x', p')$ where the functions $x'_i = X_i(x, p)$ and $p'_i = P_i(x, p)$ are homogeneous in the p_i of degrees 0 and 1, respectively.[43] Let $dT_i: (x, p) \rightarrow (x + dx, p + dp)$ be an infinitesimal homogeneous contact transformation, and let $Y_i(f)$ be the corresponding differential operator (as defined above except now there are $2n$ variables $(x, p) = (x_1, \ldots, x_n, p_1, \ldots, p_n)$ and $f = f(x, p)$. Lie discovered that a homogeneous function of the p-variables, $H_i(x, p)$, exists such that $Y_i(f) = (H_i, f)$, where the right hand side is the Poisson-Jacobi bracket (2.8). Jacobi had shown that the bracket satisfies the identity $((F, G), H) + ((G, H), F) + ((H, F), G) = 0$ for any three functions of (x, p). This identity together with the above bracket representation of $Y_i(f)$ implied that $Y_i(Y_j(f)) - Y_j(Y_i(f)) = (H_i, H_j)$. Thus if the dT_i form a group so that (3.1) holds, then also

$$(3.2) \qquad\qquad (H_i, H_j) = \sum_{k=1}^{q} c_{ijk} H_k.$$

Equation (3.2) was amazing and encouraging to Lie because it meant that the functions H_i which determine the infinitesimal contact transformations dT_i generate what he had called a "function group" in developing his invariant theory of contact transformations in 1872-73.

Both the motivation for the concept and for the theory of function groups had also come from Jacobi's "Nova methodus." Although space does not permit a detailed discussion of these matters here, it should be pointed out that, despite Lie's choice of terminology, function groups are not groups in the usual sense. Functionally independent functions $u_1, \ldots, u_r$ of the variables $x_1, \ldots, x_n, p_1, \ldots, p_n$ are said to form an "r-term group" if

$$(3.3) \qquad (u_i, u_j) = \Phi_{ij}(u_1, \ldots, u_r),$$

where (u_i, u_j) is defined in accordance with (2.8) and the Φ_{ij} are functions of r variables. In other words, the (u_i, u_j) must be functions of the u_i in order for $u_1, \ldots, u_r$ to form a function group. Any function $v = v(x, p)$ is said to belong to the group $(u_1, \ldots, u_r)$ if $v = \Phi(u_1, \ldots, u_r)$ for some function Φ. Thus by (3.3) the functions (u_j, u_j) belong to the group. More generally, if v and w belong to the group so that $v = \varphi(u_1, \ldots, u_r)$ and $w = \psi(u_1, \ldots, u_r)$, then it turns out that (v, w) belongs to the group. A function group thus possesses a closure property that might have been reminiscent in Lie's mind of the closure property of transformation groups or permutation groups and therefore inspired his choice of the word "group" for this notion. To distinguish this meaning of "group" from the usual one in cases of possible confusion, the former will be referred to as a "function group," an expression Lie himself used on a few occasions. There is no evidence that Lie perceived anything more than this remote analogy with transformation groups.[44] His only explanation upon introducing function groups was: "The [function] group concept belongs in essence to Jacobi; it is very useful to have a name for this fundamental concept" [1873a:34n.2].

One can thus appreciate the exalted sense of destiny that Lie experienced when, in the winter of 1873-74, he realized there was a significant connection between these two sorts of groups: equation (3.2) implies that the functions $H_i(x, p)$ determine a function group, and he had already obtained results on the transformation of one function group into another as part of his invariant theory of homogeneous contact transformations.[45] Lie hoped that these results would enable him to classify transformation groups up to similarity—to solve the group classification problem for these groups. He thus felt prepared to deal with this group classification problem in a way that he did not when, in 1871 and 1872, group classification problems had emerged from the geometry of tetrahedral complexes and the sphere mapping work. Although his optimism ultimately proved excessive in the sense that he never resolved the problem for $n > 3$, in the process of attempting to do so, he began to create the apparatus of his theory of transformation groups, the theory which was occupy his attention for the remainder of his life.

Although the notion of a continuous transformation group was articulated in the period 1869-72 and its importance to geometry and differential equations perceived, the birth of the theory of these groups, as it occurred during the winter of 1873-74, thus depended critically upon Lie's involvement with, and creative reaction to, the ideas stemming from Jacobi's "Nova methodus." Without Lie's involvement with Jacobi's theory in 1872-73, the events of the winter of 1873-74 are unimaginable. On the other hand, it is equally difficult to imagine how Lie would have been involved in those events were it not for the experiences of the years 1869-72, which grew out of his youthful involvement with line geometry and his association with Klein. Indeed, as we have seen, it was the work on the sphere mapping that served to crystallize the general concept of a contact transformation in Lie's mind and that inspired him to conceive of an invariant theory of contact transformations. And it was Lie's continuing concern with his *idée fixe*—which originated with the work on tetrahedral complexes – that led to the problem which, in turn, produced equation (3.1). On a more fundamental level, it was the experiences of those early years that brought to Lie's attention the notion of a continuous transformation group and convinced him of its importance in applications to the study of geometry and differential equations. With those experiences behind him, he looked for a group-theoretic interpretation of equation (3.1). The conviction that continuous groups are important to the study of geometry and differential equations was already in Lie's mind by the end of 1872. What was lacking at that time was the belief that it was possible to develop the theory of continuous groups to the point where it could be systematically applied in these areas of mathematics. Such a systematization was tied up in Lie's mind with the possibility of being able to resolve group classification problems. It was the belief that problems of this sort are mathematically feasible that was an outgrowth of his involvement with Jacobi's theory. But without the experiences of the early years it is difficult to imagine why Lie would have been aware of, and interested in, the notion of a continuous group in the first place and why, furthermore, the resolution of group classification problems would have been deemed of great importance.

NOTES

1. Research on the project, much of it done at the Institute for Advanced Study in Princeton, has been supported by National Science Foundation grants DIR 8808646 and DMS 8610730. A full presentation of the results of the research project will be incorporated into a planned book on the origins of the theory of semisimple Lie algebras.

2. The analysis of (3) and its role in the creation of Lie's theory will be presented in "Partial Differential Equations and the Birth of Lie's Theory of Groups" [in preparation].

3. The following comparision is based upon Klein's recollections [MS1892: 22–23], which I found confirmed by my study of their papers and letters, both published and unpublished.

4. Eventually Lie's generalized notion of a curve became part of his notion of an "element union." See in this connection the discussion of Lie's concept of a contact transformation in §2, especially note 26.

5. Here I have relied on Lie's recollections [Lie MS1893:3ff.], which on this matter appear reliable. The same uniform procedure is presented in [Klein & Lie 1871:§4]. In [1870] Lie did not present the above construction of p-, g- and r-transformations by means of $\mathfrak{G}$. Instead, in each case, he (or Klein) translated his definitions into terms more familiar to his readers.

6. To see this, let $S = S(c_0)$ and let $c_0(t)$ denote the points of c_0 as t varies. Then for every point $c_0(t)$ a transformation $S_t \in \mathfrak{G}$ exists such that $S_t(p_0) = c_0(t)$ (by transitivity). Since the species $S(c_0)$ consists of curves $c = T[c_0]$, so that $c(t) = T(c_0(t)) = TS_t(p_0)$, by definition of P and commutativity of $\mathfrak{G}$, $P(c(t)) = P(TS_t)(p_0) = (TS_t)^{-1}(q_0) = T^{-1}S_t^{-1}(q_0)$. Thus $P[c] = P[T[c_0]] = T^{-1}[c_0']$, where $c_0'(t) = S_t^{-1}(q_0)$. It follows that $S' = P[S(c_0)] = S(c_0')$, since the curves $T^{-1}[c_0']$ generate $S(c_0')$ as T runs through $\mathfrak{G}$. Given a point p_1, a $T_1 \in \mathfrak{G}$ exists such that $p_1 = T_1(p_0)$ (transitivity). Consider now a curve $c = T[c_0] \in S(c_0)$. The question is: for which T does $c = T[c_0]$ pass through p_1? Now c passes through p_1 if $c(t) = T(c_0(t)) = p_1$ for some t. That is, if $TS_t(p_0) = T_1(p_0)$ for some t. Since $TS_t = T_1$ means $T = S_t^{-1}T_1$, it follows that the one-parameter family of transformations $T_t = S_t^{-1}T_1$, yield curves $c_t = T_t[c_0] \in S(c_0)$ which pass through p_1. These curves determine the surface $\Sigma(S, p_1)$; it is the locus of points $\sigma(s, t)$ where $\sigma(s,t) = c_t(s) = S_t^{-1}T_1S_s(p_0) = S_t^{-1}S_sT_1(p_0) = S_t^{-1}S_s(p_1)$. Similar considerations applied to the species $S' = P[S(c_0)] = S(c_0')$ show that the corresponding surface $\Sigma(S', p_1)$ is the locus of points $\sigma'(s,t) = S_tS_s^{-1}(p_1)$, so that $\sigma'(s,t) = \sigma(t,s)$ and therefore $\Sigma(S', p_1) = \Sigma(S, p_1)$ as asserted in Lie's theorem.

7. This letter, dated 17 Dec. 1870, is quoted in [Klein MS1916:13ff]; the original is in the archives of the Handschriftenabteilung, Niedersächsische Staats- und Universitätsbibliothek, Göttingen (Cod. Ms. F. Klein 12).

8. In the preface of his *Traité* Jordan said the material on geometrical applications (Book III, Ch.III) was made possible by the "libérales communications de M. Clebsch" [1870:viii]. Many of the examples considered there were drawn from papers by Clebsch.

9. What follows is based on Lie's recollections [MS1893:6]. Judging by the marginalia in [Klein MS1892:5], Lie reminded Klein of his role in the discovery of W-curves, which Klein then stated as a fact in his *Ges. Abh.* 1, 415.

10. Jordan made the same tacit assumption in his memoir [1869a] on groups of rigid motions. They all took it for granted that groups as they defined them always possess inverses—that closure under composition entails closure under inversion, as it does for permutation groups. As Lie first began to develop his theory (from 1874 on), he thought he could prove this for r-parameter continuous transformation groups [1876:14,19]. When he realized he could not prove it, closure under inversion was explicitly assumed, although he still conjectured a proof was possible [1879:5]. Finally he withdrew the general conjecture but showed it was true for groups comprised of certain types of transformations [1888:20-22]. Cf. Klein's addition (1893) to note 8 of his Erlangen Program [Klein 1872b:463].

11. Klein (rather than Lie) used "closed system" and "transformation cycle" interchangeably. In the margin of [Klein MS1871b:1], he asked Lie if "cycle" would be preferable to "closed system."

12. "Gegenwärtig fasse ich dies so: ... Anknüpfend an den oben aufgestellten allgemeinen Satz denken wir uns nun nicht etwa die geometrischen Gebilde gegeben und fragen nach den Transformationen, sondern wir denken uns die Systeme von Transformationen gegeben und fragen nach den geometrischen Gebilden, welche durch dieselben unverändert bleiben. Da liegen dann unter den verschiedenartigsten denkbaren Transformationszyklen die Systeme der von einem, ... oder mehreren kontinuirlich veränderlichen Parametern abhängen vertauschbaren linearen Transformationen am nächsten. Unsere Theorie der W-Kurven und Flächen, sowie weiter der Reyeschen und der linearen Komplexe, sowie des Komplexes, dessen Geraden einen festen Kegelschnitt schneiden, sind nun nichts als eine Ausführung der allgemeinen Schlußweise für die speziell vorliegenden vertauschbaren linearen Transformationen." [Klein 1916:15-16]. See note 7 above.

13. Jordan's *Traité* evidently made a considerable impression upon Klein. In [MS1892:6-7] Klein wrote regarding his and Lie's interest in applying and developing the geometrical concept of a group of substitutions: "Ursprünglich dachten wir dabei gern an zahlentheoretische Analogieen. Fernerhin gab uns in dieser Richtung viel Anregung C. Jordan's grosses Werk über die Galois'schen Gruppen, welches eben erschienen war, als wir nach Paris kamen. Es handelte sich dabei natürlich nur um eine Analogie, da in der Zahlentheorie wie in der Galois'schen Theorie nur discontinuirliche Gruppen auftreten. Aber solche Analogieen sind für die Entstehung neuer Denkweise oft viel wichtiger als direkte Ansätze." In the margin, next to the mention of the influence of Jordan's *Traité*, Klein added: "d.h. jedenfalls ich. Bin ich von vorneherein für die discontinuirlichen Gruppen prädestinirt gewesen? Jedenfalls habe ich in 1871 auch viel über continuirliche gearbeitet." In the notes to his collected works Klein repeated his recollection that Jordan's book made a great impression upon him and added that to him and Lie it appeared "as a book with seven seals" (*Ges. Abh.* 1,51), which suggests that he had not penetrated deeply into its contents but was stimulated by the overall impression it made.

14. As Bourbaki has pointed out [1972:287], Klein and Lie were concerned only with connected groups. Lie always seems to have regarded connectivity as an aspect of the continuity of a group. Thus e.g. the group $\mathfrak{G}$ of orthogonal transformations was for Lie neither continuous nor discontinuous since it consists of two separated sets—those $T \in \mathfrak{G}$ with $\det T > 0$ and those with $\det T < 0$ [Lie 1888:7].

15. "Wollten wir uns eine erschoepfinde Uebersicht ueber das Gesammtgebiet dieser Gebilde verschaffen, so haetten wir zunaechst fuer den Raum die Untersuchung zu fuehren, welche wir damals in der Ebene an die Spitze stellten: **Welche geschlossene Systeme vertauschbarer linearer Transformationen gibt es ueberhaupt im Raume?**" [MS 1871b: 1]

16. "Eine ganz andere sehr schwierige Frage scheint die Beantwortung bei einem unbestimmt gelassenen n zu sein. Cf. C. Jordan" [MS1871b:1]. It is instructive to speculate on what Klein meant by this reference to Jordan. Jordan had dealt with group classification problems in two works with which Klein was probably acquainted. One was his memoir [1869a] classifying all groups of rigid motions in space. The other was his *Traité des substitutions* [1870] where the fourth, and final, part (*Livre IV*) is devoted to the problem of constructing all transitive, solvable subgroups of permutations on n letters which are maximal

with respect to these two properties. Since Jordan's memoir [1869a] dealt with a group classification problem in a fixed dimension $n = 3$, it seems more likely he was referring primarily to the *Traité*, where the classification problem involves an indeterminate n. (Jordan showed the problem for $n = n_1$ can be resolved by utilizing the resolution of the problem for values of n less than n_1.) Although evidence suggests that Klein took a look at the *Traité* in Paris when it appeared, it also suggests he did not digest its contents. (See note 13 above.) It contained difficult mathematics of a sort that was far removed from his current geometrical interests. Nevertheless, it seems likely that out of curiosity Klein browsed through the pages of the *Traité*. If he did, he probably would have been struck by the importance Jordan evidently attached to the above mentioned group classification problem. Book IV comprises 42% of the *Traité*, and relates to a problem fundamental to Galois' theory. As Jordan explained [1870:vii,396], the solution to his group classification problem yields a classification of all "general" types of irreducible polynomials which are solvable by radicals. Thus the importance Jordan attached to the group classification problem should have been clear to a casual reader, such as Klein. Another fact that would have been clear to such a reader is that Jordan's solution of the problem involved complex mathematical reasoning which utilized the previously developed theory of groups extensively. For Klein to have stressed the difficulty of his own group classification problem for any n by a reference to Book IV of Jordan's *Traité* would thus have been quite appropriate.

17. That is, "eigentlich zu leicht" (Klein to Lie, 4 Feb. 1871; quoted on p.744 of Lie, *Ges. Abh.* 1).

18. The following presentation draws upon Lie's recollections [Lie MS1893: 7ff,11,16ff] and on Ch.8, §5 of Lie's lectures [1896]. According to Lie the developments in §5 "stammen alle von Lie und zwar aus der Zeit vom winter 1869 auf 1870" [1896:356n].

19. Two groups are similar if the equations defining the one group can be transformed into those defining the other by a change of variables and parameters. Lie and Klein came to regard similar groups as essentially the same.

20. This problem of Lie's is "Mongean" in spirit, but it does not seem that before he arrived in Paris that he had read Monge's works. He formulated it independently of knowing Monge's theory [Lie MS1893:17].

21. Lie [MS1893:7] described his solution to the above differential equation as a corollary to his discovery that the group $\mathfrak{G}$ is similar to the group $\mathfrak{G}^*$ of all translations. Cf. Lie [MS1893:17].

22. According to Engel [1899:xviii], "Klein machte nämlich Lie schon in Berlin darauf aufmerksam, dass dessen Verfahren zur Integration von Differentialgleichungen, die eine Gruppe von ∞^3 verauschbaren Transformationen gestatten, grosse Aehnlichkeit besitze mit Abels Behandlung der algebraischen Gleichungen, die man jetzt als Abelsche bezeichnet." According to Engel, Lie often and emphatically referred to this episode. In [MS1893:12] Lie wrote (to Klein): "Ich erkannte in Berlin und Paris, dass die alten Methoden bei der Behandlung von Differential auf die Verwerthung von Gruppen hinauskommen.... Du bemerktest die Analogie meiner Betrachtungen mit der **Abel**schen Gleichungen (Vertauschbarkeit)." The fact that it occurred in Berlin is also supported by [Lie MS1893:16,25-26] since there we learn the logarithmic map and the attendant transformation of $\mathfrak{G}$ into translations, occurred in Berlin shortly after a French language lesson.

23. In [MS1893:17] Lie wrote (to Klein): "Wir bemerkten wohl beide, dass (**transitive**) Gruppen von projektiven vertauschbarer Transformationen der Ebene, resp. Raumes in den Ausartungsfällen in Translationen umge[hen]...." In the nondegenerate case, exemplified by tetrahedron yielding (1.1), the transformation of the group is given by (1.5)-(1.6).

24. Parts (ii) and (iii) were first presented in Lie's papers on the sphere map since by means of this map Lie had obtained an unlimited number of partial differential equations to which Theorem 1.3 applied [1871c:§16; 1871e:§17] Apparently the results of Theorem 1.3 were obtained (possibly with Klein's help) in Paris in 1870 [Lie MS1893:27]. An analogous result for first order ordinary differential equations was added, at Lie's insistence, to his joint paper with Klein on W-curves in the plane [1871:§7].

25. In [MS1893:19] Lie wrote (to Klein): "Du hast wohl geglaubt, ich hätte mein Bischen Substitutionentheorie bei Dir gelernt. Das ist nicht der Fall. Ich habe einmal bei **Sylow** eine Vorlesung darüber gehört, freilich wenig **verstanden**."

26. There are exceptions to this. For example, the Legendre transformation takes the surface $z = 1$ into the point $(0,0,1)$. However it takes the 2-dimensional "manifold" of its surface elements $ds = (x, y, 1, 0, 0)$ into the 2-dimensional manifold of elements $dS = (0, 0, 1, y, -x)$, which consists of all planes through the point $(0,0,1)$.

Such sets of elements (along with those that constitute a genuine surface) are included in Lie's concept of an "element union." A characteristic property of contact transformations is that they take element unions into element unions. Regarding surface element unions, see Lie [1896:522ff;1890:44ff].

27. To see analytically how such reciprocities determine contact transformations, and the conditions on the functions G, see Blaschke's exposition in Klein's lectures [1926:292ff].

28. "Es lässt sich die Aufgabe stellen, alle Gruppen linearer Transformationen anzugeben" [1871c:208]. Except for some minor stylistic improvements, the text in [1871e:119] is identical.

29. Due to a printer's strike, Klein's memoir [1872a] did not appear until 1873, so that Klein's second version for publication, his Erlangen Program, actually appeared first.

30. "Lie studierte in Paris den Imschenetski sehr eifrig, bis zum Ueberdruss" [Klein MS1916:10].

31. Lie, *Ges. Abh.* 7,101-102. Lie discussed his transition from n-dimensional Plückerian to n-dimensional Grassmannian geometry in unfinished drafts probably written in late 1872 or 1873. See Lie, *Ges. Abh.* 7, 57,59,101–102,105. Lie's use of Grassmann's work was admittedly superficial. As he explained in a footnote "Sei hier gleich hervorgehoben ... , dass nur der Grundgedanke Grassmanns von seinen Nachfolgern, wie auch von mir, aufgenommen ist. So gross meine Bewunderung für die tiefen Schöpfungen Grassmanns ist, scheint es mir doch zweifelhaft, ob sein Algorithmus, welcher freilich für sein Werk eine Hauptsache ist, die beste Form der Ausdehnungslehre ist" (*Ges. Abh.* 7,101n.10).

32. Unlike most of the examples of continuous groups arising in the work of Klein and Lie, $\mathfrak{B}$ is not generated by a finite number of infinitesimal transformations; it does not define a (finite dimensional) Lie group.

33. Although Klein's essay, "Methoden der Geometrie," is no longer extant, it seems necessary to assume that Klein had already formulated his ideas there within the context of n-dimensional spaces since in his seminal paper [1871c] of October 1871, we find him considering the geometry of an $M_{n-1}^{(2)}$ in $P^n(\mathbb{C})$ and its relation to $(n-1)$-dimensional metrical geometry. In any case such a formulation occurs in the next presentation of his ideas on geometrical methods, in June of 1872 [1872a:315-325]. The transition to the n-dimensional context which

we find in Klein's [1871c] may have been encouraged by Lie's work on n-dimensional curvature theory (cf. [Klein 1871c:§3]). Lie's identification of his n-dimensional geometry with Grassmann may, in turn, have come from Klein. In the Erlangen Program [1872b:Note IV] Klein explicitly distinguished between n-dimensional geometry in the sense of Plücker and of Grassmann, so it may have been Klein, with his broader knowledge of mathematics, who made Lie aware of the affinity between the type of n-dimensional geometry he was now pursuing and Grassmann's *Ausdehnungslehre*.

34. See Dugas' history of mechanics [1950:368-374].

35. See in this connection [Hawkins 1984].

36. Except for a brief announcement of his theory in *Göttinger Nachrichten* [1874b], all of Lie's publications on his theory of transformation groups were published (in German) in the Norwegian journal, *Archiv for Mathematik og Naturvidenskab* which he edited. These publications are contained in Lie, *Ges. Abh.* 5.

37. "Dass ich Deine Programmschrift sehr schätze, weiss Du. Ich habe das Wort 'gedankenreich' eingeschaltet. Ganz sicher ist sie mir anregend gewesen. Aber anderseits ist zu bemerken, dass die Aufgabe: **alle** Gruppen zu bestimmen nicht in Deiner Schrift gestellt wurde; und wohl aus dem Grunde dass eine solche Aufgabe sich damals Dir wie mir als absurd oder unmöglich stellte. Anderseits giebt Deine Schrift auch nicht Mitteln zur Erledigung meines Problems, jedenfalls nicht andere als solche die früher bekannt waren" [Lie MS1879].

38. A much fuller account, which indicates how Jacobi's theory motivated and informed Lie's theory of function groups, is given in the paper mentioned in note 2 above. The relevant publications by Lie are [1873a,1873b,1873c,1874a].

39. For example, using the sphere map Lie showed how to generate an unlimited supply of first order partial differential equations admitting one or more commuting infinitesimal transformation, equations to which Theorem 1.3 thus applied [1871e:§17].

40. Lie, *Ges. Abh.* **7**, 98. Lie's manuscripts from 1872-73 indicate that generalizations of Theorem 1.3 were linked in his mind with continuous groups and Galois' theory (*Ges. Abh.* **7**, 27-28, 89, 92, 98).

41. Lie *Ges. Abh.* **7**, p. 50.

42. There are no extant documents recording Lie's actual reasoning at the time, so that what follows is a reconstruction based upon later

documents and recollections. The basis for the reconstruction will be presented in the paper referred to in note 2.

43. A function $F = F(x,p)$ is homogeneous in p of degree ν (ν a nonnegative integer) if it satisfies the Euler equation $\sum_{i=1}^{n} p_i \partial F/\partial p_i = \nu F$. Such functions need not be homogeneous algebraic functions.

44. He may also have perceived a vague connection with Galois' theory that does not involve groups. See in this connection the paper referred to in note 2.

45. The H_i are linearly independent but may be functionally dependent. In that case a functionally independent subset of the H_i would be the "basis" for the function group. See Lie's letters to Mayer [1873-4:590-594].

PUBLISHED REFERENCES

Abel, N., 1829. "Mémoire sur une classe particulière d'équations résolubles algébriquement," *Journal für d. reine u. angew. Math.* **4** = *Oeuvres complètes* **1**, 478–507.

Bourbaki, N., 1972. *Élements de mathématiques. Fasc. XXXVII. Groupes et Algèbres de Lie* (Chapters II–III), Paris 1972.

Du Bois-Reymond, P., 1864. *Beiträge zur Interpretation der partiellen Differentialgleichungen mit drei Variablen*, Leipzig 1864.

Clebsch, A., 1866. "Über die simultane Integration linearer partieller Differentialgleichungen," *Journal für d. reine u. angew. Math.* **65** (1866), 257–268.

Dugas, R., 1950. *Histoire de la méchanique*, Paris & Neuchâtel, 1950.

Engel, F., 1899. "Sophus Lie," *Berichte über die Verhandlungen der Königlich Sächsichen Gesellschaft der Wissenschaften zu Leipzig, Math.-phys. Cl., 1899* (1899), xi–lxi.

Galois, E., 1846. "Mémoire sur les conditions de résolubilité des équations par radicaux," *Journal des math. pures et appl.* **11** = *Oeuvres mathématiques*, pp. 33–50. [16 January 1831]

Hawkins, T., 1984. "The *Erlanger Programm* of Felix Klein: Reflections on its Place in the History of Mathematics," *Historia Mathematica* **11**, 442–470.

Imschenetsky, W., 1869. "Sur l'intégration des équations aux dérivées partielles du premiere ordre," *Archiv für Math. u. Physik* **50**, 278–474.

Jacobi, C., 1862. "Nova methodus, aequationes differentiales partiales primi ordinis inter numerum variabilium quemcumque propositas integrandi," *Journal für d. reine u. angew. Math.* **60**, 1–181 = *Gesammelte Werke* **4**, 3–189. Translated into German with commentary by G. Kowalewski as *Ostwald's Klassiker der exakten Wissenschaften, Nr. 156*, Leipzig 1906.

Jordan, C., 1869a. "Mémoire sur les groupes de mouvements," *Annali di matematiche* **2**, 167–215, 322–345 = *Oeuvres* **4**, 231–302.

Jordan, C., 1869b. "Sur une équation du 16ᵉ degré," *Journal für d. reine u. angew. Math.* **70**, 182–184 = *Oevvres* **1**, 207–209.

Jordan, C., 1869c. "Commentaire sur Galois," *Mathematische Annalen* **1**, 142–160 = *Oeuvres* **1**, 211–230.

Jordan, C., 1870. *Traité des substitutions et des équations algébriques*, Paris 1870.

Klein, F., 1869. "Zur Theorie der Liniencomplexe des ersten und zweiten Grades," *Mathematische Annalen* **2** (1870) = *Gesammelte Mathematische Abhandlungen* **1**, 53–80. [4 August 1869]

Klein, F., 1871a. "Über einen Satz aus der Theorie der Linienkomplexe, welcher dem Dupinschen Theorem analog ist," *Göttinger Nachrichten 1871* (1871) = *Gesammelte Mathematische Abhandlungen* **1**, 98–105. [4 March 1871]

Klein, F., 1871b. "Über die sogenannte Nicht-Euklidische Geometrie," *Mathematische Annalen* **4** (1871) = *Gesammelte Mathematische Abhandlungen* **1**, 254–305. [19 August 1871]

Klein, F., 1871c. "Über Liniengeometrie und metrische Geometrie," *Mathematische Annalen* **5** (1872) = *Gesammelte Mathematische Abhandlungen* **1**, 106–126. [October 1871]

Klein, F., 1872a. "Über die sogenannte Nicht-Euklidische Geometrie (Zweiter Aufsatz)," *Mathematische Annalen* **6** (1873) = *Gesammelte Mathematische Abhandlungen* **1**, 311–343. [8 June 1872]

Klein, F., 1872b. *Vergleichende Betrachtungen über neuere geometrische Forschungen. Programm zum Eintritt in die philosophische Fakultät und den Senat der k. Friedrich-Alexanders-Universität zu Erlangen*, Erlangen 1872. Reprinted with additional notes in *Mathematische Annalen* **43** (1893) = *Gesammelte Mathematische Abhandlungen* **1**, 460–497. [October 1872]

Klein, F., 1926. *Vorlesungen über höhere Geometrie. Dritte Auflage, bearbeitet und herausgegeben von W. Blaschke*, Berlin 1926.

Klein, F.& S. Lie, 1870. "Sur une certaine famille de courbes et de sur-faces," *Comptes Rendus Acad. Sci. Paris* **70** (1870), 1222–1226, 1275–1279 = Klein *Gesammelte Mathematische Abhandlungen* **1**, 415–423 = Lie *Ges. Abh.* **1**, 78–85. [6 June 1870, 13 June 1870]

Klein, F.& S. Lie, 1871. "Über diejenigen ebenen Kurven, welche durch ein geschlossenes System von einfach unendlich vielen vertauschbaren linearen Transformationen in sich übergehen," *Mathematishe Annalen* **4** (1871), 50–84 = Klein *Gesammelte Mathematische Abhandlungen* **1**, 424–459 = Lie *Ges. Abh.* **1**, 229–285. [March 1871]

Kronecker, L., 1853. "Über die algebraisch auflösbaren Gleichungen," *Monatsberichte K. Akademie der Wissenschaften Berlin, 1853*, 365–374 = *Werke* **4**, 1–11.

Kronecker, L., 1877. "Über Abelsche Gleichungen," *Monatsberichte K. Akademie der Wissenschaften Berlin, 1877*, 845–851 = *Werke* **4**, 65–71

Lie, S., 1870. "Über die Reziprocitätsverhältnisse des Reyeschen Komplexes," *Göttinger Nachrichten 1870* (1870), 53–66 = *Ges. Abh.* **1**, 68–77. [January 1870.]

Lie, S., 1871a. "Over en Classe geometriske Transformationer," *Forh. Videnskabs-Selskabet Christiania 1871* (1872), 67–109 ≈ *Ges. Abh.* **1**, 105–152. [25 March 1871] In *Ges. Abh.* edition, this paper (Lie's doctoral dissertation) is translated into German, along with the explanatory remarks Lie had added to the text.

Lie, S., 1871b. "Über diejenige Theorie eines Raumes mit beliebig vielen Dimensionen, die der Krümmungstheorie des gewöhnlichen Raumes entspricht," *Göttingen Nachrichten 1871* (1871), 191–209 = *Ges. Abh.* **1**, 215–228. [24 April 1871]

Lie, S., 1871c. "Über eine Klasse geometrischer Transformationen," *Forh. Videnskabs-Selskabet Christiania 1871* (1872), 182–245 = *Ges. Abh.* **1**, 153–210. [Summer 1871]

Lie, S., 1871d. Report in Norwegian to the Scientific Society of Christania. First published in 1899. Reprinted in Lie *Ges. Abh.* **1**, 267–270. [26 September 1871]

Lie, S., 1871e. "Über Komplexe, inbesondere Linien- und Kugelkomplexe, mit Anwendung auf die Theorie der partieller Differentialgleichungen," *Mathematische Annalen* **5** (1872), 145–208, 209–256 = *Ges. Abh.* **2**, 1–121. [10 October 1871, 15 November 1871]

Lie, S., 1871f. "Zur Theorie eines Raumes von *n* Dimensionen," *Göttinger Nachrichten 1871* (1871), 535–557 = *Ges. Abh.* **1**, 271–285. [15 November 1871.]

Lie, S., 1871g. "Zur Theorie eines Raumes von *n* Dimensionen II," *Ges. Abh.* **7**, 1–10. [3 December 1871]

Lie, S., 1872a. "Kurzes Résumé mehrerer neuer Theorien," *Forh. Videnskabs-Selskabet Christiania 1872* (1873), 24–27 = *Ges. Abh.* **3**, 1–3. [30 April 1872]

Lie, S., 1872b. "Zur Theorie partieller Differentialgleichungen erster Ordnung, inbesondere über eine Klassifikation derselben," *Göttinger Nachrichten 1872* (1872), 473–489 = *Ges. Abh.* **3**, 16–26. [11 October 1872]

Lie, S., 1872c. "Zur Theorie der Differentialprobleme," *Forh. Videnskabs-Selskabet Christiania 1872* (1873), 132–133 = *Ges. Abh.* **3**, 27–28. [14 November 1872]

Lie, S., 1873a. "Über partielle Differentialgleichungen erster Ordnung," *Forh. Videnskabs-Selskabet Christiania 1873* (1874), 16–51 = *Ges. Abh.* **3**, 32–63. [21 March 1873]

Lie, S., 1873b. "Partielle Differentialgleichungen erster Ordnung, in denen die unbekannte Funktion explizite vorkommt," *Forh. Videnskabs-Selskabet Christiania 1873* (1874), 52–85 = *Ges. Abh.* **3**, 64–95. [21 March 1873]

Lie, S., 1873c. "Zur analytischen Theorie der Berührungstransformationen," *Forh. Videnskabs-Selskabet Christiania 1873* (1874), 237–262 = *Ges. Abh.* **3**, 96–119. [August 1873]

Lie, S., 1873-4. "Lie über die Anfänge seiner Theorie der Transformationsgruppen. Aus Briefen an Adolph Mayer," *Ges. Abh.* **5**, 583–614. [Excerpts from letters received by Mayer between 3 December 1873 and the summer of 1874. Edited with commentary by F. Engel.]

Lie, S., 1874a. "Begründung einer Invariantentheorie der Berührungstransformationen," *Mathematische Annalen* **8** (1874–75), 215–303 = *Ges. Abh.* **4**, 1–96. [5 July 1874].

Lie, S., 1874b. "Über Gruppen von Transformationen," *Göttinger Nachrichten 1874* (1874), 529–542 = *Ges. Abh.* **5**, 1–8.

Lie, S., 1879. "Theorie der Transformationsgruppen I," *Mathematische Annalen* **16** (1880), 441–528 = *Ges. Abh.* **6**, 1–94. [December 1879]

Lie, S., 1888,1890,1893. *Theorie der Transformationsgruppen*, 3 vols., Leipzig.

Lie, S., 1896. *Geometrie der Berührungstransformationen.*, Leipzig 1896.

Plücker, J., 1868, 1869. *Neue Geometrie des Raumes gegründet auf die Betrachtung der geraden Linie als Raumelement,* v.1 (A. Clebsch, ed.), v.2 (F. Klein, ed.), Leipzig 1868–69.

MANUSCRIPT REFERENCES

N.B. In what follows the abbreviation NSUB is used for Niedersächsiche Staats- und Universitätsbibliothek, Göttingen.

Klein, F., MS1871a. "Inhaltsangabe zu: *Ueber raumliche Gebilde mit mehrfach unendlichen vielen linearen Transformationen in sich selbst,*" Handschriften abteilung, NSUB, Cod. Ms. F. Klein 22 G II. [3 pages. Undated. Probably written in the first half of 1871.]

Klein, F., MS1871b. Untitled and undated essay by Klein on W-configurations in space. Papers of Sophus Lie, Universitetsbiblioteket i Oslo. [22 pages. Probably written in the first half of 1871. Draws upon material on W-curves and surfaces in space in Cod. Ms. F. Klein 22 G II (Handscriftenabteilung, NSUB).]

Klein, F., MS1892. "Ueber unsere Arbeiten aus den Jahren 1870-72, von F. Kl[ein] und S. Lie. Über Lie's und meine ältere geometrische Arbeiten, von F. Klein (Erster, Borkumer, Entwurf von August 1892)," Handschriftenabteilung, NSUB, Cod. Ms. F. Klein 22 G II. [24 pages]

Klein, F., MS1916. "Mitteilungen von Felix Klein über seine Beziehungen zu Lie. (Nach Aufzeichnungen, die Engel im Sept. 1916 in Göttingen gemacht hat, wohin zu kommen ihn Klein aufgefordert hatte.)" In possession of Professor Ernst Hölder, Mainz, Federal Republic of Germany. [25 pages. Manuscript in the hand of Friedrich Engel.]

Lie, S., MS1879. Letter to F. Klein, Handschriftenabeilung, NSUB, Cod. Ms. F. Klein 10, Nr. 673. [Undated. Probably written in 1879, since it discusses [Lie 1879].]

Lie, S., MS1893. "Zur Auseinandersetzung mit Felix Klein." In possession of Professor Ernst Hölder, Mainz, Federal Republic of Germany. [41 pages. Written by Friedrich Engel and originally intended for inclusion in volume 7 of Lie's collected works. (See the forward to Lie *Ges. Abh.* 7.) Contains extensive transcribed excerpts from the Klein-Lie correspondence (1891-93) regarding [Klein MS1892] and the revised version prompted by Lie's criticisms (Lie Papers, Universitetsbiblioteket i Oslo, Ms. fol. 3839, LXVII: 11). Also contains extensive transcribed excerpts from manuscripts by Lie in which he attempted to present his views on their geometrical work in 1869-72. Many of the excerpted documents are in the form of letters to Klein that Lie apparently never sent. The Lie documents quoted by Engel are undated but were probably written in 1892-93.]

Projective and Algebraic Geometry

Leonhard Euler (1707–1783)
Courtesy of Springer-Verlag Archives

The Background to Gergonne's Treatment of Duality: Spherical Trigonometry in the late 18th Century

Karine Chemla

Pour Victor, aujourd'hui âgé de 234 jours...

During the first decades of the 19th century, duality began to be considered as a mathematical subject in itself. It was dealt with in a variety of ways, associated respectively with the names of Poncelet, Gergonne, Plücker, et al. This essay will concentrate on the approach to duality developed by Joseph Diaz Gergonne (1771-1859) as it emerged in a series of works that appeared from 1810 onward.[1] Gergonne was interested in the fact that various domains of geometry (the theory of polyhedra, spherical trigonometry, some parts of plane and solid geometry) present a common phenomenon: they can be presented in such a way that their theorems and proofs are joined in couples, the members of which correspond to one another by a systematic linguistic translation.[2] For all these domains, including spherical trigonometry, he inserted such presentations in his journal, although it was not until the mid-1820s that he actually drew explicit conclusions about the essence of this phenomenon of duality. This paper represents a preliminary inquiry into the genesis Gergonne's interest in duality as evidenced from work published in the field of spherical trigonometry. For a thorough analysis of this question, one would have to review systematically *all* the geometrical domains that Gergonne considered in this respect and determine whether such an idea of coupling theorems or proofs was present in the literature available to him. Our analysis will show that mathematicians who wrote on spherical trigonometry in the second half of the 18th century actually dealt with symmetries between statements or betweem proofs due to duality. The fact that they anticipated in some sense an approach that later interested Gergonne raises the question of their influence on Gergonne's elaborations. After briefly presenting Gergonne's approach to the phenomenon of duality, as he expounded it during the years 1825-26, we detail the various concepts that were associated with the application of duality in spherical trigonometry. This will provide us with the tools required to fulfil our present goal, namely, to analyse those writings of the 18th century on spherical trigonometry that utilize duality and to compare their treatment of this notion with Gergonne's own.

THE HISTORY OF
MODERN MATHEMATICS

I. GERGONNE'S APPROACH TO DUALITY

The journal *Annales de Mathématiques Pures et Appliquées*, launched by Gergonne in 1810, was one of the important media for the maturation of the concept of "duality."[3] The phenomenon upon which this interest focused cannot be better presented than by Gergonne himself. In a paper published in 1826, Gergonne recapitulated previous investigations, commenting on the extent to which duality pervades geometry as follows:

> ...We believe to have said enough to put out of any contestation these two points of philosophy of mathematics, that is to say: 1°) that there is a relatively noticeable part of geometry in which theorems correspond to one another exactly, two by two, as well as do the reasonings that have to be made to establish them, and this because of the very nature of extension; 2°) that this part of geometry, which would become quite extensive if one wanted to include curved lines and surfaces in it, can be developed independently from computation and from the knowledge of any of the metrical properties of the magnitudes under consideration...[4]

Earlier in this paper, Gergonne had this to say about the nature of the special correspondence between theorems to which he is alluding here:

> ...a very striking feature of this part of geometry that does not depend at all on metrical relationships between parts of the figures, is that, except for some theorems which are symmetrical of themselves, ...all the theorems there are double, that is to say, in plane geometry to each theorem always corresponds necessarily another one which can be deduced from it by merely exchanging the two words *points* and *lines* with one another, whereas in space geometry points and planes are the words that have to be exchanged with one another to pass from one theorem to its correlate.[5]

Actually, this kind of correspondence occurs not only between statements of theorems but also between their proofs, and Gergonne paid special attention to this:

> We could very well limit ourselves to proving only half of our theorems and deducing from them the other half, but we prefer to prove them both directly ...in order to point out that *there exists the same correspondence between the proofs of the two theorems of a couple*

as between their statements. We will even take care to
present analogous theorems in two columns, one in
front of the other, so as to make their correspondence
more obvious... [and] so that the proofs can be used
as checks on one another.[6]

In order to call attention to these matters, Gergonne had been patiently gathering, year after year, papers for the *Annales de Mathématiques
Pures et Appliquées* that dealt with very well-known areas of geometry.
However, the way in which those papers were then presented by Gergonne served to emphasize the common feature they all shared, namely
a linguistic correspondence between the statements of "correlative" theorems. Gergonne's presentation thus raised the problem of how to deal
with this phenomenon from a higher mathematical point of view. Within
plane and space geometry, for example, this type of correspondence between theorems had been shown to occur in the geometry of polyhedra.
In a paper published by Gergonne in 1825, he noted that:

Except for some theorems, such as for instance Euler's,
in the statement of which the number of faces and
the number of vertices enter in the same way, there is
no theorem of this kind which should not inevitably
correspond to another, which can be deduced from it
by merely exchanging the words *faces* and *vertices* with
one another...[7]

In the same year, Gergonne also published part of a memoir by Sorlin
devoted to spherical trigonometry [Gergonne 2-Sorlin], the presentation of
which sheds light on a similar kind of correspondence between formulas
or between theorems in that field.[8] Gergonne made this comment on it
at the conclusion of the paper:

... All the formulas of spherical trigonometry are double, and can be distributed in two series in such a way
that one can go from one formula of any of the two
series to its corresponding one in the other series by
merely replacing the sides by the supplements of the
angles and the angles by the supplements of the sides.[9]

Finally, in his recapitulation of 1826, wherein he presents the phenomenon of duality in its full generality, Gergonne called attention to all
these cases and more.[10] Thus we can distinguish two ingredients in his
approach that are essentially linked: a general description of the phenomenon of duality and an assessment of its extension. The comparison
of different fields enabled him to elaborate the conception of duality;[11]
but since his understanding of duality embraced the field of spherical

trigonometry, it is natural to assume, conversely, that this notion helped guide his presentation of this field. Therefore, by comparing how the authors of the 18th century, on the one hand, and Gergonne and Sorlin, on the other, actually employ duality within this specific context, we may hope to gain some added precision in our description of Gergonne's treatment of this phenomenon.

II. DUALITY IN SPHERICAL TRIGONOMETRY

A spherical triangle is defined by three points on a sphere between each pair of which one draws the shortest path—namely the arc of the great circle passing through them—to obtain its sides.[12] Six angles can be associated with this triangle: a, b, c measure the arcs forming its *sides* taken from the center of the sphere; A, B, C, its *angles*, are measured by the angles between the two tangents to the sides at each of the three vertices. The capital letters are used to name angles opposite the sides denoted by the corresponding small letters. (See fig. 1.) These notations,

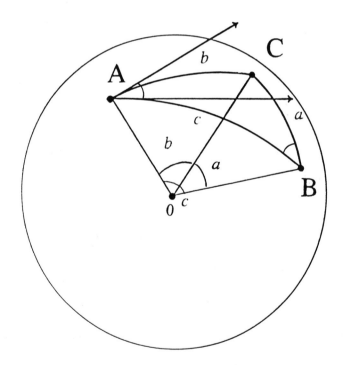

Figure 1: Euler's notations.

introduced by Euler in 1753, are today the most common ones, and are convenient for stressing associated duality phenomena, as we shall see below.[13]

One of the ways duality can be conceived is by attaching to every object a dual object that is naturally linked with it. In the framework of spherical trigonometry, this leads to the notion that to any spherical triangle T one can associate its polar or dual triangle T', the sides of which are supplements of the angles of T (namely, $\pi - A$, $\pi - B$, $\pi - C$), and whose angles are the supplements of the sides of T (namely, $\pi - a$, $\pi - b$, $\pi - c$).

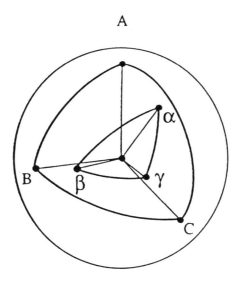

Figure 2: Construction of the polar triangle.

This triangle T' can be constructed in the following way (see fig. 2): if side BC of triangle T is located on an equator of the sphere, then the corresponding vertex α of the dual triangle will be the pole of this equator which is in the same hemisphere as the remaining vertex A of triangle T. In the same way, one gets the other vertices, β and γ, thus defining the spherical triangle T'. As α is a pole of BC, so will A be a pole of $\beta\gamma$. If the sides of the original triangle ABC have measures less than 90°, one can view this construction as follows (see fig. 3). If we extend the sides AB and AC to A' and A'' in such a way that AA' and AA'' both measure 90°, then $A'A''$ is located on the equator corresponding to the pole A. One then proceeds in the same way for B and C. By extending $A'A''$, $B'B''$, $C'C''$ until they meet one another, the polar triangle $\alpha\beta\gamma$ is constructed.[14]

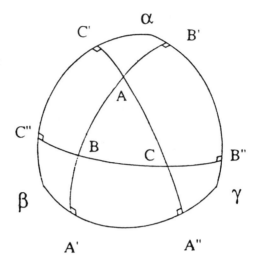

Figure 3: Construction of the polar triangle.

Moreover, the relation between a side of the polar triangle and the corresponding angle of the original triangle can be read from figure 4, which shows the projection of this situation onto the equatorial plane of A. The ray $O\beta$ is orthogonal to OA'', the intersection of our plane with

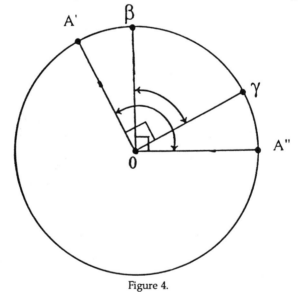

Figure 4.

the plane containing the side b; and $O\gamma$ orthogonal to OA', the intersection of our plane with the one containing the side c. The angle A of the original triangle can then be read in the equatorial plane as the measure of $A'A''$. We thus have $\beta\gamma = \beta A'' + \gamma A' - A'A'' = 2(\pi/2) - A'A''$, which is the wanted relation.[15]

The interest in this dual triangle is that every problem concerning the angles, resp. the sides, in triangle T can be converted into a problem concerning the sides, resp. the angles, of triangle T'. This leads to a reduction by half of the set of problems that have to be solved in spherical trigonometry; the other half is obtained by a transformation of the solution of a known problem to the dual triangle.

Another way of handling duality is to stress the symmetry that occurs in the set of properties attached to an object possessing a dual object. Indeed, the existence of the polar triangle T' associated with a given triangle T is linked with the fact that to any formula of spherical trigonometry one can associate a symmetrical formula. Let us take the example of formula:

(I) $$\cos a = \cos b \, \cos c + \sin b \, \sin c \, \cos A$$

This formula is valid for triangles with sides (a, b, c) and angles (A, B, C), and thus it is valid for the polar triangle as well, which gives the following relationship:

$$\cos(\pi - A) = \cos(\pi - B) \, \cos(\pi - C) + \sin(\pi - B) \, \sin(\pi - C) \, \cos(\pi - a)$$

or

(I') $$\cos A = -\cos B \, \cos C + \sin B \, \sin C \, \cos a$$

Formula (I') "looks like" formula (I); moreover, if the same transformation is applied to formula (I'), one gets formula (I) again. This is why we refer to them as being "symmetrical" with respect to one another. Actually, it was by observing this type of "symmetry" that many mathematicians first encountered duality. Formula (I') can either be considered as nothing but formula (I) applied to the triangle T', or as being a new formula, valid for the triangle T and symmetric to formula (I). This general phenomenon occurs for any formula (F) and can be represented by the following diagram, where an arrow represents a relation of symmetry, and a double bar expresses an identity:

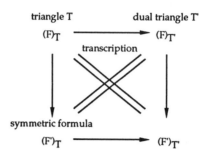

The arrows indicate that the dual formula for (F) can, in the same way, be seen either as formula (F) expressed for the polar triangle T' or as a formula (F') for triangle T, which is symmetric to (F). Both ways give the same result, as is expressed by the double bar.

Besides being related to the properties of the polar triangle and this symmetry in the set of formulas of spherical trigonometry, duality was also associated with a procedure that enabled one to transform a formula into its symmetrical counterpart. The relationship between a triangle and its polar triangle meant that the magnitudes measuring the sides in a formula were to be replaced by the supplements of the angles, and the angles by the supplements of the sides, in order to get a dual formula. Following the notations introduced by Euler, this amounts to applying a rule for rewriting the given formula which Euler states as follows: "[this formula] does not differ from the established one, except for the fact that capital and small letters have been exchanged with one another and, moreover, that all the cosines have been taken as negative."[16] An analysis of how Euler and other authors utilized and linked together all these concepts related to duality will enable us to shed some light on the specific background that shaped Gergonne's treatment of it.

III. ANALYTICAL TREATMENTS OF SPHERICAL TRIGONOMETRY

In the second half of the 18th century and the beginning of the 19th, many papers were devoted to spherical trigonometry. Among them, some dealt more specifically with an "analytical theory of spherical triangles," as Lagrange put it, and handled the question of duality in a way that has the most bearing for us. We shall, therefore, restrict ourselves to a study of those papers that adopted this approach in what follows. The fact that these papers were quite widely known is demonstrated by the number of contemporary mathematicians who cited this work. To give a hint of this, the diagram below represents the works of the authors we will be considering (in larger letters), together with those written by others who directly followed them or cited these principal papers[18]:

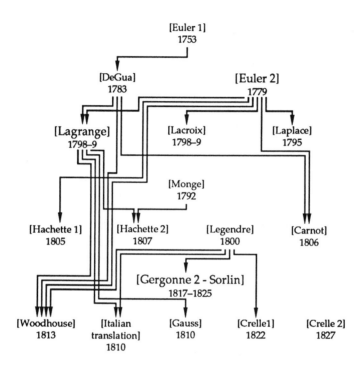

We can see from the above scheme that the analytic approach to spherical trigonometry appears to have received considerable impulse from Euler's contributions, which echoed through France in works by De Gua, Lagrange, Laplace, Legendre, Hachette, Lacroix, et al., and then diffused to other countries mainly through translations or adaptations. In the case of England, according to [Lardner], Woodhouse was responsible for introducing the analytical treatment of spherical trigonometry into British mathematics. The evolution of the analytical treatment of spherical trigonometry, which was one of the main frameworks in which duality seems to have developed, was described by Lagrange in the following terms:

> The analytical solutions of spherical triangles were only,
> at first, mere applications of algebra to geometrical

constructions. People then limited themselves to establishing some fundamental propositions geometrically and deriving all the formulae of spherical trigonometry from the equations given by those propositions. But could not this system be simplified by reducing it to a single fundamental equation? This reduction would be useful for improving the analytical theory of spherical triangles because in Analysis perfection consists in using only as few principles as possible and in deriving from those principles all the truths they contain by the very power of Analysis...[19]

Indeed, the authors of the texts we are about to analyze are less concerned with obtaining new formulas than they are with outlining a complete organization of spherical trigonometry. Their tools for this purpose consist of algebraic transformations applied to some fundamental equations which are obtained through geometry and taken as leading "principles." In this paper, we refer to such equations as "principles," "premises," or "basic formulas," without distinction. Our question then amounts to asking how does duality fit into this "architectonic" program. We shall first examine this issue in Euler's paper of 1779.[20]

IV. EULER'S PAPER OF 1779

In this work Euler adopts the same notations that he introduced in his paper of 1753. He starts by proving the following three formulas that depend on nothing more than elementary geometry:

$$(I) \qquad \cos c = \cos a \cdot \cos b + \sin a \cdot \sin b \cdot \cos C$$

$$(II) \qquad \frac{\sin C}{\sin c} = \frac{\sin A}{\sin a}$$

$$(III) \qquad \cos A = \frac{\cos a \sin b - \sin a \cos b \cos C}{\sin c}$$

Having obtained the third equation above, Euler states that: "...this equation... together with the two previous ones contains the whole spherical science, but this requires a more extensive explanation, therefore let us further develop these three equations one after the other."[21] He next proceeds to apply various algebraic transformations to these basic equations to obtain new ones. When a formula that he has just derived happens to enable him to solve some problem of spherical trigonometry, he mentions it incidentally. But even if the whole memoir provides numerous formulas for solving such problems, its principal aim is to organize the formulas rather than stress the types of problems that can be solved by them. Euler clearly hoped to create some order among these formulas by showing how, starting from the three basic ones, they can be derived algebraically one from another.

Every time he applied such transformations to obtain a formula dual to one of the basic ones, he stressed their similarity, noting that this formula can be derived from the correlated one by using a rule of rewriting.[22] In this way, he derived the dual formulas for all three basic equations. He then pointed out that these results arise from the corresponding ones by following a rule for rewriting the formulas which happens to be the same in each case: one permutes the angles and sides and then changes the signs of the cosines. Thus he arrived at the following important conclusion:

> ... as the whole of spherical trigonometry is based on the three formulas found above, the permutation between angles and sides takes place in full generality, so long as all the cosines are taken negatively.... The following noteworthy theorem proceeds from this:
>
> Theorem: *Given any spherical triangle, the angles of which are A, B, C, and the sides of which are a, b, c, another analogous spherical triangle can always be exhibited, the angles of which are the supplements from two right angles of the sides of this one, whereas its sides are the supplements from two right angles of the angles of this one.*[23]

What makes this result so noteworthy, of course, is that the "analogous" dual triangle possesses essentially all the same properties as the original one. At this point in his memoir, Euler recapitulates all the formulas which he has already obtained and presents them in tables, after which he gives their simplified form for the special case of right-angled triangles. The end of the memoir is devoted to transforming the obtained formulas into new ones that are convenient for logarithmic computations, namely formulas that state an equality between two factored expressions. He gives four such transformations.

Two points should be emphasized here, particularly because they are relevant to the work that followed Euler's own.

The first relates to the analytical framework adopted in organizing the principal results of spherical trignometry, and which naturally leads to the following question: what position within this framework should be occupied by duality? Should it be included among the initial premises or can it be deduced from the formulas that have been taken as premises? Euler proved that it can be deduced from the three basic formulas above, although this set of formulas is *not* invariant under duality. The second point is that Euler transformed not only formulas to obtain dual formulas but also their proofs. In fact, he deliberately presented dual proofs that were merely translations of the original proofs, even though he had no need to do so.

The fact that Euler's presentation is not completely symmetrical is related to the analytical organization of his paper. His three initial formulas were obtained geometrically, whereas their dual formulas were derived algebraically, and thus their derivations have no dual counterpart in Euler's paper. In fact, these algebraic derivations of the dual formulas are the only ones that are not correlated with dual computations.

As observed earlier, the phenomenon of duality can be met in two different ways: either by applying a rule for rewriting a formula to obtain another symmetrical one or by associating to each triangle a polar triangle. In this regard, it is rather interesting to observe how Euler went about introducing duality in spherical trigonometry.

First he showed that the dual formulas associated with his three basic formulas can be deduced algebraically. But since all of spherical trigonometry can be obtained from these basic formulas, he concluded that the rule for rewriting a formula so as to produce its dual is valid in full generality. For if this rule can be applied to all basic formulas and then produce valid formulas, clearly it can be applied to any formula deduced from them, and hence to any formula. Euler's "noteworthy theorem" regarding the relationship between an arbitrary triangle and its polar triangle now occurs as a natural consequence. Thus, Euler deduces from the existence of this symmetry between formulas the possibility of exhibiting a dual triangle with essentially the same properties as the original.[24] This approach is, of course, the reverse of the one followed in §II, where we deduced this symmetry from the existence of the polar triangle following the usual argument employed in modern presentations.

Thus, on the one hand, Euler's approach to duality was to deduce it from premises that do not contain it in an obvious way, for the set of initial formulas he employed is not invariant under the transformation of duality. On the other hand, whenever Euler applied a sequence of algebraic transformations to a dual formula, this sequence of transformations turned out to be dual, step by step, to another sequence which he had previously applied to the original formula.

Let us quote an example of this that will give also some of the flavor of Euler's memoir. Starting with the basic formula

(III) $\qquad$ $\cos A \sin c = \cos a \sin b - \sin a \cos b \cos C,$

Euler went on to say: "as $\sin A \sin c = \sin C \sin a$, the first member of this equation is divided by $\sin A \sin c$, whereas the last one is divided by $\sin C \sin a$, and one obtains

$$\cot A = \frac{\cos a \sin b - \sin a \cos b \cos C}{\sin a \sin C}.$$

Hence the angle A can now be obtained from the given two sides a and b with the angle C between them." In the following paragraph, he gave an algebraic derivation of the formula $\cos a \sin C = \cos A \sin B + \sin A \cos B \cos c$, which he noticed can be derived from formula (III) by using the transformation rule discussed above. He therefore applied to this new equation the operations dual to those he had applied to equation (III), that is to say: "if now the first member of the latter equation is divided by $\sin a \sin C$, and the last one by $\sin A \sin c$, this equation arises:

$$cota = \frac{\cos A \sin B + \sin A \cos B \cos c}{\sin A \sin c}$$

which is applied to find the side a from the two given angles A and B with the side c between them."

This kind of relationship between dual proofs occurs occasionally in the first part of the memoir, but it is explored systematically in the last part, where Euler exhibits formulas for logarithmic computations. There the reader is given, one after the other, dual proofs, starting from dual formulas, of dual results. Euler alludes to this process when he begins these dual computations by saying: "this is deduced in the same way from these formulas...."[25] In one particular instance Euler made a mistake in a proof that is duplicated in the dual proof. This enables us to say with certainty that the dual proof was obtained by a transformation of the original.[26] Moreover, although the rule for rewriting formulas would have allowed him to derive the dual result immediately, he, like Gergonne later, chose to write the whole proof over again, thereby laying stress on the line by line correspondence between dual proofs.

V. Subsequent developments in these two directions

In describing the evolution of analytical theories in spherical trigonometry, Lagrange pointed out that subsequent works attempted to reduce the number of basic formulas out of which one could derive the formulas of spherical trigonometry through algebraic computations.[27] De Gua, Lagrange himself, Lacroix and others proved that the entire corpus could be deduced from the following single formula[28]:

$$(A) \qquad \cos a = \cos b \cdot \cos c + \sin b \cdot \sin c \cdot \cos A.$$

Let us consider the example of Lagrange's paper of 1798-99. Starting from the above formula, he deduced algebraically all the other necessary results including:

$$(D) \qquad \cos A = -\cos B \cdot \cos C + \sin B \cdot \sin C \cdot \cos a.$$

He noticed that a rule for rewriting enables one to go from the first formula to the second one. Thus he asserted: " all the formulas which result from the formula (A) will also be true when the same changes are made." And he went on to say: "From this follows the well-known property of spherical triangles that any spherical triangle can be changed into another one, the sides and the angles of which are respectively the supplements of the angles and the sides of the former...."[29]

Thus, like Euler, Lagrange proceeds from the algebraic symmetry to deduce the existence of the polar triangle. The possibility of applying the rule for rewriting formulas in full generality is also derived in the same manner as Euler had done. But Lagrange was able to deduce the duality from a single formula rather than three, a formula which, again, does not "contain" duality in an obvious way. Unlike Euler, however, Lagrange was not interested in presenting dualized proofs of the new results. When he needed a formula dual to a formula he had already obtained, he immediately rewrote the original formula by "substituting in it the supplements of the angles A, B, C instead of the sides a, b, c, and the supplements of the opposite sides a, b, c instead of these angles."[30] Lagrange thereby improved one of the two directions opened by Euler in the analytical approach to duality in spherical trigonometry, while neglecting the other. Many of the writers who followed him tended to do the same.

Euler brought into play the idea that dual formulas can be obtained by proofs that are dual with one another. This approach can be found again in different guises in the subsequent work on spherical trigonometry undertaken by De Gua, Lacroix, and Legendre.

De Gua did not actually write down proofs that were dual to one another, but he refers to such a possibility. Starting from formulas which are dual to previous ones and led to some new formula, he states the dual result, arguing that it can be inferred by "following the same method" or by "making the same computations."[31] Lacroix, who followed Euler very closely, gave the same set of dual formulas for logarithmic computations proved by dual derivations that Euler presented in his paper.[32] Legendre, too, developed some dual proofs in the *Traité de trigonométrie*, which was appended to his widely read *Eléments de géométrie* beginning with the third edition.[33] Even though in all these texts the analytical context prevents the presentation from being completely symmetrical, and even though these authors did not emphasize the possible transformation of proofs as strongly as Euler had done, this idea pervaded much of the literature on spherical trigonometry at the end of the 18th century.

Gergonne was probably familiar with this dualistic approach to spherical trigonometry at the very least through Legendre's book, which he quoted fairly often. Moreover, some of the papers devoted to this subject which he inserted in the first volumes of the *Annales de mathématiques pures et appliquées* also stress the correlation between dual proofs as well.[34] The clearest illustration of this can be found in [Gergonne 2-Sorlin], where the fact that the proofs are symmetrical to one another is emphasized to the utmost and where the principal goal is to exhibit the whole architecture of formulas and proofs in a totally symmetrical fashion. To do so, the authors had to start from a symmetrical basis of formulas and hence sacrifice the demand for economy in the number of grounding principles.

In this paper, formula (A) is first deduced from elementary geometry. Then the dual triangle is exhibited[35], and formula (A) applied to it, leading to a new formula, the dual of (A). The authors then proceed from these two formulas, which form an invariant set under duality, and apply to them successive dual computations to obtain additional dual formulas. This methodology leads to an entirely symmetrical presentation of spherical trigonometry.[36] Euler's partially symmetrical presentation is thereby systematized, although the dual triangle can be obtained as a consequence of formula (A) there is no way to exhibit it explicitly. Thus, Gergonne was following a course that was essentially the reverse of Euler's, i.e. he begins with the existence of the dual triangle and proceeds to derive the symmetry between formulas.

Gergonne knew that the presentation of dual proofs was pointless if one were only interested in obtaining new dual results. Once the dual transformation is shown to be valid, one only has to apply it to generate dual formulas; no other proof is needed. Gergonne adopted this unnecessarily lengthy mode of presentation in order to shed light on the underlying symmetry that pervades spherical trigonometry. Moreover, this mode of presentation was identical to the one he employed in all other fields where he noticed the same phenomenon of duality. In each of these cases, he began with a set of dual propositions and proceeded by applying to them dual proofs or dual computations leading to a fully symmetrical presentation.

Thus we find that some mathematicians were interested in deriving dual formulas from the smallest number of basic formulas possible, and in deducing from them the existence of a geometrical entity responsible for this symmetry—namely the polar triangle. Those who pursued this methodological course had by and large little interest in developing Euler's approach to transformed proofs. On the other hand, there were others who stressed the fact that the whole body of spherical trigonometry can be presented in a symmetrical way. This group, however, largely neglected the task of deducing this symmetry from the minimal number of premises required.

VI. WHAT WAS NEW IN GERGONNE'S APPROACH TO DUALITY?

As we have seen, Euler had been interested in presenting parts of spherical trigonometry in a symmetrical way, thereby anticipating Gergonne's approach to duality. Does this mean that there was no qualitative difference between the two? Did Gergonne merely systematize an idea that was already clearly developed in Euler's work and which pervaded most of the subsequent litterature on spherical trigonometry? The answer to both questions is no, and the reasons why help to explain what happened during the period between Euler and Gergonne.

Euler and the mathematicians who followed him were working within the analytical tradition of spherical trigonometry. Thus, the proofs Euler gave were algebraic in nature, that is to say sequences of algebraic transformations. Such a proof can be viewed as a sequence of formulas that form the intermediate steps between the starting point and the end result. Therefore, in this context it would seem to be quite natural to proceed from a transformation of formulas to the transformation of their proofs. Nowhere in Euler's papers, nor, so far as we know, in any other works written within the analytical tradition during the second half of the 18th century, can one find *geometrical constructions* and *geometrical* proofs that are exhibited in a dual fashion.[37]

By contrast, in [Gergonne 2-Sorlin], after a first part wherein all the algebraic computations are gathered, the correlated ones following one another, the authors present a section entitled "recherches diverses" in which successive geometrical theorems and constructions correspond with one another by duality. More precisely, the first part of the paper is devoted to an analytical presentation of a set of formulas, starting from formulas (A) and (D). This has much the same flavor as the papers of the 18th century discussed above, the only difference being that the exposition is symmetrical. The final part of the paper, however, displays an entirely new feature. To pass from one paragraph to the other, one cannot, as before, merely exchange the names of the variables and the signs of the cosines. One must, in fact, transform each sentence into another one. The corresponding sentences are phrased in parallel ways so that in passing from one to the other, one only has to refer to a dictionary which associates words, elements, and constructions with one another, and then translate word by word.

For example, certain geometrical constructions on the sphere are associated with one another by duality, and the statements describing them are also phrased in parallel ways. Thus the sentence "soit d l'arc de grand cercle mené du sommet A au milieu A' du côté a" corresponds to this statement in the following proof "soit D l'angle que fait avec le côté a l'arc de grand cercle a' qui divise l'angle A en deux parties égales" (pp. 293–4). "Soient O le pôle et R le rayon sphérique du cercle circonscrit à notre triangle. On sait que, si de ce point O on abaisse des arcs perpendiculaires sur les directions de ses trois côtés, ils tomberont sur leurs milieux" is correlated with "soit présentement o le pôle et r le rayon sphérique du cercle inscrit, on sait que si de ce point on conduit des arcs de grands cercles aux trois sommets, ces arcs diviseront les angles du triangle en deux parties égales" (pp. 295, 297).

The problem "cherchons quel est, sur la sphère, le lieu des sommets A des triangles sphériques qui ont tous la base commune a et dans lesquels en outre la somme des angles est constante" becomes the problem "cherchons enfin quelle est, sur la surface de la sphère, la courbe enveloppe des bases de tous les triangles sphériques qui ont l'angle au sommet commun et même périmètre." Its solution is phrased as follows: "le lieu des sommets de tous les triangles sphériques de même base et de même somme d'angles est la circonférence d'un petit cercle de la sphère," whereas the solution to the dual problem is given as "l'enveloppe des bases de tous les triangles sphériques qui ont un angle commun et même périmètre est la circonférence d'un petit cercle de la sphère."

Thus the questions raised by duality within the analytical framework seem to have been transferred by Gergonne from an analytical to a geometrical context. Moreover, in the process of this transformation the results underwent a non-trivial alteration within the new realm of discourse. The presentation in the final part of [Gergonne 2-Sorlin] is an indication of an important advancement in dealing with propositions susceptible to a dualistic treatment. Perhaps this manner of addressing the question of duality was made possible by the analytical tradition, where the idea of a transformation of formulas was very common. This idea might have been transfered to geometry somewhat later under the impulse of the program aimed at translating the reasonings of analysis into geometrical operations.[38] Indeed, this could well have been part of Gergonne's contribution to the maturation of the concept of duality. At the very least, the foregoing analysis suggests that different steps have to be distinguished in the history of this concept.

NOTES

1. This work is part of a wider research project dealing with duality that I have undertaken together with Serge Pahaut. The results presented here have been discussed with him and have greatly benefited from his insights. For a detailed analysis of the memoirs discussed in this paper, see [Chemla, Pahaut]. It is a pleasure to thank Serge Pahaut and François de Gandt for their help in the preparation of this paper. I am also very grateful to Christian Houzel, David Rowe and the referee for their comments.

2. Gergonne does not refer to any common geometrical background to explain this common feature of different fields. In some cases, he mentions the transformation by reciprocation, largely developed by Poncelet, to justify the possible coupling of statements. In the case of spherical trigonometry, he refers to the polar triangle (see §II). But he does not pay a great deal of attention to this geometric background. He is mainly interested in the common formal property of the possible presentations. On the contrary, Poncelet mainly focused on the geometrical transformations that lead from a situation to its dual. In the following analysis, the word "duality" is taken as a descriptive term from modern mathematics, and is not, unless otherwise specified, part of the vocabulary of our authors. Indeed, it seems to have been introduced in this modern sense by Gergonne himself, and, as far as we know, was never employed in the context of 18th-century mathematics. Throughout the text, the radius of the sphere is taken to be one.

3. In a forthcoming paper, I plan to describe the various stages through which the notion of duality matured in this journal from the statement of some problems left to the reader to the double-column presentation, I shall analyse the various possible page-settings which have been used. I shall also describe the evolution of the linguistic aspect of the conception of duality. As for Gergonne himself, his

biography is to be found in [D. Struik], and the reader is referred to [A. Dahan], [L. Giard], and [M. Otero] for further information.

4. "... nous croyons en avoir dit suffisamment pour mettre hors de toute contestation ces deux points de philosophie mathématique, savoir 1°) qu'il est une partie assez notable de la géométrie dans laquelle les théorèmes se correspondent exactement, deux à deux, ainsi que les raisonnements qu'il faut faire pour les établir, et cela en vertu de la nature même de l'étendue; 2°) que cette partie de la géométrie, qui prendrait une très-grande étendue, si l'on voulait y comprendre les lignes et les surfaces courbes, peut être développée indépendamment du calcul et de la connaissance d'aucune des propriétés métriques des grandeurs que l'on considère..." [Gergonne 3], pp. 230-1.

5. "Mais un caractère extrêmement frappant de cette partie de la géométrie qui ne dépend aucunement des relations métriques entre les parties des figures, c'est qu'à l'exception de quelques théorèmes symétriques d'eux-mêmes... tous les théorèmes y sont doubles; c'est-à-dire que, dans la géométrie plane, à chaque théorème il en répond toujours nécessairement un autre qui s'en déduit en y échangeant simplement entre eux les deux mots *points* et *droites*; tandis que, dans la géométrie de l'espace, ce sont les mots *points* et *plans* qu'il faut échanger entre eux pour passer d'un théorème à son corrélatif." [Gergonne 3], p. 210.

6. "Nous pourrions fort bien nous borner à démontrer seulement une moitié de nos théorèmes, et à en déduire l'autre moitié, à l'aide de la théorie des pôles. Mais nous préférons les démontrer directement les uns et les autres ... pour avoir l'occasion de faire remarquer qu'il existe entre les démonstrations de deux théorèmes d'un même couple la même correspondance qu'entre leurs énoncés. Nous aurons même soin, afin de rendre cette correspondance plus apparente, de présenter les théorèmes analogues dans deux colonnes, en regard l'une de l'autre... de telle sorte que les démonstrations puissent se servir réciproquement de contrôle." [Gergonne 3], p. 211. The emphasis is my own.

7. "... excepté quelques théorèmes, tels, par exemple, que celui d'Euler, dans l'énoncé desquels le nombre des faces et celui des sommets figurent de la même manière, il n'est aucun théorème de ce genre auquel il ne doive inévitablement en répondre un autre, qui s'en déduit en y permutant simplement entre eux les mots *faces* et *sommets*;" [Gergonne 1], p. 157.

8. [Gergonne 2-Sorlin] is an extract of a memoir which, according to Gergonne's introduction to it, "was presented in 1819 to the Royal

Academy of Sciences in Paris, which in its session of February 22 of the same year, on the recommendation of commissioners Legendre and Delambre, was declared worthy of being inserted in the series of Savans étrangers" (p. 273). As a matter of record, the memoir was not published in this series. According to the reports of the sessions at the Academy, it was actually presented on October 27, 1817. The report on it, which appears to have been written by Delambre, can be found in the papers on the session of February 22, 1819. It can be deduced from it that Sorlin took care to present spherical trigonometry in a symmetrical way. Yet some discrepancies between sentences quoted in the report and those in the actually published excerpt lead one to assume that Gergonne rewrote it before publication, as he is well-known to have done for many other papers. The difficulty in dissociating Sorlin's contribution from Gergonne's has led us to refer to the paper simply as [Gergonne 2-Sorlin] in our abbreviations. Gergonne gives some biographical information on Sorlin in a long note at the end of the paper (p. 304–5).

9. "...toutes les formules de la trigonométrie sphérique sont doubles, et peuvent être distribuées en deux séries, de telle sorte qu'on passera d'une formule à sa correspondante dans l'autre série, en y remplaçant simplement les côtés par les supplémens d'angles et les angles par les supplémens de côtés." [Gergonne 1], p. 302. This paper is not presented in double columns, as was done later, but rather the dual parts follow one another systematically, as in the paper of 1819 on polyhedra (see [anonymous]). This paper appeared in the *Annales de mathématiques pures et appliquées*, which merely noted that the author was "un abonné." However, in the French adaptation of [A. Schoenflies], A. Tresse points out (p. 14) that the paper can be attributed to Gergonne himself. This attribution can be proven on the basis of independent evidence.

10. After having mentioned the role of duality in plane and spatial geometry (see note 5.), [Gergonne 3] went on to give other examples: "Parmi un grand nombre d'exemples que nous pourrions puiser, dans le présent recueil, de cette sorte de dualité des théorèmes qui constituent la géométrie de situation, nous nous bornerons à indiquer, comme les plus remarquables, les deux élégans théorèmes de M. Coriolis, démontrés d'abord à la page 326 du XIᵉ volume, puis à la page 69 du XIIᵉ, et l'article que nous avons nous-même publié à la page 157 du précédent volume, sur les lois générales qui régissent les polyèdres. C'est, au surplus, une suite inévitable des propriétés des pôles, polaires, plans polaires et polaires conjuguées des lignes et surfaces du second ordre, qui jouent ici un rôle assez analogue à celui

que joue le triangle supplémentaire, dans la trigonométrie sphérique où, comme il a été montré, dans tout le cours de l'intéressant mémoire de M. Sorlin (tom. XV, p. 273), les théorèmes peuvent être également répartis en deux séries parallèles, de manière à se correspondre deux à deux, avec la plus grande exactitude."

11. As he says in ([Gergonne 3, p. 231]), it was "l'esprit de détail" that has concealed this phenomenon from the eyes of other geometers.

12. [Euler 1] begins with this definition of a spherical triangle and uses the calculus of variations to derive all the necessary formulas of spherical trigonometry. On this approach, and Euler's contributions to spherical trigonometry, see [Chemla, Pahaut] and our forthcoming paper "Euler's work in spherical trigonometry: contributions and applications."

13. See [Fellmann], [Braunmühl 2], and [Braunmühl 3]. These notations were adopted by almost all subsequent writers discussed in this paper. See also [Chemla Pahaut].

14. The construction of the polar triangle is not given in the papers in the papers discussed below. Euler and some other writers mention that its vertices are the poles of the great circles supporting the sides of the original triangle, but they do not go any further into the details. The polar triangle associated with any given triangle has long been known. We find it used in a proof by an Arabic author of the 9th century (see [Debarnot 1], [Debarnot 2]). Many authors of the 17th century mention its construction, for instance Snellius (see [Braunmühl 1], [Braunmühl 3]). The polar triangle is referred to as well-known by our authors of the late 18th century, who mention it when they wish to link its existence with the symmetry of the set of formulas in spherical trigonometry.

15. With this definition of the polar triangle, we can see that the duality implied in spherical trigonometry is the same as the duality implied in ordinary projective geometry. In modern terms, suppose that we construct the projective plane as a quotient of the sphere by the following equivalence relation: xRy iff the line xy contains O, the center of the sphere, that is to say x and y are antipodal points. Then the vertex A of triangle T is the pole in the projective plane of the line determined by $\beta\gamma$ in the polar triangle. In his paper of 1826, Gergonne does not give any insight into the geometrical implications of duality. As already mentioned, his approach was to focus on the symmetry between various propositions and proofs that arise in various geometrical fields. He considered the transformation by reciprocation of the polar triangle as accounting for the existence of

such a symmetry in different subjects, but did not inquire further into the deeper reasons for this relationship. As he says ([Gergonne 3], p. 210): "C'est, au surplus, une suite inévitable des propriétés des pôles, polaires, plans polaires et polaires conjuguées des lignes et des surfaces du second ordre, qui jouent ici [in other domains that he just cited] un rôle assez analogue à celui que joue le triangle supplémentaire, dans la trigonométrie sphérique où, comme il a été montré, dans tout le cours de l'intéressant mémoire de M. Sorlin (tom. XV, pag. 273), les théorèmes peuvent être également répartis en deux séries parallèles, de manière à se correspondre deux à deux, avec la plus grande exactitude." Gergonne differed, in this respect, from Poncelet, who was mainly interested in the geometrical correspondances between dual objects. Yet he was close to Legendre, who commented as follows on the analogy between dual formulas in spherical trigonometry (*Eléments de géométrie*,12th ed., p. 401): "Cette dernière équation entre A, B, C, a, offre une analogie frappante avec la première entre a, b, c, A: et on peut rendre raison de cette analogie par la propriété des triangles polaires ou supplémentaires...."

16. To state it in modern terms, this situation occurs when the implied category is self-dual. These frameworks were the first ones where duality was noticed through symmetries of sentences.

17. It is a formula "quae a proposita non discrepat, nisi quod literae maiusculae et minusculae inter se permutentur, insuper vero omnes cosinus negative accipiantur" [Euler 2], p. 75. Some authors did not state the rule of rewriting this way, (see for instance [De Gua]).

18. An arrow between two names $A \rightarrow B$ indicates that A is quoted in B.

19. "Les résolutions analytiques des triangles sphériques n'ont d'abord été que de simples applications de l'Algèbre aux constructions géométriques. On s'est contenté ensuite d'établir par la Géométrie quelques propositions fondamentales, et l'on a tiré toutes les formules de la Trigonométrie sphérique des équations données par ces propositions .... Mais ne pourrait-on pas simplifier ce système, en le réduisant à une seule équation fondamentale? Cette réduction servirait à perfectionner la théorie analytique des triangles sphériques: car, dans l'Analyse, la perfection consiste à n'employer que le moindre nombre possible de principes, et à faire sortir de ces principes toutes les vérités qu'ils peuvent renfermer, par la seule force de l'Analyse...." [Lagrange], p. 341–2. As discussed below, Lagrange proceeded to achieve the project alluded to here. This project was presented in the same vein in the introduction by De Gua to his memoir, and it

was expressed in similar terms when S.F. Lacroix defined the branch of geometry he called "analytical geometry," a branch which "would consist of deducing the properties of extension from the fewest number of principles, by analytical means. . ." ([Lacroix 1], p. XXV). In the context of this paper, the term "analysis" will be given this meaning.

20. If Euler's first paper has a similar architectonic purpose, it is nevertheless quite different from the other following works in that, instead of taking certain geometrically deduced formulas as his first principles, he bases the theory on the calculus of variations. This he applies in defining the sides as the shortest paths between the vertices, thereby obtaining formulas which he then utilizes in organizing the solutions to problems encountered in spherical trigonometry. The solution of a problem is always presented together with the solution of the dual problem corresponding with it. But, if his notations and overall scheme reflect an interest in the symmetrical presentation of dual problems and formulas, he does not comment on this. Nor does he mention the rule for rewriting formulas and the polar triangle. Moreover, as this paper had less noticeable impact than his second paper of 1779, we prefer not to analyze it here, and refer the reader to [Chemla Pahaut] or to our forthcoming publication on Euler's contributions to spherical trigonometry.

21. "Deinde aequatio. . . cum binis praecedentibus coniuncta totam Doctrinam sphaericam complectitur, quod autem uberiorem explicationem postulat, unde singulas has tres aequationes magis euoluamus." [Euler 2], p. 74.

22. See paragraph 2 and note 17.

23. "Cum igitur tota Trigonometria sphaerica tribus aequationibus supra inventis innitatur, permutatio angulorum et laterum generaliter locum habet, si modo omnes cosinus negative accipiantur. . . . Unde sequens Theorema insigne nascitur:
Theorema: *Proposito quocunque triangulo sphaerico, cuius anguli sint A, B, C et latera a, b, c, semper aliud triangulum analogum exhiberi potest, cuius anguli sint complementa laterum illius ad duos rectos, latera vero complementa angulorum ad duos rectos.*" ([Euler 2], pp. 77–8.)

24. [Lautman], p. 62, shows that a similar process took place in topology.

25. "Haec simili modo deducitur ex his formulis. . ." ([Euler 2], p. 84).

26. This error occurred in [Euler 2], p. 82, where Euler began with the following two formulas:

$$\cos a - \cos b \cos c = \sin b \sin c \cos A$$
$$\cos b - \cos a \cos c = \sin a \sin c \cos B.$$

Dividing one by the other, he gets

$$\frac{\cos a - \cos b \cos c}{\cos b - \cos a \cos c} = \frac{\sin b \cos A}{\sin a \cos B} = \frac{\sin B \cos A}{\sin A \cos B}.$$

He then adds 1 to both sides, but inadvertently introduces an error by neglecting the denominator of the left hand side:

$$\frac{(\cos a + \cos b)(1 - \cos c)}{\cos b - \cos a \cos c} = \frac{\sin(A + B)}{\sin A \cos B}.$$

Then he subtracts 1, again neglecting the denominator of the left hand side:

$$\frac{(\cos a - \cos b)(1 - \cos c)}{\cos b - \cos a \cos c} = \frac{\sin(B - A)}{\sin A \cos B}.$$

Then he divides this equation by the previous one, so that the two neglected denominators, which are identical, compensate each other, thus producing a valid result.

In the argument, Euler starts with the following two formulas:

$$\cos A + \cos B \cos C = \sin B \sin C \cos a$$
$$\cos B + \cos A \cos C = \sin A \sin C \cos b.$$

Dividing one by the other, he obtains

$$\frac{\cos A + \cos B \cos C}{\cos B + \cos A \cos C} = \frac{\sin B \cos a}{\sin A \cos b} = \frac{\sin b \cos a}{\sin a \cos b}$$

Again he adds 1, but neglects the denominator on the left hand side:

$$\frac{(\cos A + \cos B)(1 + \cos C)}{\cos B + \cos A \cos C} = \frac{\sin(a + b)}{\sin a \cos b}.$$

Then he subtracts 1, making exactly the same mistake as before:

$$\frac{(\cos A - \cos B)(1 - \cos C)}{\cos B + \cos A \cos C} = \frac{\sin(b - a)}{\sin a \cos b}.$$

Finally, he divides this equation by the previous one, so that the two errors cancel one another out.

These arguments can be checked to be dual step by step. Andreas Speiser, who edited these papers in the *Opera omnia*, pointed out these symmetrical mistakes, but he drew no conclusions regarding Euler's working methods from them.

27. See note 19.

28. Actually, De Gua and Lagrange claim to derive spherical trigonometry from this formula alone, but they take for granted that the variables which these formulas imply correspond to precise geometrical entities to which one can therefore apply permutations. If the variables are taken to represent real numbers, one must add to this basic set of formulas those obtained by applying circular permutations to (A). In this way spherical trigonometry can be derived from three similar but independent equations, which is the approach followed by Lacroix. On this point, see [Chemla Pahaut], paragraph IV.1.b.

29. "Ainsi toutes les formules qui résultent de la formule (A) seront vraies aussi, en y faisant ces mêmes changements. Il résulte de là cette propriété connue des triangles sphériques, que tout triangle peut être changé en un autre dont les côtés et les angles soient respectivement suppléments des angles et des côtés du premier..." [Lagrange], p. 345.

30. See [Lagrange], p. 351.

31. [De Gua], pp. 317, 323.

32. Lacroix explicitly noted this connection "en opérant comme ci-dessus" (p. 61) when deriving a result dual to a previous one.

33. See the computations in paragraphs LXXVIII and LXXXI of the 12th edition, and in the "Résolution des triangles sphériques en général," pp. 405 ff.

34. A paper by Servois, "Démonstrations de quelques formules de trigonométrie sphérique," in volume II (1811), pp. 84–88, contains dual proofs of dual results that, moreover, are presented in double columns.

35. Actually the paper is formulated in terms of pyramids and their intersection with the sphere, rather than in terms of spherical triangles.

36. This symmetry in the presentation of the set of formulas and proofs is expressed in the numbering of the formulas obtained in the paper. Each formula is given a Roman numeral: the original formula in capital letters and its dual in small ones. Such a symmetrical presentation is similar to those which Gergonne gave to the various fields in which he noticed the phenomenon of duality.

37. In [Chemla Pahaut], we showed that there are some hints in Lagrange's and De Gua's papers that when interpreting auxiliary angles introduced in dual computations they tried to give correlative geometrical interpretations. But, in no case, are these corelations actual correspondences by duality.

38. See [Taton].

BIBLIOGRAPHY

[anonymous] " recherches sur les polyèdres, renfermant en particulier un commencement de solution du problème proposé à la page 256 du VII° volume des Annales", *Annales de mathématiques pures et appliquées*, tome IX, n° 10 (1819), pp. 321-44.

[A. von Braunmühl 1] "Zur Geschichte des Supplementardreiecks", *Bibliotheca Mathematica*, vol. 13 (1898) pp. 65–72.

[A. von Braunmühl 2] "Die Entwickelung der Zeichen- und Formelsprache in der Trigonometrie", *Bibliotheca Mathematica*, Third Series, 1 (1900), pp. 64-74.

[A. von Braunmühl 3] *Vorlesungen über Geschichte der Trigonometrie*, two volumes, Teubner, Leipzig, 1903.

[L. Carnot] *Mémoire sur la relation qui existe entre les distances respectives de cinq points quelconques pris dans l'espace suivi d'un Essai sur la théorie des transversales*, Paris, Courcier, 1806.

[M. Chasles] *Aperçu historique sur l'origine et le développement des Méthodes en Géométrie*, in 4°, Bruxelles, 1837.

[K. Chemla, S. Pahaut] "Préhistoires de la dualité: explorations algébriques en trigonométrie sphérique (1753-1825) " in *Sciences à l'époque de la Révolution Française*, papers by the research group REHSEIS, edited by R. Rashed, Blanchard, 1988, pp. 151-200.

[A.L. Crelle 1] German translation of the *Eléments de géométrie* by Legendre, with the treatise on trigonometry, 1822.

[A.L. Crelle 2] *Lehrbuch der Elemente des Geometrie und der ebenen und sphärische Trigonometrie*, zweiter Band, Berlin, 1827.

[A. Dahan] "un texte de philosophie mathématique de Gergonne", *Revue d'Histoire des Sciences* XXXIX/2 (1986) pp. 97–126.

[M. T. Debarnot 1] "Introduction du triangle polaire par Abū Nasr b.ᶜIrāq", *Journal for the History of Arabic Science*, vol. 2, n° 1 (1978), pp. 126–136.

[M. T. Debarnot 2] *Al Bīrūnī, Kitāb Maqālīd 'Ilm al-Hay'a, La trigonométrie sphérique chez les arabes de l'Est à la fin du 10° siècle*, Damas 1985, Institut Français de Damas.

[De Gua] "Trigonométrie sphérique, déduite très brièvement et complètement de la seule solution algébrique du plus simple de ses problèmes généraux, au moyen de diverses transformations dont les rapports des sinus et cosinus, tangentes et cotangentes, sécantes et cosécantes d'un même arc ou d'un même angle plan rendent cette solution susceptible, et comprenant quelques formules qu'on croit utiles et neuves.", *Histoire de L'Académie Royale des Sciences*, (Paris, 1783), pp. 291–343 + plate 4.

[L. Euler 1] "Principes de la trigonométrie sphérique tirés de la méthode des plus grands et des plus petits", *Mémoires de L'Académie des Sciences de Berlin*, 9 (1753), 1755, pp. 223–257. Opera Omnia, vol. XXVII of the first series, (edited by A. Speiser), Lausanne, MCMLIV, pp. 277 –308.

[L. Euler 2] "Trigonometria spaerica universa ex primis principiis breviter et dilucide derivata" *Acta Academiae Scientiarum Imperialis Petropolitanae 3*, I (1779), 1782, pp. 72-86. *Opera Omnia*, vol. XXVI of the first series, (edited by A. Speiser), Lausanne, MCMLIII, pp. 224–236.

[E. Fellmann] "Leonhard Euler -Ein Essay über Leben und Werk", in *Leonhard Euler, Beiträge zu Leben und Werk*, J.J. Burckjardt, E.A. Fellmann, W. Habicht (ed.), Birkhäuser, 1983, pp. 13–98.

[C. F. Gauss] "Zuzätze sur Geometrie der Stellung von Carnot" (1810) *Werke*, Band IV, pp. 401–6.

[J.D. Gergonne 1] "Recherche de quelques-unes des lois générales qui régissent les polyèdres", *Annales de Mathématiques Pures et Appliquées*, volume XV, (1825) pp. 157–164.

[J.D. Gergonne 2 -Sorlin] "recherches de trigonométrie sphérique", par M. Sorlin, *Annales de Mathématiques Pures et Appliquées*, volume XV, (1825, pp. 273–305): Gergonne gives a part of a memoir by Sorlin (presented to the Académie des sciences on October 1817, the 27th) (see note (8)).

[J.D. Gergonne 3] "considérations philosophiques sur les élémens de la science de l'étendue", *Annales de Mathématiques Pures et Appliquées*, volume XVI, (1826) pp. 209–231.

[J.D. Gergonne 4] "Recherches sur quelques lois générales qui régissent les lignes et surfaces algébriques de tous les ordres;", *Annales de Mathématiques Pures et Appliquées*, volume XVII, (1827) pp. 214–228.

[J.D. Gergonne 5] "remarques sur le précédent article", *Annales de Mathématiques Pures et Appliquées*, volume XVII, (1827), pp. 272–6.

[L. Giard] "La 'dialectique rationnelle' de Gergonne", *Revue d'Histoire des Sciences* XXV (1972) pp. 97–124.

[J. Hachette 1] "solution analytique de la pyramide triangulaire, comprenant la trigonométrie sphérique", in the section entitled "analyse appliquée à la géométrie" of the *Correspondance sur l'Ecole Impériale Polytechnique*, n° 8, Mai 1807, pp. 273-288 + plate I; this text is reprinted with slight modifications in his book *Surfaces du second degré*, Paris, Klostermann, 1813.

[J. Hachette 2] "solution complète de la pyramide triangulaire", in the section of descriptive geometry in *Correspondance sur l'Ecole Polytechnique*, n° 3, Pluviose an XIII, pp.41-51 + plate III. Its introduction is retaken in the *Second Supplément de géométrie descriptive* by Hachette, Firmin Didot, Paris, 1818.

[N.G. Hairetdinova] "On Spherical Trigonometry in the Medieval Near East and in Europe", *Historia Mathematica*, 13 (1986), pp. 136–46.

[Italian Translation] *Trigonometria piana e sferica des sig. Legendre colle note alla geometria ed altre giunte* (Presso Guglielmo Piatti, Firenze, 1811) This work contains the translations of the treatise of trigonometry appended to [Legendre] and of [Lagrange].

[S.F. Lacroix 1] *Traité du Calcul Différentiel et Intégral*, vol. I, Paris 1797.

[S.F. Lacroix 2] *Traité élémentaire de trigonométrie rectiligne et sphérique et d'application de l'algèbre à l'analyse*, Paris, Duprat, An VII.

[J.L. Lagrange] "solutions de quelques problèmes relatifs aux triangles sphériques avec une analyse complète de ces triangles", *Journal de l'Ecole Polytechnique*, VI cahier, tome II, Thermidor An VII, pp. 270–296 (*Oeuvres Complètes*, publiées par les soins de J. A. Serret, t.VII, section quatrième, pp. 331–359.)

[P.S. Laplace] seventh of the "Leçons données en 1795 à l'Ecole Normale de l'An III", ("réimprimé sans grands changements" in the *Journal de l'Ecole Polytechnique* VII° et VIII° cahiers, volume II, Imprimerie Nationale, juin 1812, p. 92–3).

[D. Lardner] *An Analytical Treatise on Plane and Spherical Trigonometry and the Analysis of Angular Sections* (London, 1826).

[A. Lautman] *Essai sur l'unité des mathématiques et divers écrits*. Collection 10-18, Union générale d'éditions, 1977, p. 62.

[A.M. Legendre] *Eléments de géométrie* (third edition, Firmin Didot, Paris, 1800); from this edition on, a *Traité de trigonométrie* is appended to the text.

[G. Monge] quoted in [Hachette 2], p. 279.

[N. Nielsen 1] *Géomètres Français sous la Révolution*, Copenhagen, danish edition of 1927.

[N. Nielsen 2] *Géomètres français du dix-huitième siècle*, Copenhagen-Paris, 1935.

[M. Otero] "Mathematical Logic and its Philosophy in the nineteenth century: J.D. Gergonne", paper presented to the 15th International Conference on the History of Science, a summary of the paper was included in the proceedings.

[A. Schoenflies] "Géométrie projective" (French adaptation by A. Tresse) in *Encyclopédie des sciences mathématiques*, tome III, vol. 1, fasc. 1, French edition under the direction of J. Molk, tome III (fondements de la géométrie) of the German edition written under the direction of F. Meyer.

[D. J. Struik] "Gergonne" in the *Dictionary of Scientific Biography*.

[R. Taton] *L'oeuvre scientifique de Gaspard Monge*, Paris, 1951.

[R. Woodhouse] *A Treatise on Plane and Spherical Trigonometry*, second edition, corrected, altered and enlarged, Cambridge, 1813.

Alfred Clebsch (1833–1872)
Courtesy of Springer-Verlag Archives

Algebraic Geometry in the late Nineteenth Century

J. J. Gray

Introduction

The scale of the geometric enterprise in the second half of the nine-teenth century cannot be doubted. Klein and Darboux managed to fill the pages of their journals with geometry. The Italians were also enthu-siastic geometers. But a roll-call of those who wrote regularly and at length on geometry (although not necessarily algebraic geometry) would include not just those who, like Cremona or Darboux, wrote on little less, but also a number of the most versatile and profound mathematicians of their day: Cayley in England, Lie in Norway and, of course, Hilbert. And although it lies outside the subject of this paper, the work of Poincaré, Picard, Lefschetz, and the Italians (Castelnuovo, Enriques, Segre, Severi) is the source of a great deal of two of the most important areas of modern mathematics: algebraic geometry and algebraic topology.

It will be helpful to begin by alluding briefly to some of the histori-cal literature, leaving aside treatments of individual aspects of the subject. More or less each generation has its own history of algebraic geometry. The earliest one I know of is Loria's (1888). There is one by Coolidge (1940), and in 1974 our generation was brought up to date by Dieudonné. From each of these, and of course from the essays in the *Encyclopädie der Mathematischen Wissenschaften*, one can learn a great deal of mathematics, and get some sense of the salient themes and the original interconnec-tions, but it is worth considering how these works can be used. Certainly they are reliable maps. They also make it possible for any subsequent writer to refer the reader to them for many a discussion of the technical-ities. However, they also raise problems for the historian and the reader. The high degree of technicality in these expositions, which only confirms what everyone knows who has studied algebraic geometry at all, sug-gests that it is both forbidding and narrowly special in its appeal. Yet, since algebraic geometry is one of the main branches of mathematics, to consign it to the merely technical would be to do considerable violence to the intellectual landscape.

This effect is compounded by the way these expositions participate in what I might call the "standard mode" of historical writing. This mode is close to the style of mathematics papers themselves; impersonal, and sparing in its motivational details. I do not think we wish to write history in this style so much these days, and that is for good reasons. The most trivial is that we lose sense of the mathematicians themselves; real people with all their differences, their personalities and their lives become the bloodless originators of theorems. Perhaps that merely denies us a sense of pleasure, and historians ought to be made of sterner stuff. But the standard mode also precludes our understanding how these mathematicians came to the questions they did: why did they find these problems important; why did they think to tackle them in this way. In the standard mode the theorems are supposed to speak for themselves, they are taken to be of self-evident interest. Sometimes the interest they have for us is confused with the interest they had for their discoverers, but even when this is not the case the standard mode seldom runs to a discussion of just what was interesting about the results whose history it presents. It does not always ask with the requisite force what made certain questions the obvious ones to ask.

There are many reasons why this might be the case, and I do not want to speculate on which apply in which case. But the history of algebraic geometry is not, I suggest, best served by authors who so completely share the estimate of the subject's central importance that they do not bother to discuss it. To open up debates, which were held by mathematicians in any case, about the aesthetic aspects of mathematics (questions of taste and importance) is not to admit wholesale relativism. Rather, in this cost-conscious day and age, it may be simple prudence for mathematicians to speak on behalf of their subject and for us historians, too, to say why it matters (or mattered).

This paper is concerned with the geometrical theory of curves and (to a lesser extent) surfaces. Some of the most important topics in the history of algebraic geometry, such as the approaches of Kronecker and of Dedekind and Weber, are not going to be discussed at all, because there is simply not space. The main route we shall follow, running from Abel, via Riemann, Clebsch, Brill and Noether, to Cremona and Segre, is in fact the path taken by the Italian geometrical school. We shall also explore more briefly a second route that runs from Abel, via Picard, Poincaré, Castelnuovo, and Segre to Severi. In the course of this journey, I hope to shed light on the following questions:

1) What did the various protagonists take algebraic geometry to be?
2) Why did they set themselves the tasks they did?
3) Why did they tackle them in the ways they did?

ABEL, CLEBSCH, AND RIEMANN

Nineteenth century algebraic geometry may be defined as the study by algebraic means of curves and surfaces defined by polynomial equations. Curves had long been studied algebraically; one thinks at once of Descartes and Newton. But from 1750 to 1820 all was differential geometry. The defining feature that brings curves and surfaces into this story is the way they were treated: the questions asked about them and the methods by which they were answered. These are projective and algebraic.

Curves and surfaces were studied in this way partly for their intrinsic interest, of course, but partly because of their intimate connection to one of the most important theorems of the whole of the nineteenth century: Abel's theorem on the integration of algebraic functions. So we have here a very potent mixture: a subject of intrinsic charm and subtlety which is also of great power in the theory of integration and the new, obscure, transcendental functions. This aspect of the history, the nature of this interaction and the methodological reflections that it provoked, was very dear to many mathematicians. Corrado Segre, for example, gave an interesting lecture in 1891, couched in the form of advice to his students, in which he said:

> the geometry of *algebraic* manifolds coincides, as is shown by the very name and definition of these forms, with the analysis of algebraic functions and the related transcendental functions .... This great variety of bonds between analysis and geometry and the resulting necessity of studying them both together, and not confining oneself to one alone of these two directions, the analytic or the synthetic, has always been recognized by the great mathematicians. (1904, p. 447)

The algebraic side of the story starts with Plücker and Hesse in the 1830s and 40s. Plücker found a set of formulae which apply to curves with certain singular points, namely double points or cusps. These formulae connect the degree k of the dual curve (in line coordinates) with the degree n of the original curve (in point coordinates) having d double points and r cusps: $k = n(n - 1) - 2d - 3r$. The dual formula $n = k(k - 1) - 2t - 3w$ is obtained by utilizing the t bitangents and w inflectional tangents of the dual curve.

Hesse's contribution was to introduce homogeneous coordinates in a systematic way, so as to make what happens at infinity as easy to understand as what happens anywhere else, thus making the enumeration of the ingredients in Plücker's formulae much more straightfoward. A delightful result, due to Plücker, is that a plane quartic with no singular points will have 28 bitangents, and that they can all be real. A good example of Hesse's contribution is his realization that the inflection points of the curve with homogeneous equation $f(x, y, z) = 0$ are precisely the points where the curve $f = 0$ meets the curve given by the equation obtained by equating the Hessian determinant of second partial derivatives of f to zero:

$$\begin{vmatrix} f_{xx} & f_{xy} & f_{xz} \\ f_{yx} & f_{yy} & f_{yz} \\ f_{zx} & f_{zy} & f_{zz} \end{vmatrix} = 0.$$

Enumerative questions—how many bitangents can a curve of degree n have? how many inflection points?—were widely regarded as best handled in the terms of projective geometry. Moreover, as with the example of the Hessian, algebraic projective geometry seemed to bring with it simple techniques. An enumeration of the inflection points is reduced to an exercise in Bezout's theorem, and thence to the simple rule that a curve of degree n will have $n(n - 2)$ inflection points. For some, algebraic means were the preferred means to realize geometric goals. That thinking algebraically was the preferred means for Weierstrass and others at Berlin is well known. Elsewhere I have documented how this was also the case for the school of geometers that grew up around Clebsch and was continued by Brill and Noether. But it will help us to reconsider those developments briefly here. We can best do so by following Abel's theorem.

Abel's theorem is about integrating rational functions on curves, that is to say, evaluating expressions of this form

$$\int f(x, y) dx$$

where f is a rational function of two variables (that is, a quotient of two polynomials in x and y) where x and y are connected by an equation, say $\chi(x, y) = 0$. The evaluation is to be given as a function of the upper end-point v of the integral. Much depends on how complicated χ is. If one can eliminate y entirely from the integrand, say because χ is of the form $y = g(x)$, the question reduces to mere integration, and the expectation (amounting to certainty) was that every such integral is a sum of rational functions and logarithmic functions of the upper end-point. One simply decomposes the function f into partial fractions, and picks up rational functions except where the integrand gives rise to the reciprocal of linear terms. Nowadays we would explain this using Cauchy's theory of complex integrals.

If, however, χ is of the form $y^2 = g(x)$, then matters are much more complicated. Nothing new occurs if g is of degree 2. If g is of degree of 3 or 4, then the integral is said to be elliptic. It was the triumphant achievement of Abel and Jacobi to develop an elegant theory for handling such cases, the theory of elliptic functions. First, one classifies the integrands that can arise into three types, as was first done essentially by Legendre. The upshot is that the sum of any number of these integrals, having the same integrand but different end-points, may be written as a sum of a number of rational functions of the upper end-points (these arise from the partial fractions having no simple poles) plus a sum of a number of logarithmic functions of the same (these arise from the simple poles) minus one elliptic integral (which derives, we would now say, from the topology of the basic curve $\chi = 0$ and the one-dimensional space of integrands with no poles that it supports). The variable end-point of this elliptic integral depends algebraically on the original end-points. For these "good" integrands, the sum of any number of integrals can be reduced to a single integral; the rational and logarithmic parts reduce to a constant. For this, the simplest class of integrand, the theory explains

$$u = \int^v f(x, y) dx$$

by insisting that the variables be thought of as complex, and by dealing not with u as a function of the upper end-point v, but by studying the function which gives the upper end-point v as a function of u. This is the reason the theory is said to proceed by inverting the elliptic integral.

The restriction to simple curves like $y^2 = g(x)$, where g is at most a quartic, is clearly severe, yet it proved hard to get beyond. The vital breakthrough was made by Abel in his memoir of 1826 (Abel, 1826), a paper that Legendre called his "Monumentum aere perrennius." Although written in 1826, it was not published until 1841. Abel's article presented, and presents, a number of difficulties:

1) It is not well organized as a paper. For English readers, Rowe in 1881 did the service of combing it into shape, but as Brill and Noether observed in their (1894) he did not tackle the difficult parts.

2) It presents the usual problems of reading about complex integrals before there was any clear analysis of how they depend on their paths, but that does not pose too serious a problem.

3) It is occasionally wrong, which is more serious, and it operates at such a level of generality that it was and is not clear that matters are always as Abel described.

That said, everyone who writes about the paper observes how remarkable it is that Abel could prove so much in what is virtually only an exercise in the calculus of several variables.

It is striking to observe how Abel's results compare with the elliptic case. What Abel showed is that given a sum of any number of integrals of the form $\int R(x,y)dx$, having the same rational function as integrand and with x and y related by a polynomial curve $\chi(x,y) = 0$, then these integrals can be expressed in terms of p such integrals plus certain algebraic and logarithmic terms. Furthermore the number p depends only on $\chi(x,y) = 0$ (it is, in fact, bounded above by what we call today the *genus* of the curve). More precisely, if

$$\int^{(x_1,y_1)} R(x,y)dx + \cdots + \int^{(x_m,y_m)} R(x,y)dx$$

are m such integrals with arbitrary fixed lower limits, then this sum can be written as a sum of rational functions of the $x_1, y_1, \ldots, x_m, y_m$, logarithms of such rational functions, and p integrals of the form

$$-\int^{(s_1,t_1)} R(x,y)dx, \ldots, -\int^{(s_p,t_p)} R(x,y)dx,$$

where the s_i depend algebraically on $x_1, y_1, \ldots, x_m, y_m$, and each t_i is determined from $\chi(x,y)$ as a rational function of $x_1, y_1, \ldots, x_m, y_m$, and the corresponding s_i alone. The number p depends only on $\chi(x,y) = 0$ and is independent of m, $R(x,y)$ and the endpoints of the m given integrals.

Abel's paper is not very geometric. He dealt with equations and presented results in a computational form that is not very clear. Writing of the matter in general, Jacobi said that his own usual method of making discoveries through calculation was completely impossible in this case because of the vast complexity of the calculations, and he suggested that the whole of mathematics would have to be raised to a higher level in the wake of this research. (Quoted in Koenigsberger, 1904, p. 261). If Jacobi could not cut his way through, nobody could.

Let us observe some of the problems that can arise and which were the subject of much subsequent work. There is, first of all, the question of how to determine the end-points in the sums of p integrals. Secondly, if the curve $\chi(x,y) = 0$ has singular points, then there might be integrands whose poles are at the singular points; this can yield integrands which behave like the good kind (those that actually have no poles). Finally, one would like to know where the number p comes from.

Taking the last problem first, we find that what Abel did was to observe that $\chi = 0$ implies, on differentiating, that $\chi_x dx + \chi_y dy = 0$. The good integrands are those of the form $\dfrac{\phi(x,y)dx}{\chi_y}$, where ϕ is of suitably low degree, indeed degree $n-3$ if χ is of degree n. Counting free coefficients in ϕ yields the number p, and a little calculation shows that

$$p = \frac{1}{2}(n-1)(n-2).$$

At points where $\chi_y = 0$, one considers $\dfrac{-\phi(x,y)dy}{\chi_x}$ instead, and this is satisfactory unless the curve has singular points, because necessarily $\chi_x = 0 = \chi_y$ at such points. To get around the problem of singular points, one must insist that the function ϕ vanishes at such points (and indeed vanishes to a suitably high order, determined by the nature of the singular point). This lowers the value of p accordingly.

To understand how the endpoints can be determined, Clebsch rewrote the theory in geometric terms. Indeed, the whole aim of his 1863 paper "Ueber die Anwendung der Abelschen Funktionen in der Geometrie" (Clebsch, 1863) was to develop the theory of Abelian functions so that it could be used to solve problems about curves of higher genus in the same way that elliptic functions could be used in the theory of cubic curves, as Clebsch had already shown. In Clebsch's geometric language, the conclusion of Abel's theorem is that the sum of μ integrals taken between the intersections of the curve $\chi = 0$ and a variable one $\theta = 0$ is a constant. Here θ is assumed to depend linearly on certain parameters, α in number, and the lower end-points and the upper end-points of the integrals are picked out by two of these (sets of) values. For definiteness, let us take two special cases, the elliptic one and the first of the so-called hyperelliptic ones, where χ is of the form $y^2 = g(y)$, and g is of degree 6. Abel's theorem in the elliptic case invites us to express the sum of two integrals having the same good integrand as a third such integral. We are given the first two endpoints and have to find the third. This is the addition theorem for elliptic integrals, and the answer is: join the first two points by a line and consider the third point $P = (x,y)$ where it meets χ, whose equation we are taking in the canonical form $y^2 = g(x)$, where g is of degree 3. Then the point $P' = (x,-y)$ is the one that is being sought. Plainly this determines P' algebraically as a function of the given points. Observe the role of the family of parallel lines that meet the curve again at its sole point at infinity: they form a one-parameter family of lines meeting the curve in pairs of (finite) points.

In the hyperelliptic case, let us suppose that we start with a sum of three integrals; we want $p = 2$ more so that a sum of 5 is constant. So we want three points on the curve $\chi = 0$ to determine two more. Five suggests using conics, but a conic meets a curve of degree 6 in twelve points. The solution is to look at the three parameter family of conics touching the curve $\chi = 0$ at its point in infinity. Such is the singularity there that these conics have seven-fold contact with the curve $\chi = 0$ at infinity, and so provide the family we seek.

In each case, the family of curves depended linearly on certain parameters, and varying these parameters changed some of the points of intersection with the curve but not others. It is the variable ones that are required for Abel's theorem, and which figure prominently in his presentation of it. One of Clebsch's contributions was to give the kind of geometrical analysis just sketched, and to extend it to deal with curves having certain simple kinds of singular points. The crucial step was to think of the curve $\theta(x, y) = 0$ as belonging to a family of curves $\sum a_i \theta_i(x, y) = 0$. Thinking of the variable end points as being determined by (and determining) the parameters in this way proved to be very perspicuous.

In his paper of 1863, Clebsch was also the first to state and prove the converse of Abel's theorem. He showed that if a sum of mn integrals is zero then their end points arise as the intersections of the ground curve $\chi = 0$ with a variable curve. This shows once more Clebsch's geometric view of the matter. To obtain a function with zeros at one set of these mn points and poles at the other, it is enough to take $\theta_0 = 0$ and $\theta_1 = 0$ as the curves picking out the two sets of points, and define the function $x \mapsto \dfrac{1 - \lambda}{\lambda}$ if x lies on the curve $\lambda \theta_0 + (1 - \lambda)\theta_1 = 0$. Another aim Clebsch had in mind was to elucidate Riemann's ideas about Abelian functions as put forth in the famous paper "Theorie der Abelschen Functionen" (Riemann, 1857). Clebsch believed that it was the great difficulties one encountered in studying this paper which had hitherto prevented young mathematicians from using it, and he hoped to alleviate matters by introducing an algebraic theorem of Jacobi's on functions of several variables. Clebsch's choice of methods was far more algebraic than Riemann's, and this has a significance that is worth exploring further.

Riemann took the view that any two-dimensional patch could be the domain of functions of a complex variable. His use of this naive idea was subtle, and invoked his version of Dirichlet's principle, to which Clebsch and his followers took exception. Riemann took an algebraic function $F(s, z) = 0$, of degree n in s and m in z, and having at worst double points as singularities. He thought of the function as defining a surface T spread out over the z-plane. He then cut the surface until he obtained a simply-connected piece, by means of an even number of cuts, say $2p$.

He then said that any single-valued function on T was determined by the real parts of its jumps across the cuts—this is the form in which he used Dirichlet's principle. So there is a p-complex-dimensional space of everywhere finite integrals (shown later in the paper to be identical to those exibited by Abel). This is an index theorem that relates a topological quantity to an analytical one. Clebsch and his followers also struck this part of the theory out; Clebsch preferred to define p as the number of independent one-forms.

Riemann was now able to produce his estimate of the number of analytical functions with a specified number of simple poles a given surface can support, the so-called *Riemann inequality*. A single-valued function on the surface T having d simple poles depends, by what has been said, on $d + p + 1$ constants which must obey $2p$ constraints (the vanishing of their jumps). The number of free constants r satisfies $r \geq d - p + 1$, r being the dimension of the space of functions. Thus the surface T will admit non-constant such functions with d simple poles if $d - p + 1 \geq 2$, i.e. $d \geq p + 1$. The inequality arises because it might happen for certain special sets of points that some of the jumps are already zero. Riemann's student Roch investigated when this can happen and was able to define a quantity i which turned the inequality into an equality, the celebrated Riemann-Roch theorem: $r - i = d - p + 1$ [see (Roch, 1864)].

Riemann applied his ideas to the study of curves in this way. Any function defined on the surface T is also an n-valued function of z, so the connection between n, p, and the total branch point order w is $w - 2n = 2p - 2$. Since any such function leads to an equation for the curve, all equations for the curve arise in this way, hence one obtains all the possible changes in the equation of the curve by means of rational changes of variable. Riemann also showed that the dimension of the space of inequivalent curves of genus p (where p is greater than 1) is $3p - 3$.

The paper is difficult to disentangle because several ideas are present in embryonic form. To some extent, Riemann was trying to think of rational functions on T intrinsically. They are created by the use of the Dirichlet Principle, not inherited from the ambient space. Witness also the emphasis on properties of T that are independent of its equation, and so of its embedding, such as its genus. On the other hand the starting point is still an equation for the curve.

Since it is embedded, the so-called rational functions on it, $(x, y) \mapsto \phi(x, y)$, are restrictions of functions defined on the whole of the ambient space. But until the difference between a curve (or Riemann surface) *in abstracto* and an embedded curve given by an (algebraic) equation in x and y is appreciated, the question of the existence of functions defined on a curve is inseparable from that of functions defined on the ambient space. The zeros of such a function are the intersections of the original curve with the curve $\phi(x, y) = 0$, and the poles are the locus of $\phi(x, y) = \infty$. After Riemann, one could speak of the existence of functions on a curve having prescribed poles; beforehand, like Clebsch, one spoke of curves meeting the given one. (To modern eyes, the equivalence of these viewpoints is the result of a very deep theorem and one peculiar to the single complex variable situation). That the function-theoretic viewpoint is worth pursuing was first argued for by Riemann. After his work, one began to see Abel's theorem interpreted as a condition for a curve to admit functions with specified kinds of poles.

The mathematical situation is two-fold. On the one hand it has to do with analysis. Abel, Riemann, and Roch dealt with integrands on an algebraic curve, and had discussed the number of integrands with specified properties. But in the hands of Clebsch, and Brill and Noether, what was singled out for particular attention was the geometric interpretation of an idea that Abel had had, that of sweeping out sets of points on a curve by looking at the intersection of the curve with the members of a linear family of curves.

But whether one thought analytically or geometrically, there was a disparity between the generality of Abel's original ideas and the more limited range of cases treated in detail by Riemann and Clebsch. Curves having complicated singularities were not amenable to Plücker's formulae, and were not considered by Riemann, yet not only were they there, the study of their intersections with moving families of curves was basic to Clebsch's whole approach. So the natural question that was raised was: is there a good theory of curves of arbitrary degree? In particular, is there a theory of their singular points sufficiently precise to ensure that Plücker's formulae hold in general, and to permit one to apply Abel's theorem? The answer to both questions was "No."

CREMONA TRANSFORMATIONS

The way forward following the work of Riemann and Clebsch was based on a method of reducing complicated singularities to simple ones. This necessarily took mathematicians out of the realm of projective geometry and projective transformations of curves, because such transformations cannot change the singular points of a curve. What was wanted is a transformation that is mildly but controllably destructive. Such transformations are the Cremona or birational transformations (to be defined below). This raises a conceptual point of some significance. Mumford in (Zariski, 1971, p. vii) has observed that the Italian algebraic geometers were to produce a completely birational theory of surfaces, and that such a theory seems more complicated, indeed unduly so, by comparison with the modern one. The reason they chose such a theory is because, ultimately, such an approach had worked very well for curves.

The theory of Cremona transformations was employed by a number of writers to resolve the singularities of a curve. The culmination of these efforts was the treatment given by Brill and Noether in "Ueber die algebraischen Functionen und ihre Anwendung in der Geometrie" (Brill & Noether, 1874), which showed how any curve could be birationally transformed into one having only multiple points with separated tangent directions. In such a way the Plücker formulae were secured for curves of arbitrary degree, and the algebraic definition of genus was made rigorous. I should add that the validity of their simplification of singular points was often contested and reworked by later mathematicians; the history of this topic formed the subject of a delightful Presidential address given by G. A. Bliss to the AMS in 1923. Here, let me instead make some remarks about the more geometric lines of enquiry.

The theory of curves was further expedited by the use of spaces of higher dimension. Given a curve of degree n with singular points P_i of orders α_i, Brill and Noether had considered those curves which passed through the P_i at least $\alpha_i - 1$ times. When the degree of the curves in the family was fixed, counting constants gave them families of finite dimension. So they would consider all the curves of degree, say, $n - 3$ satisfying a particular condition. The original point of view taken by Veronese was that this family could be taken to yield a novel but entirely geometrical object in a suitable space.

Consider, as Veronese did in (Veronese, 1882), the family of all quadratic curves (or conic sections). With respect to a set of axes chosen once and for all, any such family has an equation in homogeneous coordinates

$$ax^2 + bxy + cy^2 + dxz + eyz + fz^2 = 0,$$

so is specified by a set of six numbers, (a, b, c, d, e, f), or, more precisely, by their five ratios. Since not all of the six numbers can be zero, every conic section can be thought of as a point in five dimensional projective space. Now, pick a point (x, y, z) in the projective plane and consider the conics that pass through it. These are the conics for which $[a, b, c, d, e, f] \cdot [x^2, xy, y^2, xz, yz, z^2] = 0$. This as a condition on the unknowns a, b, c, d, e, f imposed by definite values of x, y, and z, so it defines a hyperplane in $\mathbf{P}^5$ and so, by duality, a point in the dual $\mathbf{P}^5$. In the usual algebraic way of handling duality, going back to Möbius, the point so obtained has coordinates $[x^2, xy, y^2, xz, yz, z^2]$ in $\mathbf{P}^5$. In this way $\mathbf{P}^2$ is mapped into $\mathbf{P}^5$, and the resulting surface, which is a quartic, is called Veronese's surface, or the Veronese embedding of $\mathbf{P}^2$ in $\mathbf{P}^5$. Any curve in $\mathbf{P}^2$ is of course carried along and embedded in this way in $\mathbf{P}^5$. The Veronese map is thus an example of a Cremona transformation of the projective plane into a surface in $\mathbf{P}^5$.

Quadratic Cremona transformations of the projective plane to itself also arise in this way. One takes the family of conics passing through three (non-collinear) points, which can be taken to be $[1, 0, 0]$, $[0, 1, 0]$, and $[0, 0, 1]$. The equation of any such is therefore of the form $ayz + bzx + cxy = 0$, so the point (x, y, z) can be mapped to this point of the dual $\mathbf{P}^2 - [yz, zx, xy]$. Thinking of this map as a map from $\mathbf{P}^2$ to $\mathbf{P}^2$, it sends points to points, except for those on the lines $x = 0$, $y = 0$, or $z = 0$. These are sent to the points $[1, 0, 0]$, $[0, 1, 0]$, and $[0, 0, 1]$ respectively, except for those very points, which are "blown up" (to use the modern term) to the lines $x = 0$, $y = 0$, and $z = 0$. So a curve through, say, the point $[1, 0, 0]$ is transformed to a curve through the line $x = 0$, and where the transform meets the line turns out to be determined by the direction the original curve was going when it passed through the point $[1, 0, 0]$.

What can be done for a family of quadratic curves can be done for other families. Let us take a curve of degree n with singular points P_i of order α_i. The family of adjoint curves, those of degree $n - 3$ which pass $\alpha_i - 1$ times through specified points P_i, has dimension say d, and in this way yields a map of $\mathbf{P}^2$ into a surface of some degree in $\mathbf{P}^d$. The condition that all the curves of the family pass through all the points P_i naturally means that their images are atypical. In fact the surface is blown up there, in the sense that these points are transformed into curves of some degree. What is more, while the original curve was singular at those points, its transform meets the images of these curves in points that are determined by the directions of its branches there, so its singularities are at least partially resolved. Thus the idea of the linear family of adjoint curves turns out to have a comprehensible geometrical meaning as an embedding of the original curve as a less singular curve in some higher dimensional projective space.

No one who passes through this branch of mathematics can resist this illustration. If one takes the family of cubics through six points not lying on a conic, the result is a surface of degree 3 lying in $\mathbf{P}^3$ which is the most general cubic surface. The six points are blown up, as the jargon has it, into lines, the 15 lines in $\mathbf{P}^2$ that join the 15 pairs of these points map into lines, as do the 6 conics through the six sets of five of those points: there is the cubic surface with its 27 lines. The lines had originally been discovered in an entirely different way by Cayley in 1849, they were subsequently enumerated by Salmon [see for example (Cayley, 1849) p. 132 and (Salmon, 1882), p. 496]. In the 1850s and 1860s the cubic surface was successfully studied from many different points of view. Indeed, the theory of algebraic surfaces was then chiefly confined to the treatment of cubic surfaces and some interesting quartic surfaces, a topic that was pursued vigorously at the time.

Cremona was not the first to study Cremona transformations, but he was the first to study them systematically, so for once the name is not inappropriate. A Cremona transformation of the projective plane is a slippery thing. It is not quite a map from $\mathbf{P}^2$ to $\mathbf{P}^2$, rather it is a map from almost all of $\mathbf{P}^2$ to almost all of $\mathbf{P}^2$. To define such a transformation, Cremona took 3 curves of the same degree, say n, whose equations are $F_i(x, y, z) = 0$, $i = 1, 2, 3$, and mapped the point $[x, y, z]$ to $[F_0(x, y, z), F_1(x, y, z), F_2(x, y, z)]$. In modern terms, this is a map at the point $[x, y, z]$ provided it is not the case that all the F_i vanish there, i.e. unless the point $[x, y, z]$ lies on all three curves; such a point is called a base point. As we saw in the examples above, the fate of the base points is also of interest. Let us call the map T. For general points P, the number of points in $T^{-1}(T(P))$ is of some interest: for an invertible map one requires that this number be 1, in other words that the curves of the form

$$\lambda_0 F_0(x, y, z) + \lambda_1 F_1(x, y, z) + \lambda_2 F_2(x, y, z) = 0$$

which pass through the point $P = [x, y, z]$ have only the base points in common. So, apart from the base points, two curves of the form $\lambda_0 F_0(x, y, z) + \lambda_1 F_1(x, y, z) + \lambda_2 F_2(x, y, z) = 0$ have only one point in common. Such a family is called a *homoloidal net*, and the corresponding map T is called a *Cremona transformation*. The curves of the net are mapped to lines in the plane. Cremona's contribution was to exhibit a large class of the plane transformations and to study how they depend on the curves $F = 0$.

Cremona transformations of higher dimensional projective spaces also exist, but they no longer share with the plane transformations the property of being generated by those of order two (the so-called quadratic transformations). For plane transformations this is a result of Rosanes, Clifford, Noether, and later but more rigorously Castelnuovo. The novel behaviour of higher dimensional Cremona transformations is connected with the greater variety of singularities a surface can have, as Noether and Cremona remarked.

Cremona himself studied how the transformations one obtains in this way depend on the family of curves F that one selects. He did not apply his transformations to simplify singularities of curves or surfaces until a simplest form had been obtained for study; that was the achievement of others, most notably Bertini.

THE ITALIAN SCHOOL

Following the work of Brill and Noether and the advent of the theory of Cremona transformations, the story passes to Italy and the powerful school of geometers that was to dominate the subject for the next 70 years. It will help us to understand why they gave their work its profoundly geometric cast, and what indeed they took geometry to be about, if we look briefly at some of the people involved, and their interconnections.

More than anyone else, the creation of the Italian school of projective and algebraic geometery is due to Cremona. Born in 1830, his first work had been undertaken in the analytical style of Chasles, but in the 1860s he moved to the more synthetic approach of von Staudt. However, in the opinion of his student and later obituarist, Bertini, Cremona's work always rested on some essentially algebraic ideas. As Cremona once wrote to Clebsch, with whom he maintained close relations until the latter's untimely death in 1872: "I am fully convinced of the mutual help which analysis and synthesis afford one another in geometry" [quoted in (Bertini, 1903), p. xiv]. Cremoma's first important papers were on the theory of cubic curves in space and on cubic surfaces; for one of these he was awarded a Steiner prize jointly with Sturm. Also in the 1860s he put forth his theory of what have come to be called Cremona transformations, regarded by Bertini as holding "the highest rank by reason of their importance in the progress of modern geometry" (Bertini, 1903, p. x).

Cremona's obituarists—Bertini, Veronese, and D'Ovidio—regarded him as a forceful personality and an exceptional teacher, who not only lectured with clarity but kept his students informed of the latest developments (for example, in 1868-69 the theory of Abelian functions as presented by Clebsch and Gordan). He was responsible for making projective geometry a first-year subject at Italian universities, and wrote a text-book on the subject which was translated into French, German, and English. He also taught courses in descriptive geometry and graphical statics, and he pioneered the teaching of projective geometry in the technical institutes. His influence also extended to the secondary schools, where he advocated the use of Euclid as a text-book. While his influence in these respects was doubtless helped by being located in Rome, after 1873, the political infighting there and his declining health quickly brought an end to his career as a creative mathematician.

Bertini was a student in Cremona's course on Abelian functions. He followed Cremona to Milan and wrote his first publication on this subject, giving a simplified proof of the invariance of genus of a curve under birational transformations. He took a job as a secondary school teacher in Rome in 1872, and when Cremona arrived the next year he hoped that Bertini would provide the nucleus of a group of disciples. But Betti obtained for him the chair in geometry at Pisa, and Bertini "defied the wrath of Luigi Cremona, who did not want to be separated from his disciple, and who only forgave this act of defiance two years later when Bertini, ever generous, offered to resign the chair in favour of his master." (Castelnuovo, 1933, p. 745). Cremona was particularly disturbed at the time by problems connected with his position in Rome which were keeping him from research.

Bertini is best remembered for two results, one classifying all the involutions of the projective plane and the other showing that a linear series yields an embedding that is smooth away from the base points. It seems that Cremona's forgiveness was not deep enough for him to respond as perhaps he should have done to this work, for Castelnuovo also tells us that Cremona confined his comments on the first of these results to observing that Bertini had missed a special case. It may have rankled with the older man, that it was Bertini and not he, who first used birational transformations to reduce complicated figures to simple ones and thus prove interesting things about them. Cremona was increasingly denied the opportunity to conduct his own research by the pressure of business.

As we have seen, it is entirely natural to study Cremona transformations defined by an r-parameter family of curves. The first to exploit the interpretation of Cremona transformations as mappings of a curve or surface into an r-dimensional projective space was Veronese. Veronese had received his initial university instruction abroad, under Fiedler at the Polytechnic in Zurich. Fiedler had him read Schröter's edition of Steiner's work, just published that year (1876), and the outcome was Veronese's remarkably thorough study (Veronese, 1876/77) of the 60 Pascal lines that arise from 6 points on a conic. Only after publishing this work did Veronese write to Cremona with the hope that he might study with him in Rome. He was admitted to the fourth year, and Cremona was suitably impressed by him, seeing to it in due course that Veronese was appointed assistant to the professor of projective and descriptive geometry at Rome (one Salvatore-Dino). In 1880-81 Veronese was in Leipzig, where he studied under Klein, who made a considerable impression on him. One consequence of this meeting was that Veronese sent what has become his best-known paper to Klein's journal, *Die Mathematische Annalen*. In this paper, (Veronese, 1882), referred to earlier, Veronese showed how to embed surfaces in spaces of higher dimension. He was certainly not the first to advocate using higher dimensional projective spaces, but the way he constructed embeddings from linear series was dramatically successful because it enriched geometrical methods. In particular, he was able to show that several familiar surfaces with complicated singularities were the images under a projection onto P^3 of a non-singular surface in some P^n.

The last Italian geometer to be described here, although he was by no means the last in this line, is Corrado Segre. It is not possible for me to survey his work here, for he must be regarded as one of the finest of the Italian geometers. As with Veronese, his choice of topics for research in later life failed to meet with the interest of others, and this may have something to do with the apparent neglect of his work today. Yet, after Cremona, he was the decisive influence on Italian geometry. This was due not only to his great originality, but also because of his tireless efforts on behalf of the work of others. The range of his erudition, well-displayed in his article "Mehrdimensionale Räume" (Segre, 1918), is quite staggering, and he put this immense learning to thoughtful use. Also interesting is the lecture he gave in 1891, from which I have already quoted. It is a testimony to the status he enjoyed in his day that this lecture was translated into English in 1904 and published in the *Bulletin of the American Mathematical Society*.

ITALIAN CONTRIBUTIONS TO ALGEBRAIC CURVE THEORY

For the Italian school, which represented the main line approach to algebraic geometry, the principal ideas in curve theory were encapsulated in the theorems of Abel and Riemann- Roch. To quote Segre again:

> The idea of this geometry, i.e. the properties of an algebraic curve which remain invariant under a rational transformation of the curve into itself, is first found in a work of analysis, in the great memoir of Riemann on the theory of abelian functions. It is here we first find fully developed the notion of *genus* ... [The Riemann-Roch theorem] gave to modern geometry one of its most important and fruitful theorems ... And all through the geometry of algebraic curves the analytic researches on algebraic functions and their integrals have had applications of the greatest importance. It is sufficient to recall in this connection Abel's theorem ... (Segre, [1904], p. 449-450).

Among the ingredients of the theory of curves were:
1) The curve itself, here called the ground curve, given by an equation;
2) Birational transformations, which change the equation but do not affect the basic properties of the ground curve cut out by linear families of curves passing through it;

3) The way these linear families give rise to embeddings of the curve, and the use in particular of Cremona transformations to reduce a curve to one having manageable singularities;

4) The three different kinds of integrals of rational functions that can arise on a curve;

5) The connection between everywhere finite integrals on the ground curve and its adjoint curves.

The theory itself had various aspects. The analytical approach, largely taken up by Weierstrass, and the so-called arithmetic approach, due to Dedekind and Weber, are not discussed in this essay. One can further differentiate the algebraic, almost formal theory of Clebsch, Brill and Noether, from the more conceptual, if murkier approach of Riemann. The leading figures who took up the ideas of Brill and Noether and did the most to advance them and give them geometric dress were undoubtedly the Italian mathematicians.

It will help us to see more clearly what the Italians meant by geometry if we look briefly at two monographs by Bertini and Segre. Speaking of these papers, Castelnuovo said they "were the start of all subsequent research by our school of geometry" [(Cremona, 1933), p. 748]. The one by Bertini [1894] was widely regarded, as it was intended to be, as the definitive presentation of the approach taken by Brill and Noether; it follows what has often been referred to as the algebraic approach. Indeed, Bertini called it "The geometry of linear series on a plane curve according to the algebraic method." He recorded his view that "one of the great merits of Clebsch was to put geometric clothes on Riemann's ideas," but he pointed out that the first mathematicians to give purely algebraic proofs by a uniform method of the principal parts of the theory of algebraic curves were Brill and Noether. One advantage of their approach, Bertini pointed out, was that it could be usefully extended to deal with spaces of higher dimension.

Bertini considered the intersection of a fixed (ground) curve $\chi = 0$ with a curve $\psi = 0$ that depends linearly on a number of parameters. As these parameters vary the curve changes and sweeps out different members of the linear series of points on the ground curve $\chi = 0$ determined by the family of curves $\psi = 0$. It is usual just to consider the variable points of intersection. Bertini, following Brill and Noether, focused his attention on those linear series that came from curves satisfying the adjoint conditions at the singular points of the ground curve $\chi = 0$; he also picked out linear series that were complete (the point groups cannot be determined by a family of curves depending on a larger number of parameters). The canonical series was defined to be the series cut out by the complete family of adjoint curves of order $m - 3$ (where the order of the ground curve $\chi = 0$ is m); it turns out to be made up of sets of $2p - 2$ points and to depend on $p - 1$ parameters (counted projectively). A series is special if it is contained inside an adjoint system. Special linear series give rise to interesting sets of points on the curve, from both the geometric and the function-theoretic point of view.

The Riemann-Roch theorem was now stated and proved by Bertini in this form:

> If there is a special complete series depending on r parameters and consisting of groups of d points, then through any group of its points there passes a family of adjoint curves of order $m - 3$ which depends on i parameters, where $i = p - 1 - d + r$, or $r - i = d - p + 1$.

Many other theorems followed about special linear series and particular kinds of curves, but I only wish to emphasize a few main points:

1) The roots of all this in Abel's theorem are clear;
2) The Riemann-Roch theorem is presented as a theorem about the dimension of families of curves meeting the given one in appropriate ways, much as Clebsch had first re-interpreted it;
3) The subject matter is plane curves, treated geometrically.

Let us compare this with the way that Segre went over the same material in his simultaneous monograph (Segre, 1894). Segre began by observing that the study of rational functions (which he equated with the study of linear series of points) had been created by Riemann and since developed by others in various directions. He regarded the Brill-Noether direction as algebraic geometry, and set himself the task of being simply geometric, not merely for its own sake but as part of an attempt, already being carried out by Castelnuovo, to study algebraic surfaces. To this end, he worked as much as possible with varieties of arbitrary dimension. He also carefully defined what he meant by a "simply infinite" or one-dimensional algebraic object or curve, and their allowable transformations—quite explicitly following Klein's "Erlanger Programm" here. A geometric property was one invariant under birational transformations, which might, of course, take one out of the plane and into a higher dimensional space. So linear series were, not quite as with Bertini, supposed to be cut out on the curve by a family of varieties (of codimension 1, as we would say). A linear series on a curve gives an embedding of the ambient space into another, and the birational geometry of the curve and the series is the same, he showed, as the projective geometry of the new ambient space.

The crucial thing about Segre's presentation of this material was his consistent way of thinking of linear series on a curve as embeddings, and so the corresponding series of point groups as cut out on the curve (embedded in some high-dimensional space) by families of hyperplanes. In this spirit he could formulate and prove both the Brill-Noether *Restsatz* and the Riemann-Roch theorem. This latter appears in Segre's paper in the following form:

> If a group of n points determines a complete series of dimension r and lies in an i-dimensional family of canonical groups, then $i = p - n + r - 1$ (i.e. $r - i = n - p + 1$).

The genus of a curve was defined by Segre as follows: if the curve admits a g_n^1 having k double points, which means an n-fold map from the curve to the Riemann sphere, then the number $p = w/2 - n + 1$ is birationally invariant (recall Riemann's formula, $2p - 2 = w - 2n$). Plane curves with higher singularities were dealt with by means of Cremona transformations, as Noether had explained.

ALGEBRAIC SURFACE THEORY

The historian who sets out to write about the theory of algebraic surfaces is confronted with an exciting, but challenging task, somewhat similar to the one faced by the original participants. For this is a largely uncharted domain within the historical literature just as it was for the geometers at the end of the 19th century. The theory of curves existed to give them some clues, while the study of surfaces of low degree had just begun. There was no question here of proving rigorously what was known imprecisely before; the question here was: "What is to be proved?" Under such circumstances, the way forward was determined by people's judgements of two factors: what should be done, and what can be done.

We can now begin to see how the theory of curves guided the Italian school in their attempts to create a theory of surfaces. They were not the first to try this, of course, and they were able to build on the work of others, notably Noether and Clebsch. Because of its importance in the birational scheme of things, the first concept Clebsch had sought to generalize to surfaces was that of genus. For curves, the genus of a non-singular curve of degree n is $\frac{1}{2}(n - 1)(n - 2)$, which is the number of adjoint curves it has of degree $n - 3$ (i.e. the number of coefficients in a polynomial in two variables of degree $n - 3$). For singular curves, one subtracts an amount determined by the singularities. So Clebsch had proposed to consider linear families of surfaces cutting a given one. The degree of those that correspond to the zeros of everywhere finite (double) integrands is, straight-fowardly, $n - 4$. Noether then showed, by an explicit calculation, that the integrands obtained from such surfaces are transformed into others of the same kind by birational transformations.

So their number defines something which Clebsch proposed be taken as the genus of these surfaces.

But even at the very start the study of surfaces soon revealed how much harder it was to be than the study of curves. For, just as the curves introduced by Abel pass through the singular points of the ground curve, so must these surfaces do the right thing by the singularities of the original surface. What are these singularities; and what are the simplest that can be admitted which allow the new theory any scope without making it too complicated to get started? Clebsch had dealt with a simple, but arguably typical case, and generalized the formula relating degree, number of singular points, and genus for curves to give a formula for the genus of the original surface. The question that arose was whether the number incorporated in this formula agreed with the genus of the surface as defined above. The genus, as given by the formula, is birationally invariant, as Noether and Zeuthen showed, but as Cayley pointed out this quantity can be negative. So it cannot always be the same as the genus defined by Clebsch. Thus two different genera for a surface were established and gradually recognized for their importance: the geometric genus p_g defined by Clebsch, and the arithmetic genus p_a given by Zeuthen's formula. It turns out that $p_g \geq p_a$, and the non-negative quantity $p = p_g - p_a$, sometimes called the irregularity of the surface, became of considerable interest, as we shall see.

Since the details of the theory of algebraic surfaces are necessarily complicated, I shall not go into them here. This is exactly the occasion when one can refer to the literature, specifically to (Baker, 1912) and Dieudonné's book and paper, from which the above material has been taken. But it will be worthwhile recording the moments when the theory of curves spawned specific aspects of the theory of surfaces. It was only in 1886 that Noether was able to publish a short note stating a Riemann-Roch theorem for surfaces. Such a result was also one of the chief goals of the Italians later on. The papers by Bertini and Segre made extensive use of the idea of linear systems of points cut out on one curve by the members of a family of curves. The generalization was to a linear family of surfaces, and to linear systems, and to linear systems of curves on a surface, adjoint systems, and so forth. The ones who took that step most decisively were Castelnuovo and Enriques in the 1890s.

As we have seen, the source for all this work is Abel's paper on integrals on an algebraic curve. The corresponding question for surfaces would seem to be about double integrals, and it was first raised in this setting by Clebsch. In addition to double integrals, a surface admits single integrals, or one-forms as we would call them. The first to study these was Picard. I find the comments made by the English mathematician H. F. Baker in this connection quite striking:

> To many, Picard's problem must have seemed an artificial and profitless one, a mere slavish imitation of Riemann's work in a field where analogous success was not to be looked for. It was, moreover, a problem of enormous labour, extending over many years, and sustained more by faith than by any striking definite results, as at least it would appear to some. Today, however, the point of view is different. For it now appears that the results obtained by the Italian school, and the results obtained by Picard are complementary to one another in a most wonderful way. (Baker, 1912), p. 6).

Indeed, Baker felt that the "union of these two diverse theories" made it possible to classify all surfaces and to give a general theoretical account of them. He went on to tell of this convergence in some detail, as did Dieudonné later, and I shall follow them.

Picard found that the integrals on a surface come in three types:
1) those which are everywhere finite;
2) those which are algebraically infinite along certain curves;
3) those which are logarithmically infinite along certain curves.

This makes sense, because the intersection of a surface with a plane is a curve, and the singularities of the integral will show up on the curve necessarily as one of these three types. The first question is, are there any integrals of the first type? Picard noticed that on a non-singular surface in $\mathbf{P}^3$ there are none. But a result of Humbert's in 1894 showed that there should be surfaces with integrals of the first type.

What of integrals of the second kind? In 1904 Severi, the next in the series of leading Italian geometers, showed how to simplify integrals of this type until they only had one curve along which they were algebraically infinite. Single curves on a surface fit into the theory of systems of curves developed by the Italians. Suppose then that the number of integrals of the first kind is q, and that of either the first or second kind is r. Using a theorem of Castelnuovo's, Severi could deduce that any surface which admitted an integral of the second kind necessarily had a non-zero value of p, and that $r - q \leq p$. The next year Severi and Castelnuovo put all this together with a theorem of Enriques to deduce that $q \geq p$, whence $q = p$ and $r = 2p$.

Dieudonné points out (Dieudonné, 1974, p. 103) that Enriques' theorem was not rigorously established. But Baker might have mentioned a different proof of all these latter results given by Poincaré in 1910. According to Dieudonné, Poincaré's proof makes a very original use of the theory of Abelian integrals. Baker does, however, repeat Severi's observation that the Castelnuovo-Severi result of 1905 amounts to a generalization of the converse of Abel's theorem.

CONCLUDING REMARKS

This essay has indicated some of the methods by which geometrical questions were pursued and indeed how questions in what might seem to be analysis were interpreted in geometrical ways. The route taken led through the theory of curves, and discoveries there were held up as a guidepost for what might be done for surfaces and functions of two variables. The importance of the interaction between geometry and analysis was something that all the protagonists stressed, not that they would have disclaimed the relevance of purely synthetic geometry. But they were all believers in the unity of mathematics; Segre even invoked this shifting unity as one of the yardsticks by which the importance of a topic could be measured. Still it is right to remember the extensive area of synthetic geometry that I have not touched upon at all; just as it is to remember the vast amount of algebraic and projective geometry that was done without regard for any applications elsewhere in mathematics. The topic is enormous, and conclusions drawn from just one look at it must naturally be tentative.

That said, I asked three questions at the beginning, and I wish to answer them now. First, what was geometry taken to be? The answer increasingly was birational geometry, with a developing emphasis on the properties a curve (or later a surface) had in whatever space it was embedded. Segre liked to emphasize the way this point of view fit in with Klein's *Erlanger Programm*, as indeed it did.

Second, why did they set themselves the tasks they did? While Riemann pursued analytic or function-theoretic ends, Clebsch began with an attempt to understand Abel's theorem and to apply Abelian functions to the study of algebraic curves. But the study of plane curves forced a consideration of their singularities, and in an attempt to make Clebsch's work rigorous and completely general Brill and Noether were led to use Cremona transformations and to introduce the idea of a linear series. In an attempt to simplify and then extend this work, Italian geometers made it more geometric and less algebraic.

Third, why did they tackle their problems in the way that they did? Certainly the methods they chose were not forced upon them by the problems themselves; the theory of fields of rational functions goes quite a long way toward solving the same problems. But that said, there are always those who simply prefer thinking about curves rather than equations in two variables; they find it suggestive of ways to proceed, questions to ask and places to look for answers. They often find it a satisfactory terminus, too. For such people the ability to see a birational transformation at work makes possible a whole methodology that would not come so readily to hand with the techniques of pure analysis (Puiseux expansions and so forth). The method of passing from one ambient space to another, and so removing the singularities of a curve (and even a surface) gave their work a generality and a simplicity it otherwise would have lacked. Finally there is that deceptively simple question of "styles" and "schools." Clebsch saw himself quite self-consciously in the tradition of German projective geometry, and it was through this tradition that he inspired Brill and Noether. Cremona was another who saw himself as a geometer, and through him and later Segre there emerged a profusion of Italian geometers (not all of them broad enough in their skills for Segre's tastes however). Personal predeliction coupled with formal training in an established tradition is a powerful combination, and as we have seen it worked very well for over half a century.

BIBLIOGRAPHY

ABEL, N.H.

1826 *Mémoire sur une propriété générale d'une classe très-etendue de fonctions transcendantes.* Memoires presentes par divers savants, t. VII, Paris 1841, Oeuvres completes, I, ed. L. Sylow and S. Lie, Christiania, 1881. 145–211.

BAKER, H.F.

1912 *On some recent advances in the theory of surfaces.* Proceedings of the London Mathematical Society, (2), 1–40.

BERTINI, E.

1894 *La geometria delle serie lineari sopra una curva piana secondo il metodo algebrico.* Annali di Matematica, XXII, 1–40.

1903 *Life and Works of L. Cremona.* Proceedings of the London Mathematical Society, (2), 1, v–xviii.

BLISS, G.A.

1923 *The reduction of singularities of plane curves by birational transformations.* Bulletin of the American Mathematical Society, 29, 161–183.

BRILL, A., NOETHER M.

1874 *Über die algebraischen Functionen und ihre Anwendung in der Geometrie.* Mathematische Annalen, 7, 269–312.

1894 *Bericht über die Entwicklung der Theorie der algebraischen Functionen in älterer und neuerer Zeit.* Jahresbericht der Deutschen Mathematiker-Vereinigung, 3, 107–566.

CASTELNUOVO, G.
1933 *Commemorazione del Socio Eugnenio Bertini.* Rendiconti della Academia dei Lincei, 17, 745–748.

CAYLEY, A.
1849 *On the triple tangent planes of surfaces of the third order.* Cambridge and Dublin Mathematical Journal, 4, 118–132, Collected Mathematical Papers, I, 1889, no. 76, 445–456.

CLEBSCH, A.
1863 *Ueber die Anwendung der Abelschen Functionen in der Geometrie.* Journal für die reine und angewandte Mathematik, 63, 189–243.

COOLIDGE, J. L.
1940 *A History of Geometrical Methods.* Oxford University Press, reprinted 1963, Dover Publications, New York

DIEUDONNÉ, J.
1974 *Cours de géométrie algébrique.* Presses Universitaires de France, Paris.

KLEIN, F.
1872 *Vergleichende Betrachtungen über neuere geometrische Forschungen (The "Erlanger Programm").* Deichert, Erlangen, Gesammelte Mathematische Abhandlungen, I, nr. XXVII, 460–497.

KOENIGSBERGER, L.
1904 *Carl Gustav Jacob Jacobi.* Teubner, Leipzig.

LORIA, G.
1888 *Il passato ed il presente delle principali teorie geometriche.* Memorie della Accademia delle Scienze, Torino, (2), 38, later published separately, 4th edition, Turin, 1930.

NOETHER, M.
1886 *Extension du théorème de Riemann-Roch aux surfaces algébriques.* Comptes rendus, 103, 734-737.

RIEMANN, B.
1857 *Theorie der Abelschen Functionen.* Journal für die reine und angewandte Mathematik, 54, 115–155, Gesammelte Mathematische Werke und Wissenschaftliche Nachlass, ed. R. Dedekind and H. Weber, Dover reprint, 1953, 88–144

ROWE, H.
1881 *Memoir on Abel's Theorem.* Philosophical Transactions of the Royal Society, 172, 713–750.

SALMON, G.
1882 *A Treatise on the Analytical Geometry of Three Dimensions.* Dublin.

SEGRE, C.

1891 *On some tendencies in geometrical investigations.* Rivista di Matematica, 1, as translated by J. W. Young and published in the Bulletin of the American Mathematical Society, 10, 442–468.

1894 *Introduzione alla geometria sopra un ente algebrico semplicement infinito.* Annali di Matematica, (2), 41–142.

1918 *Mehrdimensionale Räume.* Encyclopädie der Mathematischen Wissenschaften, III C9.

VERONESE, G.

1876 *Nuovo teoremi sull' Hexagrammum mysticum.* Memorie della Reale Accademia dei Lincei, (3), 1, 649–703.

1882 *Behandlung der projectivischen Verhältnisse der Räume von verschiedenen Dimensionen durch das Princip des Projicirens und Schneidens.* Mathematische Annalen, 19, 161–234.

ZARISKI, O.

1971 *Algebraic Surfaces,* second supplemented edition. Springer Verlag, New York.

Abel's Theorem

Niels Henrik Abel (1802–1829)
Courtesy of Springer-Verlag Archives

Abel's Theorem

Roger Cooke

In the history of mathematics there have been few instances in which as great an advance has been made by one writer at one time, and the theorem is one which every mathematician should know.

G. A. Bliss

INTRODUCTION

The present paper has been written to furnish the background for a study of the history of the Jacobi Inversion Problem. This problem arose as a result of an attempt to interpret Abel's Theorem in terms of the inverse functions that Abel and Jacobi had shown to be essential to an understanding of elliptic and hyperelliptic integrals. As background the present paper gives some of the history of the discovery of Abel's Theorem itself, as well as a discussion of the technical points that lie at the heart of the theorem. The publishing history of Abel's Theorem is too well known to require repeating in detail. As is known from biographies of Abel such as the one by Ore [1957, 150–151], Abel presented his theorem to the Paris Academy in October of 1826 and submitted a long memoir on the subject at the same time. The memoir was given to Cauchy and Legendre to read and apparently simply forgotten. Meanwhile Abel's methods were becoming known through his long papers on elliptic integrals which appeared in Crelle's *Journal für die reine und angewandte Mathematik* [1827b], [1829b] as well as a paper [1829a] on more general algebraic integrals. Abel eventually decided that his great paper was not going to appear in France and published a shorter paper on the hyperelliptic case [1828] and an outline of his general methods [1829a] in Crelle's *Journal*. These papers were sufficient to establish Abel's methods, and in particular they led Jacobi [1832] to call upon the Paris Academy to publish Abel's large memoir, which finally appeared in 1841. By this time, however, much of Abel's work had been duplicated by Broch, Minding, and Rosenhain.

In this paper we shall focus on the hyperelliptic case discussed in Abel's memoir [1828]. This is the case that attracted Jacobi's attention, and it contains most of the major ideas in the general case (though of course the extension from the hyperelliptic case to the general case is by no means obvious). Even in this special case, however, Abel's Theorem is sufficiently unusual in its formulation to leave the reader grasping for some perspective. That sensation of bewilderment was the motive for the present exposition. Mathematicians and historians of mathematics, especially the late-nineteenth-century Germans, have attempted to recover the path followed by Abel to this discovery, and in their efforts they have found many clear markers of that path. The most authoritative explanation occurs as part of the large-scale study of Brill and Noether [1894], which was originally intended as a report describing the then-current status of the subject of algebraic functions, but turned out to involve long passages of pure history. The account of Brill and Noether has been relied on by many authors in other contexts, especially in the reports that are found in the *Encyklopädie der mathematischen Wissenschaften*, including the articles by Fricke [1920], Krazer and Wirtinger [1920], and Wirtinger [1920]. Although recent Cauchy scholarship has made the Brill–Noether account incomplete, it seems to have held up fairly well where Abel is concerned and will be relied on in this paper, although part of the reason for writing this paper is to dispute an important point made by Brill and Noether.

The division of the paper is as follows. Section 1 contains a brief sketch of the earliest work on the evaluation of algebraic integrals. Since this material does not directly lead up to Abel's Theorem, no attempt is made to be comprehensive. Section 2 discusses the work of Fagnano and Euler, which Brill and Noether, following a suggestion of Jacobi, take as the starting point for the theory of elliptic integrals, and hence of nonelementary transcendental functions in general. Section 3 contains a detailed discussion of the Brill–Noether account of the history of Abel's Theorem, including the debatable points mentioned above. Section 4 discusses some technical details involved with Abel's Theorem and shows how Abel used algebraic equivalents of some concepts now associated with contour integration.

1. Early Algebraic Integrals

The only aspect of integration directly relevant to a discussion of Abel's Theorem is the fact that the integral of an algebraic function may be transcendental. For that reason we do not need to recall many facts from the seventeenth and early eighteenth centuries. We can therefore conveniently skip over most of the earliest history of integration, of which Hoffman [1965] has made a very thorough study containing an extensive bibliography. As Hoffman notes, the question of the existence and (explicit) evaluation of algebraic integrals of algebraic functions was the outstanding problem of the day. Even in the early days of the calculus

Leibniz, for example, knew that some infinite series expressions for algebraic integrals could be reduced to finite form, but he had no systematic way of determining when such a reduction was possible. Tschirnhaus had believed that all algebraic functions could be expressed as *finite* sums of radicals, so that no transcendental functions would be needed for their quadrature, and this misconception on his part was the cause of a complicated and interesting series of adventures, not directly connected with the present story, however. (See Hoffman [1965, 288–305].)

Our point of departure is the mid-eighteenth-century situation. By then, as is well-known, formulas involving algebraic and logarithmic integrals had been discovered (or recognized as the inverse of differentiation formulas); and when these functions had proved inadequate for handling integrands with the square root of a quadratic, circular functions were used. The uncertain status of complex numbers kept certain unifying principles from being recognized, especially the reason for the constant occurrence of logarithms and arctangents in the same formula, but the successes obtained through integration with these functions caused them to be recognized as the simplest, "elementary" functions. Subsequent work, then, was directed at reducing as many integrals as possible to such functions. That some functions are not so reducible was known at least by the mid-eighteenth century, as square roots of cubic and quartic polynomials occurred in more and more contexts and were recognized as not being reducible to elementary functions. The fact that *some* square roots of cubics had elementary primitives led to an investigation of the general conditions under which such reductions were possible. In fact, the search for such reductions contributed to Abel's discovery, and, I shall argue, is reflected in the famous theorem that bears his name. In a sense Abel's Theorem involves the integral analogue of his search for conditions under which an algebraic equation is solvable by radicals.

The situation in the middle of the eighteenth century is described in MacLaurin's *Treatise on Fluxions* [1742]. In the opening of Chapter III of Volume II (p. 615) MacLaurin writes:

> When it does not appear that a fluent [integral] can be assigned in a finite number of algebraic terms, we are not therefore to have recourse immediately to an infinite series. The arches of a circle, and hyperbolic areas or logarithms, cannot be assigned in algebraic terms, but have been computed with great exactness by several methods. By these, with algebraic quantities, any segments of conic sections and the arks of a parabola are easily measured; and when a fluent can be assigned by them, this is considered as the second

degree of resolution. When it does not appear that a
fluent can be measured by the areas of conic sections,
it may, however, be measured in some cases by their
arks; and this may be considered as the third degree
of resolution. If it does not appear that a fluent can be
assigned by the arks of any conic sections (the circle
included) it may however be of some use to assign the
fluent by an area or ark of some other figure that is
easily constructed or described; and it is often important that the proposed fluxion [integrand] be reduced
to a proper form, in order that the series for the fluent
may not be too complex, and that it may not converge
at too slow a rate.

Here we see already in the middle of the eighteenth century the
beginnings of an attempt to classify algebraic functions—at least those
that are integrals of algebraic functions—in terms of their complexity.
The degrees of resolution MacLaurin describes here remind one of an
unsystematic prefiguration of the concept of genus.

MacLaurin remarks, in the passage just quoted, on the possibility of
using arcs of more general figures than the conic sections. This possibility began to be exploited in earnest just ten years after the publication of
MacLaurin's *Treatise*. This is the point where we shall start our discussion
of Abel's Theorem. In 1751 Euler was given the duty of reading the collected works of Fagnano, which led him to discover an addition theorem
for elliptic integrals. It seems to have been Jacobi who first pointed out
in a letter to Fuss (Stäckel & Ahrens [1908, p. 23]) in connection with the
project of publishing Euler's collected works, that the theory of elliptic
functions arose from Euler's examination of Fagnano's work. Jacobi's remark has stood the test of time very well, and as a result this work of
Euler is regarded as a natural milestone in the development of algebraic
integrals in many accounts of the subject, for example, Brill and Noether
[1894], Krazer and Wirtinger [1920], Siegel [1956], and Ayoub [1984]. The
most thorough analysis of Fagnano's contribution is the paper by Ayoub
[1984]. Since Fagnano presents analytic proofs of his results and does not
describe the process by which he discovered them, conjectures as to the
route he followed cannot be confirmed by reading the text. Nevertheless
a plausible conjecture was made by Siegel in his report [1956], to which
the reader is referred for details.

2. FAGNANO AND EULER

Fagnano's results, as just mentioned, involve the use of arcs of more general curves than conic sections, in particular the lemniscate. As Siegel [1956] points out, Fagnano's great discovery was made in 1714, but lay dormant for 37 years, until Euler realized its implications. Fagnano published a series of papers [*Opere*, Vol. II, 293–318] on arcs of the lemniscate whose Cartesian equation is $x^2 + y^2 = a\sqrt{x^2 - y^2}$. In the first of these he showed how to construct an algebraic relation among three chords on a lemniscate, an ellipse, and an hyperbola. Letting z represent the length of the chord from the origin (C) to the point $Q = (x, y)$, of the lemniscate, Fagnano computed the length of the arc CQ subtended by this chord as

$$\int \frac{a^2}{\sqrt{a^4 - z^4}}\, dz.$$

That is, the arc length is the particular primitive of $\dfrac{a^2}{\sqrt{a^4 - z^4}}$ whose value at $z = 0$ is 0. He then considered the ellipse $2x^2 + y^2 = 2a^2$, whose semiminor axis is a and whose semimajor axis is $a\sqrt{2}$. On this ellipse he considered the arc DI measured from the point $D = (0, a\sqrt{2})$ to the point $I = (z, \sqrt{2a^2 - 2z^2})$ and showed that this arc has length

$$\int \sqrt{\frac{a^2 + z^2}{a^2 - z^2}}\, dz.$$

Again, he means the particular primitive of $\sqrt{\dfrac{a^2 + z^2}{a^2 - z^2}}$ whose value at $z = 0$ is 0. Finally he also considers the equilateral hyperbola whose Cartesian equation is $x^2 - y^2 = a^2$ and an arc measured from the vertex O of the hyperbola to a point M on the hyperbola at distance t from the origin. Fagnano computed the length of the arc MO as

$$\int \frac{t^2}{\sqrt{t^4 - a^4}}\, dt.$$

In this case the primitive has the value 0 when $t = a$. Fagnano discovered that if t and z are related by the equation

(1) $$t = a\sqrt{\frac{a^2 + z^2}{a^2 - z^2}},$$

then

$$\text{arc.}CQ = \text{arc.}DI + \text{arc.}MO - \frac{zt}{a}.$$

In other words, although the arcs themselves are transcendental functions of the variables z and t, the particular combination of them is an algebraic function when the variables are related by the algebraic expression (1).

Of still more significance was the later paper [*Opere* II, 304–313] in which Fagnano showed that the chord z of twice an arc from the center of the lemniscate is related to the chord u of the arc itself by the equation

$$z = \frac{2u\sqrt{1 - u^4}}{1 + u^4}.$$

Fagnano's discoveries were of a very particular type, and it was not clear how they fit into the general picture of integration. However, one of Fagnano's particular results on arcs of the lemniscate attracted the attention of Euler, namely the discovery that the differential equation

$$\text{(2)} \qquad \frac{dx}{\sqrt{1 - x^4}} = \frac{dy}{\sqrt{1 - y^4}}$$

has the particular solution

$$x = -\sqrt{\frac{1 - y^2}{1 + y^2}}.$$

Since Eq. (2) clearly also has the particular solution $x = y$, Euler sought to study this equation by finding its "complete integral," which was to be a single equation in x and y containing a parameter c which, for appropriate values of c, would reduce to the particular solutions already known. He remarked that the significance of Fagnano's discovery was not clear because it was of the nature of those remarks that "are not based on a systematic method" and whose "inner grounds seem to be hidden." In the abstract to Euler's first paper on the subject [1761], we find the natural comparison of Eq. (2), or rather, the more general equation

$$\frac{m\,dx}{\sqrt{1 - x^4}} = \frac{n\,dy}{\sqrt{1 - y^4}}$$

with the equation for the inverse sine

$$\text{(3)} \qquad \frac{m\,dx}{\sqrt{1 - x^2}} = \frac{n\,dy}{\sqrt{1 - y^2}}.$$

In this abstract Euler (or whoever wrote the abstract) points out that the integrals occurring in (3) cannot be expressed algebraically, yet the sines of commensurable angles can be compared algebraically, and

> it seems even more remarkable, since an integral of the form $\dfrac{dz}{\sqrt{1-z^4}}$ cannot be expressed either by angles or logarithms ... that nevertheless, in regard to the differential equation in question, it is possible to exhibit the relation between x and y algebraically.

In the passage just quoted Euler was pointing out that even when an integral such as

$$(4) \qquad \int \frac{1}{\sqrt{X}}\, dx,$$

X being a polynomial ($X = p(x)$ in modern notation), cannot be expressed in finite form, it is nevertheless sometimes possible to find the complete integral of the differential equation

$$(5) \qquad \int \frac{1}{\sqrt{X}}\, dx + \int \frac{1}{\sqrt{Y}}\, dy = 0,$$

(where again $X = p(x)$ and $Y = p(y)$, p being a polynomial) in closed algebraic form as (in modern notation)

$$(6) \qquad P(x, y; c) = 0,$$

where $P(x, y; c)$ is a family of polynomials in the variables x and y whose coefficients depend on the parameter c. (We shall discuss below the way in which Euler obtained $P(x, y; c)$.) For a proper understanding of Abel's Theorem it is essential to understand the connection between the one-parameter family of polynomials in Eq. (6) and the indefinite integral (which we shall denote $\psi(x)$) in Eq. (4). On the one hand, as Euler pointed out, the general integral of Eq. (5) is Eq. (6). On the other hand, it is obvious that this general integral is

$$(7) \qquad \psi(x) + \psi(y) = K,$$

for an arbitrary constant K. It appears, then, that the transcendental relation (7) between x and y is equivalent to the algebraic relation (6). Equivalence of this sort is actually encountered quite frequently. For example, the transcendental relation between x and y expressed by the equation

(8) $4f^2(x) - 4f(x) + f(y) = 0,$

where $f(x) = \sin^2(x)$, is certainly implied by the algebraic relation

(9) $y = 2x;$

and conversely if x and y are continuous functions of other variables such that $y = 2x$ at some point where $x \neq 0$, then Eq. (8) implies Eq. (9) in a neighborhood of the point. From this point of view Euler's discovery amounts to the observation that there may be an algebraic relation between $\psi(x)$ and $\psi(y)$ when x and y are commensurable, even when ψ is a transcendental function.

In the particular case considered in his 1761 papers Euler showed that for any integers m and n the complete integral of the differential equation
(10)
$$\frac{m\,dx}{\sqrt{A + 2Bx + Cx^2 + 2Dx^3 + Ex^4}} = \frac{n\,dy}{\sqrt{A + 2By + Cy^2 + 2Dy^3 + Ey^4}}$$

can be expressed as a polynomial in x and y whose coefficients depend on an arbitrary parameter. He claimed, however, that not only the method but also the result was limited to square roots of quartics, since the square root of a sixth-degree polynomial could be a cubic polynomial. In particular, if the polynomial in (10) is increased to the sixth degree, one might be faced with
$$\frac{m\,dx}{1 + x^3} = \frac{n\,dy}{1 + y^3},$$
whose general integral is

(11) $\dfrac{m}{3} \ln \dfrac{x+1}{\sqrt{x^2 - x + 1}} + \dfrac{m}{\sqrt{3}} \arctan \dfrac{2}{\sqrt{3}} (x - \dfrac{1}{2}) =$

$\dfrac{n}{3} \ln \dfrac{y+1}{\sqrt{y^2 - y + 1}} + \dfrac{n}{\sqrt{3}} \arctan \dfrac{2}{\sqrt{3}} (y - \dfrac{1}{2}) + C.$

In this integral, as Euler remarked, both terms contain both a logarithm and an arctangent, between which no algebraic relationship exists. This remark of Euler's seems to show that the uncertain status of complex numbers prevented Euler from seeing the true generality to which his results could be extended. (The general integral (11) can be expressed without arctangents by use of complex numbers.) Of course, even if

Euler had noticed this fact, the method he was using might not have been adequate to obtain such results. Let us now examine this method in more detail, since, as Brill and Noether* have pointed out, Abel's work really starts from Euler's. (According to Brill and Noether [1894, p. 209], Abel applied to Euler's case some algebra not known in Euler's day. The details will be discussed below.) Euler's most systematic exposition of his method occurs in his *Institutiones Calculi Integralis*, in §612. To make it clear what Euler achieved, we first give a schematic representation of his general procedure.

We start with a polynomial equation in two variables, say

(12) $$p(x, y) = 0.$$

This equation leads immediately to the exact differential equation

$$M(x, y)\, dx + N(x, y)\, dy = 0$$

where $M(x, y) = \dfrac{\partial p}{\partial x}$ and $N(x, y) = \dfrac{\partial p}{\partial y}$. But Eq. (12) can be solved (locally) for either variable, giving

$$x = \theta(y), \quad y = \phi(x),$$

so that if $X = M(x, \phi(x))$ and $Y = N(\theta(y), y)$, we have another exact (variables separated) equation

(13) $$X\, dx + Y\, dy = 0,$$

whose solution is obviously $\int X\, dx + \int Y\, dy = C$. The functions X and Y are algebraic functions, and (12) can be regarded as a new, non-obvious integral of (13). In fact, if the number of coefficients in the two equations is just right, (12) can be regarded as the complete (general) integral of (13). The case where $p(x, y)$ is symmetric in x and y leads to a case where $X = f(x)$ and $Y = f(y)$, f being an algebraic function.

Euler applied a slight variation of the scheme just described to a particular equation in §612 of his *Institutiones Calculi Integralis*. The equation is

(14) $$0 = \alpha + \gamma(x^2 + y^2) + 2\delta xy + \zeta x^2 y^2.$$

---

* The foreword to the paper by Brill and Noether [1894] mentions that the early portion of the paper was written largely by Brill. The historical arguments summarized here are therefore probably due to that author.

This equation can be solved as a quadratic in either variable, giving

$$(\zeta x^2 + \gamma)y + \delta x = \pm\sqrt{A + Cx^2 + Ex^4},$$
$$(\zeta y^2 + \gamma)x + \delta y = \pm\sqrt{A + Cy^2 + Ey^4},$$

where $A = -\alpha\gamma$, $C = \delta^2 - \gamma^2 - \alpha\zeta$, and $E = -\zeta\gamma$. (I have omitted an inessential factor m used by Euler, since it soon drops out of the discussion anyway.) Since the exact differential equation derived from (14) is

$$[(\zeta y^2 + \gamma)x + \delta y]\,dx + [(\zeta x^2 + \gamma)y + \delta x]\,dy = 0,$$

we have the variables-separated equation

(15)
$$\frac{dx}{\sqrt{A + Cx^2 + Ex^4}} + \frac{dy}{\sqrt{A + Cy^2 + Ey^4}} = 0.$$

One can see already an enormous number of important special cases in (15). For example $A = 1$, $C = -1$, $E = 0$ makes (14) into an algebraic relation between the sines of two angles that satisfy a certain linear relation; or, if you look at it the other way, (15) gives a linear relation between the arcsines of two algebraically related numbers. The case $A = 1$, $C = 0$, $E = -1$ gives analogous relations between arcs on the lemniscate; and the case $A = 1$, $C = -(k^2 + 1)$, $E = k^2$ gives an addition theorem for elliptic integrals of the first kind (so named later by Legendre).

Equation (14) can be written in terms of the constants A, C, and E occurring in (15) and an arbitrary constant $k = \gamma^2$ as

(14′) $$0 = -A + k(x^2 + y^2) + 2\sqrt{AE + Ck + k^2}\,xy - Ex^2y^2.$$

Equation (14′), which represents a one-parameter family of curves, may be regarded as the complete integral of Eq. (15). Euler then (p. 455) applies this result to solve his Problem 79:

Let $\Pi : z$ denote the function of z given by

$$\Pi : z = \int \frac{dz}{\sqrt{A + Cz^2 + Ez^4}}$$

so chosen that it equals 0 when $z = 0$. Find the relations among functions of this type.

Solution

Let x and y satisfy

$$\frac{dx}{\sqrt{A + Cx^2 + Ex^4}} + \frac{dy}{\sqrt{A + Cy^2 + Ey^4}} = 0$$

and $y = b$ when $x = 0$. Then

$$\Pi : x + \Pi : y = \Pi : b.$$

To summarize the remainder of Euler's argument it will be convenient to use our own notation for functions. We shall therefore replace Euler's $\Pi : z$ by $\psi(z)$, so that Eq. (7) becomes ($K = \psi(b)$)

$$\psi(x) + \psi(y) = \psi(b).$$

Since $X = A + Cx^2 + Ex^4$ is an even function and the indefinite integral $\psi(z)$ starts from the point $z = 0$, we have $\psi(-b) = -\psi(b)$, and so

(7') $\psi(x) + \psi(y) + \psi(-b) = 0.$

Equation (7') is symmetric in the variables x, y, and $-b$; and if the parameter c in Eq. (6) (i.e., k in Eq. (14')) is replaced by its expression in terms of b, the resulting equation in x, y, and b, namely

(16) $\quad 0 = A^2(b^4 + x^4 + y^4) - 2A^2(b^2x^2 + b^2y^2 + x^2y^2) - 4ACb^2x^2y^2 -$
$$2AEb^2x^2y^2(b^2 + x^2 + y^2) + E^2b^4x^4y^4,$$

is also symmetric. Strictly speaking, Eq. (16) does not exactly imply Eq. (7'). It may be necessary to change the sign of the square root in one term in order to get Eq. (7'). Abel's Theorem takes account of this minor technical point by multiplying each term by an appropriate factor ± 1 (see Eq. (17) below).

It is clear that the relation (16) is an order of magnitude more sophisticated than the multitudinous particular integration formulas that apply in the rare cases when an integral such as $\psi(x)$ reduces to an elementary integral. The method does have its limitations, however. For instance, there is not any algorithm for choosing the initial polynomial in (12) so as to obtain a given integrand in (13). In fact the relation between integrand and integral is very indirect. (Brill and Noether [1894, p. 207] believed that Euler must have obtained the particular integral (14) for the equation (15) by trial and error.) Furthermore, as Euler's remark on the limitations of his method shows, he was not able to exploit relations like (16) in a general context. In fact fifteen years later he was again writing numerous papers [1777abc] on cases where particular algebraic integrals could be evaluated in terms of elementary functions. Brill and Noether, who traced Abel's ideas to Eq. (16), attributed the lack of further progress to the need for purely algebraic advances [1894, p. 209]. Although Brill and Noether do not go into details, it seems clear that a still-primitive understanding of complex numbers had retarded the full understanding of the significance of symmetric polynomials in Euler's day. We now turn to the analysis of Abel's approach to the subject, as conjectured by Brill and Noether.

3. THE DISCOVERY OF ABEL'S THEOREM.

The result that came to be called (at Jacobi's suggestion) Abel's Theorem is such a large advance on everything that came before it that even Abel's contemporaries, to whom the results leading up to it were familiar, must have found it puzzling. To put the whole matter in perspective we need to state the theorem informally and obtain a schema that summarizes Abel's technique in a way analogous to the summary of Euler's technique given above. This kind of information will make it possible to compare Euler's result with Abel's and see the extent to which Abel was influenced by Euler. We begin by giving the commonly-accepted informal statement of Abel's Theorem.

THEOREM. *For any given algebraic differential the sum of an arbitrary number of integrals can be reduced to a fixed number of integrals plus an elementary function, and the fixed number is determined by the irrationalities of the integrand.*

It should be noted that the concept of integral as used by both Euler and Abel is applicable to any exact differential in several variables and means a function (of several variables) whose differential is the given exact differential form. (In particular, it obviously does *not* mean a contour integral or a Riemann integral.) Because the notion is an algebraic one, the difficulties caused by singularities of the integrand cannot be systematically handled.

The version of Abel's Theorem stated above, however, does not give any clue as to the underlying techniques or the general insight into such integrals that the discovery of the theorem reflects. Such insight comes only from examining the proof Abel gave to the theorem. As a basis for discussion of this proof, we shall paraphrase Abel's statement of the fundamental theorem (Theorem VI of [1828]), from which he derives Abel's Theorem, for the special subcase that applies to the theorem of Euler just discussed.

THEOREM. *Let $\psi(x)$ denote the function*

$$\int \frac{dx}{\sqrt{A + Cx^2 + Ex^4}},$$

and let $\varphi_1(x)$ and $\varphi_2(x)$ be any two polynomials such that

$$\varphi_1(x)\varphi_2(x) = A + Cx^2 + Ex^4.$$

Then there is a constant K such that for any for polynomials $\theta_1(x)$, $\theta_2(x)$ the relation

(17) $\varepsilon_1 \psi(x_1) + \cdots + \varepsilon_\mu \psi(x_\mu) = K$

holds with some choice of $\varepsilon_j = \pm 1$, $j = 1, 2, \ldots, \mu$, where $x_1, \ldots, x_\mu$ are the roots of the equation

(18) $\theta_1^2(x)\varphi_1(x) - \theta_2^2(x)\varphi_2(x) = 0.$

(The constant K remains the same if the coefficients of θ_1 and θ_2 vary by small amounts.)

It is clear that the point of view has shifted when we turn from Euler's theorem to Abel's. Although Euler's equation (7′) is just Abel's (17) with $K = 0$ and $\mu = 3$ (and the coefficients ε that were ignored by Euler), Euler's condition (16) on the upper limits of integration is quite different from the corresponding condition (18) of Abel. Euler, it will be noted, gives the condition on the arguments of $\psi(x)$ as a symmetric polynomial in the arguments. In a sense, of course, Abel, does the same thing, since the expression on the left-hand side of Eq. (18) can be regarded as a symmetric polynomial in the variables x_j whose coefficients depend on x as well as the usual coefficients of the four polynomials. Abel's language, however, makes it clear that he regards the roots x_j as functions of the coefficients of θ_1 and θ_2. There is no hint of this point of in Euler's writing. The fact that the expression on the left-hand side of Eq. (17) is constant is a statement about functions of these variables. Now there are both documentary reasons, based on what is known of Abel's education, and mathematical reasons, based on a similarity of approach and content, for believing that a direct connection can be made between the work of Euler and Abel's Theorem. Brill and Noether attempt to make this connection. They go far beyond a simple connection, however, conjecturing the line of reasoning Abel must have followed all the way to the discovery of his theorem. It is this part of their argument that I wish to present and consider how plausible it is that Abel actually proceeded in this way.

The argument presented by Brill and Noether [1894, 209–212] begins with a discussion of the work of Euler just discussed, of which Abel was certainly aware, since Holmboe [1839, p. V] states explicitly that he taught Abel from Euler's *Institutiones Calculi Integralis*. Since the exact words used by Brill and Noether are important to my argument, I give both the original and a translation at this point. Before doing so, I should explain their notation. They use X, Y, and B to denote respectively $A + Cx^2 + Ex^4$, $A + Cy^2 + Ey^4$ and $A + Cb^2 + Eb^4$, and analogously for divisors of the polynomial $A + Cx^2 + Ex^4$, which they denote by X', X'', Y', Y'', etc. After discussing Euler's work Brill and Noether present their account of a likely road to Abel's discoveries.

5. Mit dem Vorstehenden haben wir den Gedankenkreis und die Probleme umgrenzt, die nun ABEL's Erfindungskraft herausforderten. Die Brücke aber, die von EULER zu ABEL hinüberführt, glauben wir in einer gewissen Umgestaltung zu erkennen, welche ABEL mit der zwischen x, y, b bestehenden algebraischen Gleichung [Eq. (16)] vornahm.

EULER hatte diese Gleichung bereits in eine hinsichtlich dieser drei Größen symmetrische Gestalt gebracht. Nun waren bekanntlich die Fortschritte, welche die Theorie der Gleichungen seit EULER gemacht hatte, für ABEL nich verloren gegangen. Ihm lag der Gedanke nahe, jene zwischen den symmetrischen Funktionen von x, y, b bestehende Relation zur Bildung derjenigen Gleichung dritten Grades zu verwenden, die für eine dieser Größen, etwa x, besteht, *welcher dann aber auch y und b genügen müssen.* Man erhält sie sogleich, indem man die identische Gleichung, deren Wurzeln x, y, b sind, mit Hilfe jener Relation umformt.

Diese Gleichung nun ist es, die der Kernpunkt der Entdeckung ABEL's gewesen zu sein scheint. Sie läßt sich in die Form bringen:

(1) $$vx.\sqrt{X'} - v'x.\sqrt{X''} = 0,$$

wo $X'X'' = X$ ist, und vx, $v'x$ ganze Funktionen von x sind, von solchem Grad, daß (1), auf rationale Form gebracht, den dritten Grad nicht übersteigt. Die Koeffizienten in v, v' hängen von y und b, oder vielmehr von zwei symmetrischen Funktionen dieser Größen ab, und sind durch y, b bestimmt. Man kann aber auch die (zwei) wesentlichen Koeffizienten in v, v' (v is vom ersten Grad, v' konstant, wenn X' vom ersten,

X'' vom dritten Grad ist) als gegebene Größen ansehen; sie
sind mit y, b durch die Relationen verbunden:

$$(2) \qquad \begin{cases} vy.\sqrt{Y'} - v'y.\sqrt{Y''} = 0 \\ vb.\sqrt{B'} - v'b.\sqrt{B''} = 0. \end{cases}$$

Jene Gleichung (1) besteht nun zunächst auch im Falle beliebig vieler elliptischer Integrale. Denn die Zahl der Integrale, welche mit einander verglichen werden, vermehrt
sich einfach dadurch, daß man den Grad der Funktionen
vx, $v'x$ erhöht, wobei zugleich die Zahl der willkürlich an-
nehmbaren Größen in (1) entsprechend zunimmt. Aber eines
bleibt: die Eigenschaft der Integralsumme, durch eine algebraische und logarithmische Funktion der Koeffizienten von
v, v' ausdrückbar zu sein. Denn wie im einfachsten Falle
vermöge der Gleichungen (1), (2) die irrationale Differen-
tialsumme durch ein rationales Differential symmetrischer
Funktionen von x, y, b darstellbar ist, so läßt sich hier wieder
die Summe durch symmetrische Funktionen der neuen oberen
Grenzen rational ausdrücken, die man erhält, wenn man die
Summanden unter Benutzung der Gleichungen (1), (2) und
ihrer Differentiale umformt, indem man die symmetrischen
Funktionen der Grenzen durch die Koeffizienten ersetzt.

6. *Alle diese Schlüße sind aber auch*, wie man ohne Weiteres
einsieht, *durchaus nicht an den Grad von X gebunden*; auch
der Zähler rx der Integranden braucht nur allgemein eine
rationale Funktion zu sein. So beseitigt der Gedanke ABEL's,
die Gleichung (1) an Stelle der zwischen den oberen Grenzen
bestehenden Relation zum Ausgangspunkt zu wählen, die
Schwierigkeiten, die sich EULER gegenübergestellt hatten, als
er zu den hyperelliptischen Integralen aufsteigen wollte.

A translation of this passage now follows.

5. In the preceding paragraphs we have outlined the circle
of ideas and the problems that now challenged Abel's inge-
nuity. We believe, however, that we have found the bridge
that leads from Euler to Abel in a certain transformation that
Abel applied to the algebraic equation that holds among x,
y, and b, [Eq. (16)].
Euler had already brought this equation into a form that was
symmetric with respect to these three quantities. Now, as is
well-known, the progress that had been made in the theory
of equations since the time of Euler had not been lost on Abel.

It occurred to him to apply the relation that holds between the symmetric functions of x, y, and b to form the cubic equation that holds for one of these quantities, say x, *which, however, must then be satisfied by y and b also.* It is obtained immediately by transforming the identity whose roots are x, y, and b using the former relation.

Now it is this equation that seems to have been the essence of Abel's discovery. It can be brought into the form

$$(1) \qquad vx.\sqrt{X'} - v'x.\sqrt{X''} = 0,$$

where $X'X'' = X$ and vx and $v'x$ are polynomials in x whose degrees are such that Eq. (1), when rationalized, is of degree at most three. The coefficients in v and v' depend on y and b, or rather, on two symmetric functions of these quantities, and are determined by y and b. One can, however, also regard the (two) essential coefficients in v and v' (v is linear and v' constant if X' is of first degree and X'' is of third degree) as independent variables. They are related to y and b by the relations

$$(2) \qquad \begin{cases} vy.\sqrt{Y'} - v'y.\sqrt{Y''} = 0 \\ vb.\sqrt{B'} - v'b.\sqrt{B''} = 0. \end{cases}$$

Now Equation (1) holds also in the case of arbitrarily many elliptic integrals; for the number of integrals being equated to one another increases only when the degree of the functions vx and $v'x$ is increased, and the number of quantities that can be chosen arbitrarily in (1) increases correspondingly. However, one thing remains constant, namely the property that the sum of the integrals is expressible by an algebraico-logarithmic function of the coefficients of v and v'; for, just as in the simplest case, where the irrational differential form can be represented by means of (1) and (2) via a rational differential of symmetric functions in x, y, and b, so again in this case the sum can be expressed rationally in terms of symmetric functions of the new upper limits of integration obtained by transforming the summands using equations (1) and (2) and their differentials, replacing the symmetric functions of the upper limits by the coefficients.

6. *All these conclusions, however, are in no way dependent on the degree of X,* as can be seen immediately; even the numerator

rx of the integrands in general need only be a rational function. In this way Abel's idea of choosing Eq. (1) rather than the relation on the upper limits of integration as a starting point eliminates the difficulties Euler had encountered when he tried to proceed to hyperelliptic integrals.

As always, Brill and Noether write very clearly here. The picture they present may be summarized as follows: Abel noticed that the algebraic relation (16) given by Euler could be written in terms of the elementary symmetric functions of x, y and b. (For aesthetic reasons I shall henceforth use z instead of b.) The elementary symmetric functions are $P = x + y + z$, $Q = yz + zx + xy$, and $R = xyz$. The identity equation satisfied by these variables is

$$p(w) = w^3 - Pw^2 + Qw - R = 0.$$

One can also solve the biquadratic equation (16) for x in terms of y and b $(= z)$. The variable x will be symmetric in these variables. When this value of x is substituted in P, Q, and R, the result will be a polynomial $p(w)$ whose coefficients are symmetric functions of y and z and whose zeros will be at x, y, and z. In fact it is true that Abel frequently uses the fact that a root of a polynomial can be expressed rationally in terms of the coefficients and the other roots. Another way of phrasing this statement is to say that Euler's relation (16) can be interpreted as a relation on P, Q, and R, namely

$$(19) \qquad (ER^2 - AP^2)^2 + 4AQ(ER^2 - AP^2) + 4AR(2AP - CR) = 0,$$

and this relation can be used to eliminate one of the variables P, Q, and R from the polynomial $p(w)$.

Likewise the differential form

$$(20) \qquad \frac{dx}{\sqrt{A + Cx^2 + Ex^4}} + \frac{dy}{\sqrt{A + Cy^2 + Ey^4}} + \frac{dz}{\sqrt{A + Cz^2 + Ez^4}},$$

being symmetric in the variables (and, because of Eq. (16), actually expressible rationally in these variables), can be expressed as a differential in the variables P, Q, and R, guaranteed to be integrable in terms of elementary functions. Here then is the main germ of Abel's idea: given three "upper limits of integration" x, y, and z, regard them as functions of the symmetric polynomials P, Q, and R, expressing the differential (20) in terms of the latter. According to Brill and Noether, the polynomial having the given roots x, y, and z, whose coefficients are P, Q, and R, is expressible in the form $\theta_1^2(x)\varphi_1(x) - \theta_2^2(x)\varphi_2(x)$. One is still left wondering how Abel thought of expressing the polynomial in terms of the functions

θ_i and φ_i mentioned in the theorem, however. Brill and Noether attempt to answer this question also, but at this point their reasoning becomes much less convincing, and, in fact, very misleading. They state that the equation $p(x)$ $(= x^3 - Px^2 + Qx - R) = 0$ can be brought into the form

$$(21) \qquad \theta_1(x)\sqrt{\varphi_1(x)} - \theta_2(x)\sqrt{\varphi_2(x)} = 0,$$

where $\varphi_1(x)\varphi_2(x) = A + Cx^2 + Ex^4$. (I am denoting vx by $\theta_1(x)$, etc., so that Eq. (21) is Brill and Noether's Equation (2). The comments made after this equation, especially the comments regarding the degrees of the polynomials θ_i and φ_i, show that they are claiming that $p(w)$ can be represented in the form

$$(22) \quad p(w) = w^3 - Pw^2 + Qw - R = (\alpha w + \beta)^2(w + \omega) - \gamma^2(w^2 - \omega_1^2)(w - \omega)$$

where $\omega, -\omega, \omega_1$, and $-\omega_1$ are the zeros of the polynomial $A + Cx^2 + Ex^4$. To make Eq. (22) a plausible route to the discovery of Abel's Theorem it would be desirable to show not only that this representation of $p(w)$ is possible, but also that it arises naturally. Since there are four coefficients to be matched by adjusting the three constants α, β, and γ, it is obvious that such a representation is not possible for all polynomials $p(w)$. Some restriction on the coefficients P, Q, and R is needed. The reader is left to infer that Euler's relation (19) must be the required relation, and that it leads naturally to the representation (22). If such were the case, we would have a very clear idea of how Abel came to discover his theorem. Now it is certainly true that Abel sought a representation of the type (22), as the statement of his theorem quoted above shows. However, I have not found anything intrinsic about the function $p(w)$ or about Eq. (19) that leads naturally to the representation (22). In fact even when Eq. (19) holds, it is not possible in general to represent $p(w)$ in the form (22) claimed by Brill and Noether. The polynomial $p(w)$ admits a representation in the form (22) if and only if the following condition on its coefficients holds:

$$(23) \quad 4ER\omega^3 + (EQ^2 - 2EPR + A)\omega^2 + 2CR\omega - (AP^2 - 2AQ + ER^2) = 0.$$

If ω in Eq. (23) is replaced by $-\omega$, ω_1, and $-\omega_1$, and the four resulting equations are multiplied, the result is a condition on the coefficients P, Q, and R that is necessary and sufficient for a representation in the form (22) with ω some root of the $A + Cx^2 + Ex^4$. This corollary of Eq. (23) may be thought of as Abel's algebraic relation on x, y and z in order for the equation

$$\psi(x) + \psi(y) + \psi(z) = K$$

to hold, in contrast to the relation (19) given by Euler. Of course Abel nowhere wrote or suggested such a relation; we mention it here only to show the contrast between his technique and that of Euler. *The relations* (19) *and* (23) *are not equivalent.* For instance in Fagnano's integral for the arc length of a lemniscate, where $A = 1$, $C = 0$, and $E = -1$, the values $y = -x$, $z = 0$, which give $P = 0 = R$, $Q = -x^2$, where x is arbitrary, fall under Euler's theorem; but a representation (22) is possible only for the eight values $x = \pm\sqrt{\pm 1 \pm \sqrt{2}}$.

It is significant that in the places where Abel does apply the theorem paraphrased here to elliptic integrals he never seeks a representation in the form (22). In a letter to Legendre [*Œuvres*, Vol. II, 271–279] dated 25 November 1828 he considers such a representation involving the square root of a quartic. For the particular quartic we have been discussing this representation would be

$$(24) \quad (fx)^2 - (\varphi x)^2 (A + Cx^2 + Ex^4) = B(x^2 - x_1^2)(x^2 - x_2^2) \cdots (x^2 - x_\mu^2).$$

(The mathematics discussed in this letter was published the following year in Crelle's *Journal*, cf. [1829b].) He proceeds similarly in the fragments of his early work published in T. II of his *Œuvres*, for example in No. XV. Equation (24) makes it unlikely that Abel started from the factorization for the minimal polynomial claimed by Brill and Noether.

On the other hand, Brill and Noether are quite convincing when they argue that Abel must somehow have been led to his result by studying the polynomial $p(w)$. The problem is to explain how Abel came to seek the particular felicitous decomposition of that polynomial that occurs in Abel's Theorem. With the explanation given by Brill and Noether ruled out of court, we must look elsewhere for Abel's motivation. Judging from the fragments published in T. II of Abel's *Œuvres*, it is clear that Abel was already in possession of the general techniques used in obtaining Abel's Theorem before he journeyed outside Norway. For that reason it is not obvious that he first proved Abel's Theorem in the elliptic special case and then generalized it. In trying to conjecture the route of discovery we must first give a schematic description of Abel's method analogous to that given for Euler's method, then try to combine this description with the basic principles that Abel invoked constantly and the problems he was continually absorbed in, so as to formulate a possible route to Abel's Theorem. In this connection it is worthwhile to take note of two general problems on which Abel labored, namely to find all algebraic equations that can be solved algebraically, and to find all integrals of algebraic functions that can be evaluated using algebraic functions and logarithms. In the first problem it is an important principle that a *simultaneous* solution of

two algebraic equations (if unique) can be expressed rationally in terms
of the coefficients of the two equations (by simply finding the greatest
common divisor of the two polynomials). In applying this principle to
the second problem, one encounters the difficulty that the simultaneous
solutions of two equations are simply a finite set of points, when what is
needed is a *variable* point. Abel apparently discovered that this obstacle
is overcome by making the coefficients of one of the equations into variables and regarding the common solutions as functions of these variables.
These principles lead to the following schematic representation of Abel's
technique.

Given an integral $\int g(x)\,dx$ containing one or more algebraic irrationalities, find (if possible) a rational function $f(x,y)$ and a polynomial
$p(x,y)$ such that $g(x) = f(x,y)$, where $y = y(x)$ is a function defined by
the equation $p(x,y) = 0$. The addition properties of the integral $\int g(x)\,dx$
can then be stated in the following terms: For any polynomial $q(x,y)$,
if $(x_1,y_1),\ldots,(x_\mu,y_\mu)$ are the simultaneous solutions of the equations
$p(x,y) = 0 = q(x,y)$, then

$$\int f(x_1,y_1)\,dx_1 + \cdots + \int f(x_\mu,y_\mu)\,dx_\mu = \int dE,$$

where dE is a rational exact differential in the coefficients of the polynomial $q(x,y)$, hence expressible in terms of algebraic and logarithmic
functions. By varying the polynomial $q(x,y)$ one can obtain many particular theorems about such integrals.

It is doubtful that a direct and obvious path from Euler's technique
to this technique can be found. The two techniques have certain elements in common, but they differ in many important respects. For instance in Euler's technique the algebraic equation leading to the differ-
ential $\dfrac{dx}{\sqrt{A + Cx^2 + Ex^4}}$ is only indirectly related to the differential itself,
whereas in the Abel technique just described the differential is the starting
point, and leads immediately to the algebraic relation:

$$y^2 - (A + Cx^2 + Ex^4) = 0.$$

Since the technique involves finding rational expressions for the simultaneous solutions of this equation and an arbitrary second equation, one is
led to postulate a second equation, which may be assumed of first degree
in y. (The original equation in y is monic, and so may be divided out of
whatever polynomial is postulated; the remainder will certainly then be
of first degree in y.) Thus we consider also the equation

$$p_2(x)y - p_1(x) = 0.$$

The two equations lead immediately to a third:

$$p_1^2(x) - p_2^2(x)(A + Cx^2 + Ex^4) = 0.$$

If by chance p_1 has a common factor φ_1 with $A + Cx^2 + Ex^4$, we can cancel that factor and obtain

$$\theta_1^2 \varphi_1 - \theta_2^2 \varphi_2 = 0,$$

where $\theta_1 = \dfrac{p_1}{\varphi_1}$, $\theta_2 = p_2$, and $\varphi_1 \varphi_2 = A + Cx^2 + Ex^4$. This is essentially the technique developed by Abel (in a more general context) and published in Fragment X of Volume II of his *Œuvres*. According to Holmboe, Abel had written this fragment before leaving Norway. Thus Abel's form of the equation is obtained quite naturally, though it is not generally a cubic equation, as claimed by Brill and Noether, since φ_1 is usually constant.

The preceding considerations seem to me a more plausible route to the discovery of Abel's Theorem than those given by Brill and Noether, but of course the matter cannot really be settled. However Abel came to look for the particular decomposition, it was a fortunate decision, for this form of the polynomial suggests the following well-known integration formula (cf. Euler's *Institutiones Calculi Differentialis*, Chapter VII, p. 189):

$$\int \frac{y\,dx - x\,dy}{y^2 - x^2} = \log\left(\frac{y + x}{y - x}\right).$$

This formula was Abel's constant friend in many situations. In seeking to discover all cases in which an integral

$$\int \frac{\rho\,dx}{\sqrt{R}}$$

can be expressed as an elementary function [1826a], Abel showed that, when such a reduction is possible, this integration formula is the key. It is therefore certain that there is an intimate connection between Abel's interest in integration in finite terms and his addition theorem for algebraic integrals, though the exact nature of that connection is not certain. Abel was a genius, and not everything he did can be reduced to obvious steps.

What seems cogent in Brill and Noether's interpretation is the fact that Euler's relation (16), when written in the form (19), suggests regarding two of the symmetric polynomials, say P and R, as independent variables. The third polynomial Q is then determined by Eq. (19), and so x, y, and z are determined. It is then guaranteed that the differential form (20) is exact and is the differential of an elementary function. This differential form can be written as

$$F(P, R)\, dP + G(P, R)\, dR.$$

(In the work he published Abel did not actually write out the differential in terms of the new independent variables. They are always implicit in what he says. In the fragment X of T. II of his *Œuvres* he does write out such a differential, but the independent variables are the coefficients of the polynomials I have labeled θ_1 and θ_2, and the symmetric polynomials P, Q, and R are regarded as functions of these variables.) The introduction of the polynomial whose roots are the limits of integration does seem to be strongly suggested by the symmetry of Euler's equation, as Brill and Noether state. The importance of representing this polynomial in the particular form (22), considered from a purely technical standpoint, cannot be overestimated, as can be seen in Abel's paper on the hyperelliptic case [1828]. If we let $F(x)$ be the minimal polynomial for the values of the upper limits of integration we are interested in and regard the coefficients as variables, so that x becomes a function of the coefficients, the values of x being determined from the equation $F(x) = 0$, then we have the total differential equation

$$F'(x)\, dx + \delta F(x) = 0,$$

where $\delta F(x)$ denotes the differential of F on the coefficients. Then the differential

$$\frac{dx}{\sqrt{A + Cx^2 + Ex^4}}$$

becomes

(25)
$$\frac{-\delta F(x)}{F'(x)\sqrt{A + Cx^2 + Ex^4}}.$$

Equation (25) suggests that the way to get rid of the square root is to get a factor of this square root into $\delta F(x)$; and, as Abel shows, the way to do this is to assume that $F(x)$ has the representation (22).

To summarize this section, the path to Abel's theorem outlined by Brill and Noether is in general correct; but on one important point, Abel's discovery of the way to eliminate the irrationality, Brill and Noether seem to have made a mistake. I do not claim that the alternative route I have sketched here is proved to be the one followed by Abel, only that it is a plausible explanation for what he might have done. Now I would like to analyze the statement and proof of Abel's Theorem, since it connects with many areas of analysis that were being developed by the French mathematicians of the period, especially Cauchy.

4. The Content of Abel's Theorem.

In 1828 Abel published a memoir [1828] in Band 3 of Crelle's *Journal für die reine und angewandte Mathematik* in which he showed how the sum of any number of hyperelliptic integrals could be reduced to an elementary function plus the sum of a fixed number of such integrals whose upper limits could be expressed as the roots of a polynomial having a certain rather ambiguous relationship to the integrand. In the preceding section I discussed a plausible route by which Abel may have arrived at this discovery. In the present section I wish to examine his proofs to show their connections with the contemporaneous work of other mathematicians, especially that of Cauchy. For some purposes it might be better to examine the long paper [1826b], which contains his ideas in more generality and with complete explanations. (The paper owes its great length to Abel's insecurity about the foundations of analysis, based on the bitter experience of errors he had fallen into by using insufficiently grounded methods.) However, since the longer paper did not appear until 1841, and by then some of it had been rediscovered by Minding and Rosenhain, it was primarily the hyperelliptic case that introduced Abel's new methods to the world.

Abel states very clearly in his introduction that he intends to demonstrate an analogue of the addition theorem for elliptic integrals, which he rightly took to be well-known. He says,

> I have in mind the functions that can be regarded as *integrals of arbitrary algebraic differentials*. While one cannot express the sum of an arbitrarily given number of such functions by a single function of the same type, as in the case of elliptic functions, in all cases one can at least express such a sum by the sum of a fixed number of other functions of the same type as the original ones upon adding a certain algebraico-logarithmic expression. We shall demonstrate this property in a subsequent issue of this journal. At present I shall consider a special case, which includes the elliptic functions, namely the case of functions subsumed in the formula

$$\psi x = \int \frac{r \, dx}{\sqrt{R}},$$

> R being an arbitrary rational and entire function [a polynomial], and r a rational function.

Abel's basic theorem, from which he planned to derive the result announced in the introduction, was obtained in a somewhat roundabout way. He considered a more special integral

$$(26) \qquad\qquad \psi x = \int \frac{fx.dx}{(x-\alpha)\sqrt{\varphi x}},$$

where fx and φx are arbitrary "entire functions" and α is an arbitrary "constant quantity," i.e., a complex number. (Since Abel later uses the degree of φx in the discussion, there is no doubt that he meant the term "entire" to refer to a polynomial, although he usually refers to polynomials as "rational integral" functions, in accordance with accepted nineteenth-century usage.) Already one sees the Cauchy kernel appearing in the integral, and some of what Abel proved about the integral (26) amounts to the Cauchy integral theorem. However, it should be kept in mind that to Abel this integral is what we would call a "Newton integral," i.e., an arbitrary (complex-valued) function whose derivative is a given complete differential. The Cauchy integral as we know it did not yet exist. By this time Cauchy had already introduced a preliminary version of a contour integral and had discovered the concept of residue, but only for simple poles (cf. Brill and Noether [1894, 165–168]). Although Abel knew of this work, he did not refer to it, even indirectly, in the present work, undoubtedly because his ideas on Abel's Theorem were already in more or less complete form before he began buying Cauchy's works in Paris. Absolutely no knowledge of contour integrals or the residue theorem is even remotely hinted at in the present paper, even though their implicit presence is obvious to a modern mathematician. Abel, it seems, had found his own way of doing things before coming to Paris and does not seem to have been influenced in this particular by the intimate acquaintance he had with the works of Cauchy. In this connection see the biography of Abel by Ore, [1957, 153–154]. One should also remember Abel's famous comment in a letter to Holmboe [Œuvres, Vol. II, p. 259] that

> Cauchy is crazy, and there is no way of communicating with him, even if he does happen to be just now the one who knows how mathematics ought to be handled. What he does is excellent, but very unclear. At first I understood almost none of it, but now I see it more clearly. He is publishing a series of memoirs under the title *Exercices de mathématiques*. I am buying them and reading them avidly. Nine issues have appeared so far this year . . . .

However, as just stated, Abel nowhere uses Cauchy's contour integrals, and Cauchy himself said—when a work by Abel was finally obtruded on his notice—that Abel's work had very little connection with his own. (The work in question, to be sure, was on approximation of integrals, and so Cauchy's remark should not be interpreted as applying to the whole of Abel's work.)

Abel's approach to the integral (26) seems very indirect at first, as well as full of ambiguity. Without giving any motivation he supposes the polynomial φx factored as $\varphi x = \varphi_1 x . \varphi_2 x$, an arbitrary factorization. He adds more arbitrariness by considering two more arbitrary polynomials $\theta x = a_0 + a_1 x + \cdots + a_n x^n$ and $\theta_1 x = c_0 + c_1 x + \cdots + c_m x^m$ and forming the function $F x = (\theta x)^2 \varphi_1 x - (\theta_1 x)^2 \varphi_2 x$. (The preceding sections were intended to supply some perspective on Abel's reasons for taking this approach, which, studied out of context, seems bizarre in the highest degree.) The equation $F x = 0$ then defines a set of functions $x_j(a_0, \ldots, a_n; c_0, \ldots, c_m)$, $j = 1, 2, \ldots, \mu$, where $F x = A(x - x_1) \cdots (x - x_\mu)$. (Abel explicitly stated that the x_j were to be considered as functions of the coefficients a_k and c_l, but he did not write out the functional form just given; that form has been included for the sake of clarity.) Each of the functions x_j is a solution of the total differential equation

$$F'x.dx + \delta F x = 0,$$

where $\delta F x$ denotes the differential of F regarded as a function of the a_k and the c_l. It is not difficult to obtain a very explicit formula for $\delta F x$, but Abel had no need to do so. Indeed the formulas he actually derived contained implicitly terms that were zero, but were retained anyway in order to preserve symmetry in the formulas.

The equation $F x = 0$ leads immediately to the two equations

$$\theta x \sqrt{\varphi_1 x} = \varepsilon \theta_1 x \sqrt{\varphi_2 x},$$

where $\varepsilon = \pm 1$. From this equation and a similar one, Abel writes the differential equation

(27) $$F'x.dx = 2\varepsilon(\theta x.\delta\theta_1 x - \theta_1 x.\delta\theta x)\sqrt{\varphi x}.$$

Multiplying Eq. (27) by $\varepsilon \dfrac{f x}{\sqrt{\varphi x}} \dfrac{1}{F'x} \dfrac{1}{x - \alpha}$, Abel changes the exact differential equation (27) into

(28) $$\varepsilon \frac{f x.dx}{(x - \alpha)\sqrt{\varphi x}} = \frac{\lambda x}{(x - \alpha)F'x},$$

where λx denotes the differential $2fx(\theta x.\delta\theta_1 x - \theta_1 x.\delta\theta x)$, whose coefficients are polynomials in x. Abel wanted to show that the differential on the right-hand side of (28) is an exact differential. To do so he used a principle that we recognize nowadays as a special case of the residue theorem. The principle can be expressed as the formula

$$(29) \qquad \sum_{j=1}^{\mu} \frac{g(x_j)}{F'(x_j)} = \Pi\left(\frac{g(x)}{F(x)}\right),$$

where, $x_1, \ldots, x_\mu$ are the zeros of $F(x)$ and for any rational function $h(x)$ the symbol $\Pi(h(x))$ denotes the coefficient of $\frac{1}{x}$ in the power series (Laurent) expansion of $h(x)$ for large values of x. The notation Π for this coefficient is Abel's, though we have modified his notation for functions slightly by introducing parentheses. The relation (29) is clearly a consequence of the residue theorem, since the left-hand side is the sum of the residues of the function $\frac{g(x)}{F(x)}$ and the right-hand side is, as we know, $1/2\pi i$ times the integral of $\frac{g(x)}{F(x)}$ around a large circle with center at the origin. Not having the residue theorem, Abel based his proof of this result (which he needed only for polynomials $g(x)$) on the known formula

$$\sum_{j=1}^{\mu} \frac{1}{(x_j - \alpha)F'x_j} = -\frac{1}{F\alpha}$$

whose proof is an immediate algebraic fact, not given by Abel. (The proof amounts to the remark that the polynomial $G(\alpha) = 1 + \sum_{j=1}^{\mu} \frac{F\alpha}{(x_j - \alpha)F'x_j} = 1 - \sum_{j=1}^{\mu} \frac{Fx_j - F\alpha}{(x_j - \alpha)F'x_j}$ has zeros at $\alpha = x_j$, $j = 1, \ldots, \mu$, yet is of degree $\mu - 1$. Hence it must be identically zero.) Abel applied this rule to each polynomial coefficient in the differential δFx; but, as we shall see, a small complication arises.

Writing $\lambda x = \lambda\alpha + (x - \alpha)\lambda_1 x$, where $\lambda_1 x$ is a polynomial, Abel has

$$\sum_{j=1}^{\mu} \frac{\lambda x_j}{(x_j - \alpha)F'x_j} = -\frac{\lambda\alpha}{F\alpha} + \sum_{j=1}^{\mu} \frac{\lambda_1 x_j}{F'x_j},$$

from which he was able to find a primitive for the differential form λx. The term $\dfrac{\lambda \alpha}{F \alpha}$ is easily handled, since α is a constant independent of the a_k and c_l. The indefinite integral formula (cf. the remarks above)

$$\int \frac{p\,dq - q\,dp}{p^2 m - q^2 n} = \frac{1}{2\sqrt{mn}} \log \frac{p\sqrt{m} + q\sqrt{n}}{p\sqrt{m} - q\sqrt{n}}$$

is immediately applicable and yields the function

$$G = \frac{f\alpha}{\sqrt{\varphi \alpha}} \log \frac{\theta \alpha \sqrt{\varphi_1 \alpha} + \theta_1 \alpha \sqrt{\varphi_2 \alpha}}{\theta \alpha \sqrt{\varphi_1 \alpha} - \theta_1 \alpha \sqrt{\varphi_2 \alpha}}$$

as a primitive whose differential on the variables a_k and c_l is $-\dfrac{\lambda \alpha}{F \alpha}$. As for the term $\displaystyle\sum_{j=1}^{\mu} \frac{\lambda_1 x_j}{F' x_j}$, which is equal, according to the formula given by Abel, to $\Pi \dfrac{\lambda_1 x}{F x}$, Abel notes that since $\lambda_1 x = \dfrac{\lambda x - \lambda \alpha}{x - \alpha}$, it must be equal to $\Pi \dfrac{\lambda x}{(x - \alpha) F x}$, since $\Pi \dfrac{\lambda \alpha}{(x - \alpha) F x} = 0$ (the expansion of $\dfrac{1}{(x - \alpha) F(x)}$ in powers of $\dfrac{1}{x}$ starts with $\left(\dfrac{1}{x}\right)^{\mu+1}$). Such being the case, the rule $\delta \Pi = \Pi \delta$, which Abel did not mention or justify (though the justification is easy), leads to a primitive for the remaining term, namely

$$H = \Pi \frac{f x}{(x - \alpha)\sqrt{\varphi x}} \log \frac{\theta x \sqrt{\varphi_1 x} + \theta_1 x \sqrt{\varphi_2 x}}{\theta x \sqrt{\varphi_1 x} - \theta_1 x \sqrt{\varphi_2 x}}.$$

Thus, if ψx denotes any indefinite integral of $\dfrac{f x . dx}{(x - \alpha)\sqrt{\varphi x}}$, and x_j are the functions of a_k and c_l determined by the equation $F x = 0$, it follows that

$$\varepsilon_1 \psi x_1 + \varepsilon_2 \psi x_2 + \cdots \varepsilon_\mu \psi x_\mu + G - H$$

is a function of a_k and g_l whose differential is zero. This function is therefore constant. Abel expressed this result as the following equation:

$$(30) \quad \varepsilon_1 \psi x_1 + \varepsilon_2 \psi x_2 + \cdots \varepsilon_\mu \psi x_\mu = C - \frac{f\alpha}{\sqrt{\varphi \alpha}} \log \frac{\theta \alpha \sqrt{\varphi_1 \alpha} + \theta_1 \alpha \sqrt{\varphi_2 \alpha}}{\theta \alpha \sqrt{\varphi_1 \alpha} - \theta_1 \alpha \sqrt{\varphi_2 \alpha}}$$

$$+ \Pi \frac{f x}{(x - \alpha)\sqrt{\varphi x}} \log \frac{\theta x \sqrt{\varphi_1 x} + \theta_1 x \sqrt{\varphi_2 x}}{\theta x \sqrt{\varphi_1 x} - \theta_1 x \sqrt{\varphi_2 x}}.$$

The choice of values for ε_j, (either $+1$ or -1) is dictated by the choice of square roots $\sqrt{\varphi_1 x_j}$ and $\sqrt{\varphi_2 x_j}$. Formula (30) is Theorem I of the paper.

Abel remarked (Theorem II) that the proof just given assumes that the x_j are all distinct, although the final formula is valid without this restriction. If Fx has the factorization $Fx = A(x-x_1)^{m_1}(x-x_2)^{m_2}\cdots(x-x_\mu)^{m_\mu}$, formula (30) becomes

(31)

$$\varepsilon_1 m_1 \psi x_1 + \varepsilon_2 m_2 \psi x_2 + \cdots \varepsilon_\mu m_\mu \psi x_\mu = C - \frac{f\alpha}{\sqrt{\varphi\alpha}} \log \frac{\theta\alpha\sqrt{\varphi_1\alpha} + \theta_1\alpha\sqrt{\varphi_2\alpha}}{\theta\alpha\sqrt{\varphi_1\alpha} - \theta_1\alpha\sqrt{\varphi_2\alpha}}$$

$$+ \Pi \frac{fx}{(x-\alpha)\sqrt{\varphi x}} \log \frac{\theta x\sqrt{\varphi_1 x} + \theta_1 x\sqrt{\varphi_2 x}}{\theta x\sqrt{\varphi_1 x} - \theta_1 x\sqrt{\varphi_2 x}}.$$

The fundamental formula (31) is in convenient form for wide application, after suitable small adjustments are made. The main idea is to cause one or both of the logarithmic terms to disappear by suitable auxiliary assumptions. For example, if $f\alpha = 0$, the first logarithmic term will disappear. Since α is an arbitrary constant, this is easily done. As Abel put it, one need only consider $(x-\alpha)fx$ instead of fx. The formula (31) then becomes

$$\varepsilon_1 m_1 \psi x_1 + \varepsilon_2 m_2 \psi x_2 + \cdots \varepsilon_\mu m_\mu \psi x_\mu = C + \Pi \frac{1}{\sqrt{\varphi x}} \log \frac{\theta x\sqrt{\varphi_1 x} + \theta_1 x\sqrt{\varphi_2 x}}{\theta x\sqrt{\varphi_1 x} - \theta_1 x\sqrt{\varphi_2 x}}.$$

where now $\psi x = \int \frac{fx.dx}{\sqrt{\varphi x}}$. (This is Theorem III of Abel's paper.) If, in addition, one assumes that φx is a polynomial of degree ν and fx a polynomial of degree at most $\frac{\nu-2}{2}$, the second term also vanishes (since the expansion of the logarithmic term in powers of $\frac{1}{x}$ begins with $(\frac{1}{x})^2$), and so $\varepsilon_1 m_1 \psi x_1 + \varepsilon_2 m_2 \psi x_2 + \cdots \varepsilon_\mu m_\mu \psi x_\mu$ is constant (Theorem VI). More generally (Theorem IV) the second logarithmic term disappears in (31), leaving only the first logarithmic term, when $(fx)^2$ is of lower degree than φx (in this theorem ψx is once again $\int \frac{fx.dx}{(x-\alpha)\sqrt{\varphi x}}$). Of more importance for the actual application of the formula was a generalization that seems to foreshadow Cauchy's formulas for the derivatives of an analytic function. Letting $\psi x = \int \frac{dx}{(x-\alpha)^k\sqrt{\varphi x}}$, Abel deduces (Theorem V) that

$$\varepsilon_1 m_1 \psi x_1 + \varepsilon_2 m_2 \psi x_2 + \cdots \varepsilon_\mu m_\mu \psi x_\mu =$$

$$= C - \frac{1}{1\cdot 2\cdots(k-1)} \cdot \frac{d^{k-1}}{d\alpha^{k-1}}\left(\frac{1}{\sqrt{\varphi\alpha}} \log \frac{\theta\alpha\sqrt{\varphi_1\alpha} + \theta_1\alpha\sqrt{\varphi_2\alpha}}{\theta\alpha\sqrt{\varphi_1\alpha} - \theta_1\alpha\sqrt{\varphi_2\alpha}}\right).$$

This last formula makes it possible to replace the polynomial fx by a rational function rx through the simple device of a partial-fractions expansion (Theorem VIII).

We come at last to the question of the principal application of formula (31). It is intended, as Abel showed in the beginning, to handle integrals of the form

$$\psi x = \int \frac{r\,dx}{\sqrt{\varphi x}}.$$

where rx is a rational function and φx is a polynomial of degree $2\nu - 1$ or ν. In order to apply the theorem to a sum (or difference), which Abel denotes

$$\psi x_1 + \psi x_2 + \cdots + \psi x_{\mu_1} - \psi x_1' - \psi x_2' - \cdots - \psi x_{\mu_2}',$$

it is necessary to choose the functions θx and $\theta_1 x$, i.e., the coefficients a_k and c_l, in such a way that the values x_j will be roots of the polynomial $Fx = (\theta x)^2 \varphi_1 x - (\theta_1 x)^2 \varphi_2 x$. This can be done, since the equations

$$\theta x_i \sqrt{\varphi_1 x_i} = \theta_1 x_i \sqrt{\varphi_2 x_i}$$

and

$$\theta x_j' \sqrt{\varphi_1 x_j'} = -\theta_1 x_j' \sqrt{\varphi_2 x_j'}$$

are linear in these coefficients. There are $m + n + 2$ such coefficients and it might seem that there could be $m + n + 2$ such equations. However, at most $m + n + 1$ of these equations can be independent, since the roots of cFx are the same as the roots of Fx. Assuming that there really are $m + n + 1$ independent equations, we may regard Fx as determined. The equation $Fx = 0$ then has x_j and x_j' among its roots. The question is: what *other* roots does this equation have? These remaining roots will yield the fixed number of indefinite integrals to which the sum of any number of indefinite integrals can be reduced. A simple computation shows that the number must be at least $\nu - 1$ when the degree of φx is $2\nu - 1$ or 2ν. Thus Abel had his principal theorem (Theorem VIII):

(10) $\psi x_1 + \psi x_2 + \cdots + \psi x_{\mu_1} - \psi x_1' - \psi x_2' - \cdots - \psi x_{\mu_2}' =$

$$= v + \varepsilon_1 \psi y_1 + \varepsilon_2 \psi y_2 + \cdots + \varepsilon_{\nu-1} \psi y_{\nu-1},$$

v being algebraico-logarithmic and $\varepsilon_1, \varepsilon_2, \ldots, \varepsilon_{\nu-1}$ equal to $+1$ or -1. Here the y_j are determined algebraically in terms of the x_j, x_j', $\sqrt{\varphi_1 x_j}$, etc., as the roots of the polynomial equation

$$\frac{Fy}{(y - x_1) \cdots (y - x_{\mu_1})(y - x_1') \cdots (y - x_{\mu_2}')} = 0.$$

and the ε_j are determined from the equations

$$\theta y_j \sqrt{\varphi_1 y_j} = \varepsilon_j \theta_1 y_j \sqrt{\varphi_2 y_j}.$$

This concludes our summary of Abel's paper [1828]. By understanding the motivation for the steps we can also gain an intuitive feeling for Abel's approach to algebraic integrals as expressed in the general Abel Theorem. In inexact form the theorem goes somewhat as follows: Consider any integral of the form $\psi(x) = \int f(x,y)\,dx$, where $f(x,y)$ is a rational function of x and y and y is regarded as a function of x determined by an equation $p(x,y) = 0$. In general $\psi(x)$ is not an elementary function—reducing such integrals to elementary functions was a major preoccupation of mathematicians both before and after Abel and one of Abel's preoccupations as well—but it is possible to find a finite number of upper limits such that when the integrals are combined by addition or subtraction the result is an elementary function. This is achieved by restricting x by means of a second polynomial relation $q(x,y) = 0$. This second restriction allows y to be expressed as a rational function $\varphi(x)$, and thereby leads to an equation of the form $0 = F(x) = A(x)p(x, \varphi(x))$, whose roots $x_1, \ldots, x_\mu$ are the allowable upper limits of integration. Herein is the essence of Abel's Theorem. Seen in this light, it appears to be a natural extension of one of his primary preoccupations: To find all cases in which integrals of this type are expressible in terms of logarithms. Assuming that problem solved (as Abel claimed in [1826a]), Abel's Theorem answers the next logical question: What can be said about integrals that are *not* expressible in terms of logarithms?

The paper just summarized is important not only for the intrinsic value of the results it contained but also because of the further work it inspired. Abel's use of an algebraic principle equivalent (for rational functions) to the residue theorem, which was bound to be noticed, certainly provided a motivation—even if there had been no other—to apply the Cauchy integral and put Abel's great insights in their proper context, in what we now call analytic function theory. A second source of future work arising from this paper, for which the Cauchy integral would eventually prove to be absolutely essential, was the Jacobi Inversion Problem. The fact that the number of integrals to which a sum of integrals could be reduced is in general larger than one introduces a certain indeterminacy into Abel's Theorem. This indeterminacy was pointed out by Jacobi, and the work arising from it will be the subject of future papers by the present author.

BIBLIOGRAPHY

ABEL, N. H. (1802–1829)

1826a Sur l'intégration de la formule différentielle $\frac{\rho\,dx}{\sqrt{R}}$, R et ρ étant des fonctions entières. *Journal für die reine und angewandte Mathematik*, 1 = *Œuvres*, T. I, 104–144.

1826b Mémoire sur une propriété générale d'une classe très étendue de fonctions transcendantes. *Mémoires présentés par divers savants*, VII, (1841) = *Œuvres*, T. I, 145–211.

1827a Recherche de la quantité qui satisfait à la fois à deux équations algébriques données. *Annales de mathématiques pures et appliquées*, XVII = *Œuvres*, T. 1. 212–218.

1827b Recherches sur les fonctions elliptiques. *Journal für die reine und angewandte Mathematik*, 2, 3 = *Œuvres*, T. I, 262–388.

1828 Remarques sur quelques propriétés générales d'une certaine sorte de fonctions transcendantes. *Journal für die reine und angewandte Mathematik*, 3 = *Œuvres*, T. I, 444–456.

1829a Démonstration d'une propriété générale d'une certain classe de fonctions transcendantes. *Journal für die reine und angewandte Mathematik*, 4 = *Œuvres*, T. I, 515–517.

1829b Précis d'une théorie des fonctions elliptiques. *Journal für die reine und angewandte Mathematik*, 4 = *Œuvres*, T. I, 518–617.

Œuvres *Œuvres Complètes de Niels Henrik Abel*, I, II, Second Edition. Christiania: Grøndahl & Son (1881). Johnson Reprint Corporation, New York, 1965.

AYOUB, Raymond

1984 The lemniscate and Fagnano's contributions to elliptic integrals, *Archive for the History of the Exact Sciences*, 29, No. 2, 131–149.

BRILL, A. (1842–1935) and NOETHER, M. (1844–1921)

1894 Die Entwicklung der Theorie der algebraischen Functionen in älterer und neuerer Zeit. *Jahresbericht der deutschen Mathematiker-Vereinigung*, III, (1892–93), I–XXII, 109–566.

EULER, L. (1707–1784)

1755 *Institutiones Calculi Differentialis*, Academiae Imperialis Scientiarum Petropolitanae.

1761 De integratione aequationis differentialis

$$\frac{m\,dx}{\sqrt{(1 - x^4)}} = \frac{n\,dy}{\sqrt{(1 - y^4)}}$$

Novi Commentarii Academiae Scientiarum Imperialis Petropolitanae, **VI**, 7–9, 37–57.

1768 *Institutionum Calculi Integralis Volumen Primum*, Academiae Imperialis Scientiarum Petropolitanae.

1777a Specimen integrationis abstrusissimae hac formula

$$\int \frac{\partial x}{(1 + x)\sqrt[4]{(2xx - 1)}}$$

contentae, *Nova Acta Acad. Imp. Scient.*, **IX**, 98–117.

1777b Evolutio formulae integralis

$$\int \frac{\partial z(3 + zz)}{(1 + zz)\sqrt[4]{(1 + 6zz + z^4)}}$$

per logarithmos et arcus circulares, *Nova Acta Acad. Imp. Scient.*, **IX**, 127–131.

1777c Integratio succincta formulae integralis

$$\int \frac{\partial z}{(3 \pm zz)\sqrt[3]{(1 \pm 3zz)}}$$

Nova Acta Acad. Imp. Scient., **X**, 20–26.

FAGNANO, G. (1682–1766)
  Opere *Opere Matematiche del Marchese Giulio Carlo De' Toschi di Fagnano*, I–III, Milano–Roma–Napoli, Società Editrice Dante Alighieri di Albrighi, Segati e C., 1911–1912.

FRICKE, R. (1861–1930)
  1920 Elliptische Funktionen. *Encyklopädie der Mathematischen Wissenschaften*, **II B 3**, 177–348.

HOFFMAN, Jos. E.
  1965 Aus der Frühzeit der Infinitesimalmethoden: Auseinandersetzung um die algebraische Quadratur algebraischer Kurven in

der zweiten Hälfte des 17. Jahrhunderts, *Archive for the History of the Exact Sciences*, **2**, No. 4, 271–343.

HOLMBOE, B. (1795–1850)

1839 Notices sur la vie de l'auteur. *Œuvres Complètes de N. H. Abel*, T. I, First edition. V–XVI. Christiania: Grøndahl.

JACOBI, C. G. J. (1804–1851)

1832 Considerationes generales de transcendentibus abelianis. *Journal für die reine und angewandte Mathematik*, **9**, 394–403 = *Werke*, Bd. 2, 5–16.

Werke *C.G.J. Jacobi's Gesammelte Werke*, I-VIII, Berlin 1881–1891. Chelsea Publishing Company, 1969.

KRAZER, A. (1858–1926) and WIRTINGER, W. (1865–1945)

1920 Abelsche Funktionen und allgemeine Thetafunktionen. *Encyclopädie der mathematischen Wissenschaften*, **II B 7**, 604–873.

MACLAURIN, Colin (1698–1746)

1742 *Treatise on Fluxions*, Vols. I and II, T. W. and T. Ruddimans, Edinburgh.

ORE, Øystein (1899–1968)

1957 *NIELS HENRIK ABEL, Mathematician Extraordinary*. University of Minnesota Press.

SIEGEL, C. L.

1956 Zur Vorgeschichte des Eulerschen Additionstheorems, (Beitrag, eingereicht am 29.9.1956) = *Sammelband Leonhard Euler*, Akademie-Verlag, Berlin 1959, 315–317 = *Gesammelte Abhandlungen*, Band III (1966), 249–251.

STÄCKEL, P. (1862–1919) AND AHRENS, W. (1872–1927)

1908 *Der Briefwechsel zwischen C. G. J. Jacobi und P. H. von Fuss über die Herausgabe der Werke Leonhard Eulers*, Leipzig.

WIRTINGER, W. (1865–1945)

1920 Algebraische Funktionen und ihre Integrale. *Encyklopädie der Mathematischen Wissenschaften*, **II B 2**, 115–175.

Number Theory

Richard Dedekind (1831–1916)
Courtesy of Springer-Verlag Archives

Heinrich Weber (1842–1913)
Courtesy of Springer-Verlag Archives

Heinrich Weber and the Emergence of Class Field Theory

Günther Frei

Dedicated to my inspiring teacher Professor B.L. van der Waerden

Table of Contents

1. Historical Concepts and Approaches to Class Field Theory
2. Euler's ζ-function
3. Dirichlet's L-series
4. Dedekind's ζ-function
5. Weber's Complex Multiplication, L-series and Class Field Theory
6. Dedekind's Contribution towards Class Field Theory
7. Chronological Table to Heinrich Weber

1. HISTORICAL CONCEPTS
   AND APPROACHES TO CLASS FIELD THEORY

1.1 Historical Concepts and Definitions of a Class Field

1. The modern concept of a *class field* is due to Teiji *Takagi* (1875-1960), who established in a long and fundamental paper in 1920 (see [Ta-1920]) class field theory as the theory of all abelian algebraic extensions **K**, **K/k**, with respect to an algebraic number field **k** as a base field. This general point of view however emerged only in the wake of extensive developments that began some 60, or even 140 years earlier.

2. The concept of a class field first arises in the work of Leopold *Kronecker* (1823-1891) on elliptic functions with complex multiplication applied to algebraic number theory. Kronecker uses the terminology "species associated with a field k" (instead of "class field of k"), which means the smallest field **K** over **k** in which every ideal in **k** becomes principal (see [Kr-1882]; Werke II, p. 322) and refers in particular to the situation where the base field **k** is an imaginary quadratic field, $\mathbf{k} = \mathbb{Q}(\sqrt{-m})$, $m > 0$, $m \in \mathbb{N}$ and **K** is the so called absolute class field to **k**, namely the field generated by the singular moduli $j(\omega)$ of **k**,

$K = k(j(\omega)) = \mathbb{Q}(\sqrt{-m}, j(\omega))$. It has the property, as Kronecker discovered, that the Galois group $G = Gal(K,k)$ is isomorphic to the class group C_k in k and that the structure of the class group in k is reflected by arithmetic properties in K (see [Kr-1882], in particular §19, pp. 65–68 (Werke II, pp. 322–324) and [Kr-1857], [Kr-1862], in particular Werke IV, p. 214 and [Kr-1877], in particular Werke IV, pp. 70-71). Furthermore, it is an unramified extension over k, a fact that Kronecker uncovered empirically and for which he had no proof (see [Kr-1862], Werke IV, p. 213). We shall henceforth call $K = \mathbb{Q}(\sqrt{-m}, j(\omega))$ the *Kronecker class field* of $k = \mathbb{Q}(\sqrt{-m})$.

3. When Heinrich *Weber* (1842-1913) introduced the term *class field* for the first time in his book on elliptic functions and algebraic numbers "*Elliptische Funktionen und algebraische Zahlen*" in 1891 he still only meant the Kronecker class field (see [We-1891], p.439). Only in 1896, in an important series of three papers on number groups in algebraic number fields (see [We-1897]), did Weber enlarge the concept of a class field to fields K associated with a congruence class group in k (see [We-1897], II, p. 87 (§1, Voraussetzung 4) and p. 97 (§4)), but only in the second edition of his *Lehrbuch der Algebra* (see [We-1908], Viertes Buch) was the term *class field* used to designate a general class field. This general definition is related to the decomposition law of primes in k with respect to K and is based on an observation going back to Fermat and Euler that prime numbers p represented by certain norm forms, in this case binary quadratic forms, belong to certain congruence classes modulo a number m, i.e. can be described by congruence conditions modulo m; e.g. a prime $p \neq 2$ is a sum of two squares $p = x^2 + y^2$ with integers $x, y \in \mathbb{Z}$ if and only if $p \equiv 1 \bmod 4$. Following Weber's point of view, class field theory becomes the theory of norm forms of algebraic field extensions K/k, that is the theory of diophantine equations in k arising as norm forms with respect to an abelian extension K.

4. David *Hilbert* (1862-1943) in his *Zahlbericht* of 1897 gives still another definition of a class field (see [Hi-1897], §58, p. 279). For Hilbert the class field K of k designates an unramified abelian extension K over k, i.e. an abelian extension whose relative discriminant $d(K/k)$ is equal to 1. It has the property that certain ideal classes in k become principal in K. Hence K is related to certain ideal classes in k. A particular case is the *absolute class field* H, or *Hilbert class field* to k, as we say today, which has the property that its Galois group $Gal(H/k)$ is isomorphic to the ideal class group C_k in k. This extension H/k is unramified, abelian and every ideal in k becomes principal in H, properties that were proved later by Philipp *Furtwängler* (1869-1940) in 1907 and by Emil *Artin* (1898-1962) and *Furtwängler* in 1930 (see [Fu-1907] and [Fu-1930]).

5. A larger view was later taken by *Hilbert*, certainly under the influence of Weber's work (Hilbert became Weber's successor in Göttingen in 1895 (see [Fr-1985], Brief 97, p. 115)) in his theory of relatively quadratic number fields, where Hilbert developed class field theory for these particular fields, but having in mind a general theory of relatively abelian extensions (see [Hi-1899a], in particular p. 94 (Werke I, p. 369), [Hi-1899b], in particular p. 1 (Werke I, p. 370) and [Hi-1898], in particular Werke I, p. 484 and pp. 508-509).

1.2 Historical Approaches to Class Field Theory

There are four aspects of number theory that finally merged into class field theory, each one having its own independent development.

1. Genus theory and representation of primes
   by norm forms.

The first aspect goes back to Pierre de *Fermat* (1601-1665) and Leonhard *Euler* (1707-1783), who noticed that primes p represented by norm forms, in this case *binary quadratic forms*, are characterized by the fact that they split completely in the extension field K/k belonging to this norm form, e. g. $p = x^2 + y^2 = (x + iy)(x - iy)$ in $\mathbb{Q}(i) = \mathbb{Q}(\sqrt{-1})$ with the norm form $x^2 + y^2$. These primes can be characterized by congruence conditions, e.g. $p = x^2 + y^2$ is solvable with integers $x, y \in \mathbb{Z}$ if and only if $p \equiv 1 \bmod 4$.

This aspect was developed by Leonhard *Euler* (convenient numbers, 1778) and Carl Friedrich *Gauss* (1801) for quadratic forms (see [Fr-1983], in particular Theorem 3.15, p. 39 and [Fr-1979], in particular Theorem 3.8, p. 41), by Ernst Eduard *Kummer* (1810-1893) for cyclotomic fields (see [Ku-1859], in particular §14, pp. 116-128) by *Hilbert* for Kummer extensions (see [Hi-1897], Capitel XXXII) and by *Takagi* for cyclic extensions (see [Ta-1920], in particular Capitel II, pp. 48-62). The clue to this is the *norm residue symbol* and the relation to the *reciprocity law*. This work has been studied to some extent in [Fr-1979] and [Fr-1983].

2. Explicit construction of abelian extensions.

The second aspect, the construction of a class field by means of singular values of certain transcendental functions, was initiated by *Kronecker* and further developed by *Weber* for fields $k = \mathbb{Q}$ and $k = \mathbb{Q}(\sqrt{-m})$ by means of the exponential functions and the elliptic functions (with complex multiplication) and by Erich *Hecke* (1887-1947) in his doctoral thesis (1910) and in his habilitation thesis (1912) for real quadratic fields and the corresponding biquadratic imaginary fields by means of Hilbert's modular functions (see [He-1912] and [He-1913]). This approach will be studied in more detail in a separate paper.

3. Non-ramified extensions.

A third aspect, the study of non-ramified extensions was developed by *Hilbert* (see [Hi-1897], §58, p. 279). Hilbert seems to have been motivated by the analogy with the existence theorem of a finite integral on a closed Riemann surface (see [Hi-1900], Problem 12). Another motivation was the search for a field K/k, such that all ideals in k become principal in K (see [Hi-1897], §58, p. 279).

Already *Kronecker* had conjectured that the absolute class field over $\mathbb{Q}(\sqrt{-m})$, $m \in \mathbb{N}$, is non-ramified (see [Kr-1862], Werke IV, p. 213) and *Weber* (1896) was very well aware of this important property (see [We-1897], I, p. 435 and III, p. 25). We plan to take up these developments in a sequel to this paper.

4. Analytic approach.
Primes in arithmetic progressions.

In this paper we shall study the analytic approach pursued by *Euler* (1737, ζ-function), *Dirichlet* (1837, *L*-functions), *Dedekind* (1871, ζ-function of a field) and *Weber* (1896, *L*-functions for congruence groups). This approach was also the one taken by Weber in his series of papers of 1896/7 (see [We-1897]), where he introduced the general notion of a class field. Weber shows, in fact, that the analytic part yields the *first fundamental inequality* of class field theory, $h \leq \nu$ (see Theorem 28(a) or [We-1897], II, §1, p. 88, Satz I). This played an essential part in Teiji *Takagi's* proof of the main theorems, the other being the *second fundamental inequality* $\nu \leq h$ obtained by Takagi from the genus theory (see [Ta-1920], §12, in particular pp. 48-50).

5. *Weber* was already well aware of these four aspects of class field theory (see [We-1897]; in particular, as to 1. see I, p. 434 and p. 362 ff., II, pp. 95-96 and p. 99; as to 2. see III; as to 3. see I, p. 435 and III, p. 25 and as to 4. see I, II, III).

2. EULER'S ζ-FUNCTION

2.1 Early History

1. In his treatise *Quaestiones super Geometriam Euclidis* about 1360, *Nicole Oresme* (1323-1382) already distinguished certain types of series by giving a criterion for convergence and divergence which he used, among other things, to establish the divergenece of the *harmonic series* (see [Cl-1968]):

THEOREM 1.

$$1 + \frac{1}{2} + \frac{1}{3} + \frac{1}{4} + \ldots = \sum_{n=1}^{\infty} \frac{1}{n}$$

is divergent.

2. Pietro *Mengoli* (1625-1686) reproved this theorem. He followed up the work of Cataldi in his *Novae Quadraturae Arithmeticae*, published in Bologna in 1650, where he considered infinite algorithms and infinite series. He then posed the following

PROBLEM 2. *Show that*

$$\sum_{n=1}^{\infty} \frac{1}{n^2}$$

has a finite value v and determine this value v.

3. Approximate values for this value $v = \zeta(2)$ were given by:

John *Wallis* (1616-1703) in his *Arithmetica Infinitorum* in 1655

Daniel *Bernoulli* (1700-1782) in a letter to Goldbach in 1728

Christian *Goldbach* (1690-1764) in his answer to Daniel Bernoulli in 1728

James *Sterling* (1692-1770) in 1720

Leonhard *Euler* (1707-1783) in his article *De Summatione Innumerabilium Progressionum* of 1731 and more precisely in *Invertio Summae Cuiusque Seriei ex Dato Termino Generali* of 1736.

4. Finally *Euler* succeeded in giving the explicit value for $v = \zeta(2)$ in a remarkable paper *De Summis Serierum Reciprocarum* of 1735 that contributed much to his fame:

THEOREM 3.

$$\zeta(2) = \frac{\pi^2}{6} \qquad \zeta(4) = \frac{\pi^4}{90}$$

2.2 THE EULER ζ-FUNCTION

Euler introduced the ζ-function in 1737 in his paper *Variae Observationes circa Series Infinitas*, published in 1744 (see [Eu], series (1), Vol.14, pp. 216-244):

DEFINITION 4. $\zeta(s) = \sum_{n=1}^{\infty} \frac{1}{n^s} \qquad s \in \mathbb{R}$

Remarks.

1. $\zeta(k)$ for $k \in \mathbb{N}$, $k \geq 2$ was already considered by Jacob *Bernoulli* (1654-1705); e. g. he showed that

$$\left(\sum_{n \equiv 1(2)} \frac{1}{n^k} \right) \Big/ \left(\sum_{n \equiv 0(2)} \frac{1}{n^k} \right) = 2^k - 1$$

2. $\zeta(s)$ converges absolutely if $s > 1$.

In 1737 Euler proved the following results:

THEOREM 5. $\zeta(s) = \sum\limits_{n=1}^{\infty} \dfrac{1}{n^s} = \prod\limits_{p} \dfrac{1}{1 - \dfrac{1}{p^s}}$ $\qquad$ (over all primes p) $s > 1$.

PROOF. By means of the fundamental theorem of arithmetic (first proved by Gauss in his *Disquisitiones Arithmeticae* of 1801).

COROLLARY 6. *There are infinitely many prime numbers.*

More precisely, Euler obtained a stronger result:

THEOREM 7. $\sum\limits_{p} \dfrac{1}{p}$ *is divergent, i. e.* $\lim\limits_{s \to 1+} \sum\limits_{p} \dfrac{1}{p^s} = \infty$.

PROOF.

$$\log \zeta(s) = \log \left(\sum_{n=1}^{\infty} \frac{1}{n^s} \right) = -\sum_{p} \log \left(1 - \frac{1}{p^s} \right)$$

$$= \sum_{p} \left(\frac{1}{p^s} + \frac{1}{2} \left(\frac{1}{p^s} \right)^2 + \frac{1}{3} \left(\frac{1}{p^s} \right)^3 + \cdots \right)$$

$$= \sum_{p} \frac{1}{p^s} + \sum_{p} \left(\frac{1}{2} \left(\frac{1}{p^s} \right)^2 + \cdots + \frac{1}{m} \left(\frac{1}{p^s} \right)^m + \cdots \right) = \sum_{p} \frac{1}{p^s} + A$$

where $A = \sum\limits_{p} \left(\sum\limits_{m=2}^{\infty} \dfrac{1}{m} \left(\dfrac{1}{p^s} \right)^m \right)$ remains bounded when $s \to 1^+$. Since $\lim\limits_{s \to 1^+} \log \zeta(s) = \infty$, one has that the series $\sum\limits_{p} \dfrac{1}{p}$ is divergent.

Remarks. Euler also found the following two properties.

1. After having first computed $\zeta(2k)$ in 1735 (see [Eu], E. 41), Euler derived in 1739 the connection with the Bernoulli numbers (see [Eu], E. 130):

$$\zeta(2k) = \sum_{n=1}^{\infty} \frac{1}{n^{2k}} = (-1)^{k-1} \frac{(2\pi)^{2k}}{2(2k)!} B_{2k}$$

Euler defined the *Bernoulli numbers* B_{2k} as follows.
Definition:

$$\frac{t}{e^t - 1} = 1 - \frac{t}{2} + B_2 \frac{t^2}{2!} + \cdots = \sum_{k=0}^{\infty} B_k \frac{t^k}{k!}$$

$$B_0 = 1, \; B_1 = -\frac{1}{2}, \; B_2 = \frac{1}{6}, \; B_3 = 0, \; B_4 = -\frac{1}{30}, \; B_6 = \frac{1}{42}, \; B_8 = -\frac{1}{30},$$
$$B_{10} = \frac{5}{66}, \; \cdots .$$

Hence

$$\zeta(2) = \frac{(2\pi)^2}{2 \cdot 2} \cdot \frac{1}{6} = \frac{\pi^6}{6}, \qquad \zeta(4) = \frac{(2\pi)^4}{2 \cdot 4!} \cdot \frac{1}{30} = \frac{\pi^4}{90}.$$

The Bernoulli numbers were introduced by Jacob Bernoulli in his book *Ars Conjectandi*, published in 1713.

In 1735 Euler also computed $\zeta(2k + 1)$ for $k \in \mathbf{N}$ without finding an explicit formula.

2. In 1749 Euler found the *functional equation* (see [Eu], Series (1), 15, pp. 70–90; E. 352)

$$\zeta(1 - s) = 2(2\pi)^{-s} \cos \frac{\pi s}{2} \Gamma(s)\zeta(s) \qquad \text{for } s \in \mathbf{R}$$

It was proven by Bernhard *Riemann* (1826-1866) in 1859.

3. *Riemann* introduced $\zeta(s)$ as a complex function, $s \in \mathbf{C}$ and showed:

THEOREM 8.

(a) $\zeta(s)$ can be extended to the whole complex plane $\mathbf{C}$.

(b) $\zeta(s)$ is a meromorphic function with a single pole at $s = 1$ which is simple and has residue 1.

4. *Euler* also conjectured (see [Eu], *Opuscula Analytica* 2, 1783):

   If $(a, m) = 1$, there are infinitely many primes p with $p \equiv a \pmod{m}$.

3. DIRICHLET'S L-SERIES

3.1 Dirichlet's Theorem

Peter Gustav Lejeune *Dirichlet* (1805-1859) built upon Euler's ideas in a fundamental article *Beweis des Satzes, dass jede unbegrenzte arithmetische Progression, deren erstes Glied und Differenz ganze Zahlen ohne gemeinschaftlichen Factor sind, unendlich viele Primzahlen enthält* (see [Di-1837] and also [Di-1838] and [DD-1863], Supplement VI). There he showed:

THEOREM 9. *Let $m \in \mathbf{N}$ and $a \in \mathbf{Z}$ and $(a, m) = 1$. Then*

1.

$$\lim_{s \to 1^+} \sum_{p \equiv a(m)} \frac{1}{p^s} = \infty$$

(*over all primes p congruent to a modulo m*), i.e.

$$\sum_{p \equiv a(m)} \frac{1}{p}$$

is divergent.

2. *There are infinitely many primes* p, *with* $p \equiv a \pmod{m}$

PROOF: Several new ideas had to be introduced by Dirichlet.

1. First Dirichlet had to single out the primes $p \equiv a \pmod{m}$. This was done by means of characters modulo m. Let us consider the case $m = q = \text{prime} \neq 2$ (see [Di-1837], §1). Let g be a primitive root modulo q and ω a $(q-1)$st root of unity. If $\gamma(n)$ is the index of n modulo q with respect to g, i.e. $\gamma(n)$ is the smallest non-negative integer such that $g^{\gamma(n)} \equiv n \pmod{q}$, hence $0 \le \gamma(n) \le q - 2$, then define a *character*

DEFINITION 10.

$$\chi_\omega(n) = \begin{cases} \omega^{\gamma(n)}, & \text{if } (n, q) = 1 \\ 0, & \text{otherwise.} \end{cases}$$

One has the properties

PROPOSITION 11. (a) $\chi_\omega(n + q) = \chi_\omega(n)$.

(b) $\chi_\omega(rs) = \chi_\omega(r)\chi_\omega(s)$ *that is,* χ_ω *is a character modulo* q, *a homomorphism* $\chi_\omega : \mathbf{Z}/q\mathbf{Z} \longrightarrow \mathbf{C}$.

By multiplying with the conjugate complex character $\overline{\chi}_\omega$ and summing over all characters modulo q, Dirichlet derives the characteristic function (see [Di-1837], §6, p. 332)

(c) $\dfrac{1}{q-1} \sum_\omega \overline{\chi}_\omega(a)\chi_\omega(p) = \begin{cases} 1 & \text{if } p \equiv a \pmod{q} \\ 0 & \text{otherwise.} \end{cases}$

2. Next Dirichlet introduced the L-series for each character χ modulo q (see [Di-1837], §1):

DEFINITION 12. $L(s, \chi) = \displaystyle\sum_{n=1}^{\infty} \dfrac{\chi(n)}{n^s}$ *for* $s \in \mathbf{R}$ *and* $s > 1$.

These have the *Euler product property* (see [Di-1837], §1):

PROPOSITION 13. $L(s, \chi) = \displaystyle\prod_p \left(1 - \dfrac{\chi(p)}{p^s}\right)^{-1}$ *for* $s > 1$.

3. By means of Proposition 11 as in the proof of Theorem 7, Dirichlet also proved

$$\frac{1}{q-1} \sum_{\omega} \overline{\chi}_{\omega}(a) \log L\left(s, \chi_{\omega}\right)$$

$$= \frac{1}{q-1} \sum_{\omega} \overline{\chi}_{\omega}(a) \left(-\sum_{p} \log \left(1 - \frac{\chi_{\omega}(p)}{p^s}\right)\right)$$

$$= \frac{1}{q-1} \sum_{\omega} \overline{\chi}_{\omega}(a) \sum_{p} \sum_{m=1}^{\infty} \frac{\chi_{\omega}(p^m)}{mp^{ms}}$$

$$= \sum_{p \equiv a(q)} \frac{1}{p^s} + \sum_{m=2}^{\infty} \sum_{p^m \equiv a(q)} \frac{1}{mp^{ms}}$$

$$= \sum_{p \equiv a(q)} \frac{1}{p^s} + A$$

where A remains bounded if $s \to 1^+$.

4. For the principal character $\omega = 1$, Dirichlet obtained (see [Di-1837], §2) the

PROPOSITION 14. (a) $L(s, \chi_1) = \prod_{p|q} \left(1 - \frac{1}{p^s}\right) \zeta(s) = \left(1 - \frac{1}{q^s}\right) \zeta(s)$

and hence by Theorem 8

(b) $L(s, \chi_1)$ is a meromorphic function with a single pole at $s = 1$ which is simple and has residue $\left(1 - \frac{1}{q}\right) = \frac{\phi(q)}{q}$, where ϕ denotes the Euler ϕ-function.

Hence Dirichlet had to prove that $\log L(1, \chi_{\omega})$ is finite if $\omega \neq 1$, i.e. that (see [Di-1837], §4)

THEOREM 15. $L(1, \chi_{\omega}) \neq 0$ if $\omega \neq 1$

The proof required another new idea, namely the creation of a fundamental connection with the class number of quadratic fields.

(5.1) First Dirichlet showed that $L(1, \chi_{\omega}) \neq 0$ if $\omega \neq 1, -1$, i.e. if χ_{ω} is not a real character (see [Di-1837], §5, p. 329).

(5.2) In the case where $\omega \neq 1$, but χ_{ω} real, i.e. $\omega = -1$, Dirichlet demonstrated that the corresponding character χ_{ω} is a quadratic character $\left(\frac{D}{-}\right)$ for some D, related to $m = q$; that is the field $\mathbb{Q}(\sqrt{D})$ is class field to the quadratic character $\chi = \chi_D = \left(\frac{D}{-}\right)$

or to the congruence group $\mathbf{H}_D$ generated by all primes p with $p \nmid D$ and $\left(\dfrac{D}{p}\right) = 1$, where $\left(\dfrac{D}{p}\right)$ is the Kronecker symbol. He showed further that the corresponding L-series has the property (see [Di-1837], §4 and §11 and [Di-1838]):

(5.3)

$$L(1, \chi_D) = c(D)h(D)$$

where $c(D)$ is a non-zero constant depending on the discriminant and on the units of the field $\mathbb{Q}(\sqrt{D})$ and $h(D)$ is the class number of that field (or of the corresponding quadratic forms with discriminant D). Hence $h(D) \geq 1$ and Theorem 15 follows (see [Di-1837] §4, p. 326).

Remarks.

1. The research on the *class number formula* (i.e. formula (5.3)) was precisely motivated by the study of the L-series (see [DD-1871], §135, p. 347 or [DD-1863], p. 384).

2.

$$c(D) = \begin{cases} \dfrac{2\pi}{\rho} \cdot \dfrac{1}{\sqrt{|D|}} & \text{if } D < 0 \\[2ex] \log \varepsilon \dfrac{1}{\sqrt{D}} & \text{if } D > 0. \end{cases}$$

where ρ denotes the number of roots of unity and ε stands for the fundamental unit in the field $\mathbf{k} = \mathbb{Q}(\sqrt{D})$. This is *Dirichlet's class number formula* (see [Di-1837], §4, p. 327 and [Di-1838] and also [DD-1871], §167, p. 484), already known to Gauss.

4. DEDEKIND'S ζ-FUNCTION

4.1 The class number formula

In order to compute the class number of a field $\mathbf{k}$ and in order to be able to follow the ideas of Dirichlet, Dedekind introduced in the Xth Supplement of the second edition of Dirichlet's *Vorlesungen über Zahlentheorie* (see [DD-1871], §167), and more explicitly in the XIth Supplement of the third edition (see [DD - 1879], §178, p. 578) the ζ-function for a field $\mathbf{k}$:

DEFINITION 16. $\zeta_{\mathbf{k}}(s) = \sum_{\mathcal{A}} \dfrac{1}{N(\mathcal{A})^s}$ *for $\Re s > 1$ over all integral ideals $\mathcal{A}$ in $\mathbf{k}$,*

where $N(\mathcal{A}) = [\mathfrak{o}(\mathbf{k}) : \mathcal{A}]$ denotes the norm of $\mathcal{A}$ and $\mathfrak{o}(\mathbf{k})$ stands for the ring of integers in $\mathbf{k}$.

Following Dirichlet, Dedekind had to evaluate $\lim\limits_{s \to 1^+} (s - 1)\zeta_{\mathbf{k}}(s)$ in two different ways. To that end Dedekind established first the class number formula (see [DD - 1871], §167)

THEOREM 17.

$$h(\mathbf{k}) = \frac{\rho\sqrt{|d(\mathbf{k})|}}{2^{r+t}\pi^t R(\mathbf{k})} \lim_{s\to 1+} (s-1)\zeta_{\mathbf{k}}(s)$$

and applied this formula to derive *Kummer's class number fomula* for cyclotomic fields and Dirichlet's class number formula for quadratic fields (or quadratic forms) (see [DD-1871],§168, or for more details [DD - 1879], §179 and §180).

The notations mean the following:
$h(\mathbf{k})$ = the class number in $\mathbf{k}$, ρ = number of roots of unity in $\mathbf{k}$, $d(\mathbf{k})$ = discriminant of $\mathbf{k}$, $R(\mathbf{k})$ = regulator of $\mathbf{k}$, r = number of real and t = number of pairs of conjugate complex imbeddings in **C**.

In order to compute the class number $h(\mathbf{k})$ Dedekind studied the distribution of integral ideals in a given ideal class A and established the following limit formula (see the second edition of Dirichlet's *Vorlesungen über Zahlentheorie* ([DD -1871], §167), or for a more detailed and precise treatment see the third or fourth edition ([DD - 1879], §178 or [DD - 1894], §184), which is based on a theorem on the number of ideal classes in orders of a field, proved by Dedekind in 1877 (see [De-1877], in particular pp. 118-119 and §§10-13) and which caused Dedekind to enlarge his treatment of 1871 considerably for the later editions of Dirichlet's *Vorlesungen über Zahlentheorie*).

THEOREM 18. (a) *Let* A *be an ideal class in* k *and* $\mathcal{A}$ *an integral ideal in* k. *Let* $T(A, x; \mathcal{A})$ *denote the number of integral ideals in* A *divisible by* $\mathcal{A}$, *whose norm is smaller than* x.
Then

$$\lim_{x\to\infty} \frac{T(A,x;\mathcal{A}))}{x} = \frac{2^{r+t}\pi^t R(\mathbf{k})}{N(\mathcal{A})\rho\sqrt{|d(\mathbf{k})|}}.$$

(b) *If* A = o(k) *one has*

$$\lim_{x\to\infty} \frac{T(A,x;o(\mathbf{k}))}{x} = \frac{2^{r+t}\pi^t R(\mathbf{k})}{\rho\sqrt{|d(\mathbf{k})|}} = g$$

which is a number independent of the class A, *i.e. the integral ideals are equally distributed in the ideal classes.*

This last result played an important rôle in Weber's introduction of class fields (compare with [We-1897], II, §1, p. 84).

4.2 Properties of Dedekind's ζ-function

Dedekind showed that the following properties hold (see [DD-1879], §178, p. 578).

THEOREM 19. *(a) $\zeta_k(s)$ has a simple pole at $s = 1$.*
(b) $\zeta_k(s)$ is absolutely convergent if $\Re s > 1$.

REMARKS.
(a) *Landau* proved in 1903 that $\zeta_k(s)$ can be analytically extended to
$$\Re s > 1 - \frac{1}{[k : \mathbb{Q}]} \text{ (see [La-1903])}.$$
(b) *Hecke* proved in 1917 that $\zeta_k(s)$ is a meromorphic function in the complex plane $\mathbb{C}$ with a single pole which is simple at $s = 1$ and that $\zeta_k(s)$ satisfies a functional equation (see [He-1917]).
(c) According to Theorem 17 the residue of $\zeta_k(s)$ at $s = 1$ is

$$res\zeta_k(s = 1) = \lim_{s \to 1^+} (s - 1)\zeta_k(s) = \frac{2^{r+t}\pi^t R(k)h(k)}{\rho\sqrt{|d(k)|}}.$$

Dedekind then derived the following *product formula* (see [DD-1879], §178, p. 579):

THEOREM 20. $\zeta_k(s) = \prod_{\mathcal{P}} \left(1 - \frac{1}{N(\mathcal{P})^s}\right)^{-1}$ *over all primes ideals $\mathcal{P}$ in k.*

PROOF: This is a consequence of Dedekind's theorem that in any algebraic number field k every ideal in k is the unique product of prime ideals in k.

The first treatment of Dedekind's ideal theory appeared in the Xth Supplement of the second edition of Dirichlet's *Vorlesungen über Zahlentheorie* (see [DD - 1871], §159 - §170), which was enlarged considerably in the third edition where it appeared as a new Supplement XI (see [DD - 1879], §159- §181).

In addition Dedekind indicated in a letter to Frobenius of 8 July 1896 (see [De] II, p. 435) the

THEOREM 21. *If $k/\mathbb{Q}$ is abelian, then*

$$\zeta_k(s) = \prod_{\chi \in \widehat{Gal(k/\mathbb{Q})}} L(s, \chi) = \zeta(s) \prod_{\chi \neq \chi_1} L(s, \chi).$$

REMARKS.

(a) This formula was first mentioned by Dedekind in the case where **k** is a *p*-cyclotomic field for a prime *p* (see [DD-1879], p. 596 and [DD-1894], p. 625) and in the case of a quadratic field $\mathbf{k} = \mathbb{Q}(\sqrt{d})$, where it was already known to Dirichlet (see [Di-1838]).

(b) The formula was generalized to the case **K/k** abelian by *Takagi* in 1920.

(c) That $\dfrac{\zeta_{\mathbf{K}}(s)}{\zeta_{\mathbf{k}}(s)}$ is an entire function if K/k is normal was proved by H. *Aramata* in 1933, based on Artin's work on Artin's *L*-functions. The proof was simplified by R. *Brauer* (1947) and had an important influence on the proof of *Siegel's* conjecture on a generalization of Siegel's limit formula.

5. WEBER'S COMPLEX MULTIPLICATION,
   L-SERIES AND CLASS FIELD THEORY

5.1 Introduction

1. Weber's main concern was analysis, in particular the application of analysis to number theory and physics. After he obtained his doctorate in Heidelberg on 19 February 1863, Weber went to Königsberg where he was deeply influenced by *Franz Neumann* (Bessel functions and applications to physics) and Friedrich Julius *Richelot* (1808-1875) (algebraic functions), who continued to perpetuate the tradition established ther by Carl Gustav Jacob *Jacobi* (1804-1851). Weber's first publication, *"Zur Theorie der singulären Lösungen der partiellen Differentialgleichungen"* (On the theory of singular solutions of partial differential equations), appeared in 1866 and was based on Jacobi's ideas as presented in Richelot's seminar. Weber turned next to Riemann's theory of abelian integrals. His aim was to develop this theory in a purely algebraic way. His main concern was the addition theorem of abelian integrals (1874) and the inversion problem for such integrals (1869/70). These were central problems of analysis at that time, as is demonstrated by *Carl Neumann's* book on Riemann's theory of abelian integrals (1865) and the book by Alfred *Clebsch* and Paul *Gordan* on abelian functions (1866).

2. Weber first became interested in number theory by continuing Jacobi's work on applications of θ-functions to this field. In 1882 he and Dedekind published a fundamental treatise on algebraic functions of one variable which gives a purely arithmetic and algebraic treatment of Riemann's theory of algebraic functions (without referring to the notions of continuity or limit or to geometric intuition). It was based instead on Dedekind's notion of ideals and modules, first set forth by Dedekind in 1871 in the second edition of Dirichlet's *Vorlesungen über Zahlentheorie* (see [DD-1871], Supplement X) which advanced up to the theorem of

Riemann and Roch. Weber and Dedekind had already edited Riemann's work in 1876, and from that time onward they remained in intimate scientific contact with one another. There is no doubt that Dedekind had a considerable influence on Weber's work, as Weber acknowledges, for instance, in the introduction to his book *Elliptische Funktionen und algebraische Zahlen* (November 1890). It would be an interesting task to study the mutual influence of the two close friends (see also Section 6).

3. In 1886 Weber succeeded in proving Kronecker's conjecture that any abelian algebraic field over the rationals $\mathbf{Q}$ is a subfield of a cyclotomic field. It was in Weber's book *"Elliptische Funktionen and algebraische Zahlen"* (see [We-1891]), which built on Dedekind's third edition of Dirichlet's lectures on number theory (see [DD-1879]), that the term *class field* was introduced for the first time (in connection with the now so-called absolute class field of an imaginary quadratic number field) (see [We-1891], p. 439).

4. Weber's book first presents a full treatment of the theory of elliptic functions, as created by Jacobi and Weierstrass, and links it with the then modern theories of invariants created by Cayley and Sylvester and Galois theory in order to treat the theory of transformation and division equations created by Abel and Galois. Most important is the third part on complex multiplication where for the first time a full treatment of results by Abel, Hermite and Kronecker is given. Together with the fundamental series of three papers *"Über Zahlgruppen in algebraischen Körpern"* (see [We-1897]), where a program for a general class field theory is proposed and initiated, these ideas found a final treatment as Volume III of the second edition of Weber's *Lehrbuch der Algebra* (see [We-1908]). In that edition a whole book (Book 4) is devoted to class field theory as it was known at that time.

We shall now study more closely how Weber came to develop his notion of a class field.

5.2 Complex Multiplication

We begin with a summary of what was known to Weber around 1891, prior to his series of papers of 1896/97, as presented in the third part of his book *"Elliptische Funktionen und algebraische Zahlen"* (see [We-1891], in particular §§86-120, pp. 325-498). Here he is building on investigations by Abel, Hermite and Kronecker on complex multiplication. We plan to study this theme in more detail in a separate paper.

If F is an elliptic function with periods ω_1, ω_2 we set $\omega = \dfrac{\omega_1}{\omega_2}$. We may suppose that ω has positive imaginary part, by interchanging ω_1 and ω_2 if necessary. F admits complex multiplication, i. e. $F(\alpha x)$ is a rational function of $F(x)$ for some algebraic integer α which is not a rational integer, if and only if ω is an integer in a quadratic imaginary field $\mathbf{k} = \mathbb{Q}(\sqrt{-d})$ with discriminant $-d$. The set of complex multipliers α of an elliptic function F with complex multiplication forms an order $\mathfrak{o}_f$ in $\mathbf{k}$. Suppose $\mathfrak{o}_f$ has class number $h = h_f$ and that $C_1, \ldots, C_h$ are the ideal classes in $\mathfrak{o}_f$. Each ideal class in $\mathfrak{o}_f$ determines a unique isomorphism class of elliptic curves and hence a unique value $j(C_i)$ of the modular function j, called a *singular modulus* following *Kronecker*. In fact, if $\mathcal{N} = (\nu_1, \nu_2)$ and $\mathcal{M} = (\mu_1, \mu_2)$ are two ideals in $\mathfrak{o}_f$ and if N and M are elliptic functions having ν_1, ν_2 respectively μ_1, μ_2 as periods with $\mathfrak{o}_f$ as complex multipliers, then $\mathcal{N}$ and $\mathcal{M}$ are equivalent if and only if the elliptic curves associated with N and M are isomorphic. Hence if $\nu = \dfrac{\nu_1}{\nu_2}$ and $\mu = \dfrac{\mu_1}{\mu_2}$ then

$$j(\nu) = j\left(\frac{a + b\nu}{c + d\nu}\right) = j(\mu)$$

with $\mu = \dfrac{a + b\nu}{c + d\nu}$ for certain $a, b, c, d \in \mathbb{Z}$ and $ad - bc = 1$, where

$$j(\nu) = 2^6 3^3 \frac{g_2^3}{g_2^3 - 27 g_3^2}$$

with

$$g_2 = 60 \sum_{(x,y) \in \mathbb{Z}^2 \backslash (0,0)} \frac{1}{(x\nu_1 + y\nu_2)^4} ; \qquad g_3 = 140 \sum_{(x,y) \in \mathbb{Z}^2 \backslash (0,0)} \frac{1}{(x\nu_1 + y\nu_2)^6}$$

(see [We-1891], pp. 124, 247, 330). Then Kronecker proved (see [We-1891], pp. 330, 332, 438, 447) the

THEOREM 22. (a) $j(C_i)$ *are algebraic integers for* $i = 1, \ldots, h$. *They are called the* class invariants.

(b) $j(C_i)$ *are roots of a monic polynomial* $H_n(x)$ *with coefficients in* $\mathbf{k}$, *called the* class equation.

(c) $H_n(x)$ *is irreducible over* $\mathbf{k}$ *and hence the* $j(C_i)$ *are conjugates over* $\mathbf{k}$.

(d) $\mathbf{K} = \mathbf{k}(j(C_i))$ *is independent of the class* C_i *and* K/k *is an abelian extension of degree* h *and hence a solvable extension.*

If $\mathfrak{o}_f$ is the principal order, $\mathbf{K}$ is called by Weber the *class field* of discriminant $-d$ or of the field $\mathbf{k} = \mathbb{Q}(\sqrt{-d})$. We call it a Kronecker class field.

5.3 Weber's L-Series and Congruence Groups and Weber's Notion of a Class Field

In a series of three papers, *Über Zahlgruppen in algebraischen Körpern* (see [We-1897]), *Weber* studied certain sets of fields containing the set of Kronecker class fields. As a matter of fact, the starting point for his investigations, as Weber mentions in his introduction, was (the proof of) Kronecker's conjecture that the Kronecker class field K over $\mathbb{Q}(\sqrt{-d})$ is not ramified (see also [We-1891], p. 453). Weber observed that these fields, which are in fact class fields in the modern sense, must have interesting properties, which among other things make it possible that Dirichlet's analytic methods for primes in arithmetic progressions can be applied. The key question is, in fact, the determination of the degree of these fields or the proof of the irreducibility of the corresponding defining equation, which corresponds to the class equation in the case of a Kronecker class field. This goes back to Kronecker (see [Kr-1862], Werke IV, p. 214) and it is solved by means of Dirichlet's method, as it is applied to cyclotomic fields and by the resulting Dirichlet class number formula. Weber's main concern was to gain some information on the primes dividing the relative discriminant of a Weber class field K over $k = \mathbb{Q}(\sqrt{-d})$ and thus on the above conjecture of Kronecker (compare also [We-1891], p. 453).

We shall now describe these three papers making use of Weber's more detailed exposition in Book 4, Volume III of the second edition of his *Lehrbuch der Algebra* (see [We-1908]).

Part I of *Über Zahlgruppen in algebraischen Körpern* is concerned with the presentation and index computation of ideal groups, number groups, orders and genera.

In *Part II* Weber attaches fields to certain ideal or number groups as follows:

1. Let k be a base field of degree n (over $\mathbb{Q}$) and A an ideal group in k satisfying the following properties:
(1) A contains all prime ideals in k except for a finite set $S = \{\mathcal{P}_1, ..., \mathcal{P}_s\}$, i.e. $A = A_{\mathcal{M}}$, where $\mathcal{M} = \mathcal{P}_1 \cdots \mathcal{P}_s$ and $A_{\mathcal{M}}$ is the group of all fractional ideals in k prime to $\mathcal{M}$, or else $A_{\mathcal{M}}$ is the group of fractional ideals generated by all primes not in S.
(2) Let $H_{\mathcal{M}}$ be a subgroup of $A_{\mathcal{M}}$ such that
   (a) $H_{\mathcal{M}}$ contains only principal ideals of k,
   (b) The index $[A_{\mathcal{M}} : H_{\mathcal{M}}] = h$ is finite.
In view of the methods of Dirichlet and Dedekind, Weber adds the condition:

(3) Let $\mathcal{A}$ be an integral ideal in $A_{\mathcal{M}}$ and $T(t, \mathcal{A}, \mathcal{C})$ the number of integral ideals in a class $\mathcal{C} \in C_{\mathcal{M}} = A_{\mathcal{M}}/H_{\mathcal{M}}$ divisible by $\mathcal{A}$ and with norm smaller than t. Then

$$T = T(t, \mathcal{A}, \mathcal{C}) = \frac{gt}{N(\mathcal{A})} + Mt^{1-\frac{1}{n}}$$

for a positive constant g and a constant M, where

$$N(\mathcal{A}) := [\mathbf{o}_k : \mathcal{A}]$$

denotes the *norm* from $\mathbf{k}$ to $\mathbb{Q}$ and $\mathbf{o}_k$ the integers in $\mathbf{k}$.
This means

$$\lim_{t \to \infty} \frac{T}{t} = \frac{g}{N(\mathcal{A})}$$

for a positive constant g which does not depend on $\mathcal{A}$ nor on the class $\mathcal{C}$; that is Weber assumes the analogue of the Dirichlet-Dedekind theorem that all integral ideals in $\mathbf{k}$ are equally distributed among the classes of $C_{\mathcal{M}}$.

From these three conditions Weber is able to obtain *Dirichlet's class number formula* (p. 85), whereby we set $\mathbf{o}_{\mathcal{M}} = A_{\mathcal{M}} \cap \mathbf{o}_k$:

THEOREM 23.
$$\lim_{s \to 1^+} (s - 1) \sum_{\mathcal{A} \in \mathbf{o}_{\mathcal{M}}} \frac{1}{N(\mathcal{A})^s} = gh$$

2. Next Weber introduces his L-series

DEFINITION 24. *Let χ be a character of $C_{\mathcal{M}} = A_{\mathcal{M}}/H_{\mathcal{M}}$ and extend χ following Dirichlet from $C_{\mathcal{M}}$ to $A_{\mathcal{M}}$ by the convention:*

$$\text{if } \mathcal{A} \in \mathcal{C} \in C_{\mathcal{M}}, \text{ then } \chi(\mathcal{A}) := \chi(\mathcal{C}).$$

Then define the *Weber L-series*:

$$L(s, \chi, k) = \sum_{\mathcal{A} \in \mathbf{o}_{\mathcal{M}}} \frac{\chi(\mathcal{A})}{N(\mathcal{A})^s}$$

One has again the product representation:

THEOREM 25.
$$L(s, \chi, k) = \prod_{\mathcal{P} \in A_{\mathcal{M}}} \left(1 - \frac{\chi(\mathcal{P})}{N(\mathcal{P})^s}\right)^{-1}$$

and the

THEOREM 26. (a) $L(s, \chi, k)$ are analytic if $\Re s > 1$ for any $\chi \in \widehat{C_{\mathcal{M}}}$.
(b) $L(1, \chi, \mathbf{k})$ is finite if $\chi \neq \chi_1$, where χ_1 is the principal character.
(c) $\lim\limits_{s \to 1+} (s - 1)L(s, \chi_1, \mathbf{k}) = gh$.

The last property is a consequence of Theorem 23.
    Since

$$
(s - 1) \prod_{\chi \in \widehat{C_{\mathcal{M}}}} L(s, \chi, k) = \prod_{\mathcal{P} \in A_{\mathcal{M}}} \frac{s - 1}{(1 - N(\mathcal{P})^{-fs})^e}
$$

which is finite for $s = 1$, where f is the order of the class of $\mathcal{P}$ and $h = ef$,
one has to distinguish between prime ideals of degree 1, i. e. for which
$N(\mathcal{P}) = p$, and prime ideals for which $N(\mathcal{P}) = p^i$ for some $i > 1$, where
p is the corresponding prime number.
    Since the product over the primes of degree $i > 1$, as well as the
product over the primes of degree 1 which do not belong to the principal
class, i.e. for which $f > 1$, only yield finite constants, one gets

$$
(s - 1) \prod_{\chi \in \widehat{C_{\mathcal{M}}}} L(s, \chi, k) = B \prod_{\mathcal{P} \in H_{\mathcal{M}}} \frac{s - 1}{(1 - N(\mathcal{P})^{-s})^h} ,
$$

where B is a finite constant different from 0 and where the product ranges
only over the principal prime ideals $\mathcal{P}$ in $H_{\mathcal{M}}$ of degree 1. Hence in order
to show that $L(s, \chi, k) \neq 0$ for any character $\chi \neq \chi_1$, one has to show that

$$
\prod_{\mathcal{P} \in H_{\mathcal{M}}} \frac{s - 1}{(1 - N(\mathcal{P})^{-s})^h} \neq 0
$$

where $\mathcal{P}$ rans over all principal ideals in $H_{\mathcal{M}}$ of degree 1.
    3. At this stage Weber makes an additional and very fundamental
assumption (p. 87, Voraussetzung 4):
    (4) Suppose there exists a field $\mathbf{K}$ over $\mathbf{k}$ with the properties
        (4.1) $\nu = [\mathbf{K} : \mathbf{k}] \leq h$.
        (4.2) Every prime ideal $\mathcal{P} \in H_{\mathcal{M}}$ of degree 1 splits completely in $\mathbf{K}$
            into prime ideals $\mathbf{P}_i$ of degree 1, i.e.

$$
\mathcal{P}O_{\mathbf{K}} = \mathbf{P}_1 \cdots \mathbf{P}_\nu
$$

with $f(\mathbf{P}_i/\mathcal{P}) = 1$, where $O_{\mathbf{K}}$ are the integers in $\mathbf{K}$ and where
    $f(\mathbf{P}_i/\mathcal{P})$ denotes the inertia degree of $\mathbf{P}_i$ in $\mathbf{K}/\mathbf{k}$.
These conditions lead to:

DEFINITION 27. *A field* **K** *is a Weber class field to* k, *if* **K** *satisfies (4.1) and (4.2)*.

Only in the second edition of his *Lehrbuch der Algebra*, Vol. III (see [We-1908], §164, p. 607) does Weber call such a field a class field. In the paper of 1897 ([We-1897]) the term *class field* still only denotes a Kronecker class field (see [We-1897], III, p. 8).

Under these hypotheses Weber is now able to prove (pp. 88-89):

THEOREM 28. *(a)* $\nu = h$, *i. e.* **K** *is of degree h over* k.

(b) $L(1, \chi, \mathbf{k}) \neq 0, \infty$ *if* $\chi \neq \chi_1$.

(c) $\lim\limits_{s \to 1+} (s - 1) L(s, \chi_1, \mathbf{k}) \neq 0, \infty$.

(d) Every ideal class $C \in C_{\mathcal{M}}$ *contains infinitely many prime ideals of degree 1.*

These properties follow now from the decomposition law for **K**/k or what amounts to the same, from the fact proved by Takagi in 1920, that the product

$$\prod_{\chi \in \widehat{C_{\mathcal{M}}}} L(s, \chi, k)$$

is essentially equal to the Dedekind ζ -function of the field **K**.

4. Weber gave several examples of ideal and number groups satisfying the conditions (1)—(4).

Let k be a number field of degree n over **Q** and $m \in \mathbb{N}$ a natural number and let A_m be the group of all ideals in k prime to m. Then Weber considers congruence classes which contain whole number classes modulo m, i. e. if $\alpha \in H_m$ and $\beta \equiv \alpha$ modulo m, then also $\beta \in H_m$. Weber calls such a group H_m a *harmonic number group* (p. 90).

Of course, there are two classical examples of congruence groups satisfying the conditions (1)—(4). They are already present in Dirichlet's proof, namely those for which a Dirichlet class number formula was known. In fact, these two examples served Weber as a guide, but Weber does not mention them explicitly. They belong to the case $\mathbf{k} = \mathbf{Q}$, $m \in \mathbb{N}$ and are harmonic number groups, namely

$$A_m = \{a \in \mathbb{Z} : (a, m) = 1\}$$

and

(a) $H_m = \{\alpha \in A_m : \alpha \equiv 1 \bmod m\}$, to which the cyclotomic field

$\mathbb{Q}(\zeta_m)$, $\zeta_m = e^{\frac{2\pi i}{m}}$, is the corresponding Weber class field of degree $\phi(m) = [A_m : H_m]$, and

(b) $H_m = \{\alpha \in A_m : \left(\dfrac{m}{\alpha}\right) = 1\}$ to which $\mathbb{Q}(\sqrt{m})$ is the corresponding Weber class field, or

$$H_m = \{\alpha \in A_m : \left(\dfrac{-m}{\alpha}\right) = 1\}$$

to which $\mathbb{Q}(\sqrt{-m})$ is the corresponding Weber class field.

The three examples of Weber class fields given by Weber all belong to harmonic number groups in an imaginary quadratic number field $\mathbb{Q}(\sqrt{-m})$. They can be constructed by means of the theory of elliptic functions.

Let $A_f = \mathbf{o}_f$ be an order in k. $\mathbf{o}_f$ can be characterized as follows.

$\mathbf{o}_f = \{\alpha \in \mathbf{o}_k : \alpha \equiv r \bmod f$ for a rational number $r \in \mathbb{Q}$ with $(r, f) = 1\}$

that is, as the ring of all algebraic integers α in k which are congruent to a rational number r modulo f. f is called the *conductor* of $\mathbf{o}_f$ (see [We-1908], Vol. III, p. 351).

5. A first example is obtained by taking as A_f all ideals in $\mathbf{o}_f$ (prime to f) and as H_f all principal ideals in $\mathbf{o}_f$, i. e.,

$$H_f = \{(\alpha) = \alpha\mathbf{o}_f : \alpha \in A_f\}$$

so that $C_f = \mathbf{o}_f/H_f$ is the group of ideal classes in the order $\mathbf{o}_f$ in k (p. 94). For that reason Weber calls these groups *order groups*.

If $f = 1$ one gets the ideal classes in the principal order $\mathbf{o}_k$, that is the ideal classes in k.

By Kronecker's theory there corresponds a Weber class field to C_f, namely $\mathbf{K} = k(j(C_i))$ obtained from k by adjoining the singular moduli of $\mathbf{o}_f$. Weber calls such a field a *class field of the order* $\mathbf{o}_f$ or an *order field* (p. 98 or I, p. 433).

If $f = 1$, K becomes what we called the Kronecker class field of k. If k has discriminant d then $\mathbf{o}_f$ has discriminant df^2, where f is the conductor of $\mathbf{o}_f$, and Theorem 28 (d) implies (p. 95)

THEOREM 29. *Every primitive quadratic form of discriminant df^2 represents infinitely many prime numbers.*

6. A second example is obtained by taking A_{fm} to be all the ideals in $\mathbf{o}_f$ prime to a given modulus m and by replacing H_f by the principal genus H'_{fm} of A_{fm} with respect to the modulus $m \in \mathbb{N}$, i. e.,

$$H'_{fm} = \{\mathcal{A} \subseteq \mathbf{o}_f : N(\mathcal{A}) \equiv N(\alpha) \text{ modulo } m \text{ for some } \alpha \in \mathbf{o}_f\}.$$

$C'_{fm} = A_{fm}/H'_{fm}$ is then the genus group of order $g = 2^t$ for some $t \in \mathbf{N}$ with respect to m. Weber shows that there corresponds a Weber class field $\mathbf{K}'$ to C'_{fm} for an appropriate m which is obtained by adjoining t real square roots to $\mathbf{k}$, from which he derives the remarkable property (p. 99):

THEOREM 30. $\mathbf{K}'$ *is the maximal extension contained in* $\mathbf{K}$ *which is abelian over* $\mathbf{Q}$.

Today we call $\mathbf{K}'$ the (relative) *genus field* of k (see for instance [Ha-1951]).

As a consequence of Theorem 28 (d) Weber also showed (p. 97):

THEOREM 31. (a) *Each ideal class in* $\mathbf{o}_f$ *contains infinitely many prime ideals of degree 1 belonging to a given non-empty genus modulo* m,

or else in terms of a Dirichlet theorem:

(b) *Every primitive quadratic form represents infinitely many prime numbers belonging to a given non-empty genus modulo* m.

7. In the third example Weber starts with the ideal group $A_{f\mathcal{M}}$ in $\mathbf{o}_f$ for an ideal modulus $\mathcal{M}$ and defines $H''_{f\mathcal{M}}$ as follows. $H''_{f\mathcal{M}}$ contains all numbers which are congruent to a unit in $\mathbf{o}_f$ modulo $\mathcal{M}$ (p. 97).

Part III of Weber's study is devoted to the construction of the corresponding Weber class field. It leads to extension fields $\mathbf{T}$ of $\mathbf{K}$ obtained by adjoining singular moduli of a division of elliptic functions. That is why Weber calls these fields *division fields*. They are abelian with respect to $\mathbf{K}$. Application of Theorem 28 (d) to this situation yields a generalization of Dirichlet's theorem:

THEOREM 32. *There are infinitely many algebraic integers* π *in* $\mathbf{o}_f$ *whose norm is a prime number (hence* $(\pi) = \pi \mathbf{o}_k$ *are prime ideals in* k *of degree 1) which are congruent to a given* $\alpha \in \mathbf{o}_f$ *modulo* $\mathcal{M}$.

6. DEDEKIND'S CONTRIBUTION TO CLASS FIELD THEORY

Dedekind's posthumous works, edited only in 1932, 16 years after his death, by Robert Fricke, Emmy Noether and Öystein Ore, contain many important unpublished documents relating to the groundwork for a theory of class fields.

In a paper on the irreducibility of cyclotomic polynomials, Dedekind uses a correspondence between prime numbers in a congruence group and the elements of the Galois group of the mth cyclotomic field $\mathbf{Q}(\zeta_m)$, which is a special case of Artin's reciprocity law (see [De], XL (1894)).

In another paper on group characters of number classes Dedekind considers general ray classes and congruence groups and their conductors and characters as well as groups and their conductors associated with a given character (see [De], XLI (1896)), and in a letter to Frobenius of 8 July 1896 Dedekind introduces the ζ-function or the L-funtion corresponding to a class division (see [De], XLV).

In a still further paper on discriminants of cyclotomic fields, Dedekind starts again from the ray class group modulo m and its subgroups, associates to them the subfields of the mth cyclotomic field corresponding to them as class fields and proves the conductor-discriminant theorem for that case (see [De], XLII).

In two letters written in July 1882 to Frobenius, Dedekind even studied certain properties of the Frobenius substitution (see [De], XLV).

As Emmy Noether points out in her commentary on Dedekind's paper [De], XLI, it is unfortunately no longer possible to determine to what extent these unpublished investigations on class field theory undertaken by Dedekind between 1894 and 1896 influenced the work of Weber, since the correspondence between Dedekind and Weber from that time period on class field theory is no longer extant.

7. CHRONOLOGICAL TABLE TO HEINRICH WEBER

7.1 LIFE

Born	5.5.1842	Heidelberg	(father: G. Weber, historian)
Studies	1860	Heidelberg	(Leipzig 1 year)
Doctorate	1863	Heidelberg	
		Königsberg	(F. Neumann, F.J. Richelot)
			(Math. Physics, Alg. Functions)
Habilitation	1866	Heidelberg	
A.o. Professor	1869	Heidelberg	
Professor	1870	ETH Zürich (SS)	
	1875	Königsberg (WS)	Rector (Minkowski, Hilbert)
	1883	TH Charlottenburg	(Berlin)
	1884	Marburg	Rector
	1892	Göttingen	
	1895	Strassburg	Rector
Died	17.5.1913	Strassburg	

WORK

	1862 *Kronecker*: Complex Multiplication, Singular Moduli, Non-ramification
1870 ETH *Zürich*	
	1871 *Dirichlet-Dedekind* Zahlentheorie, 2. Aufl. (ζ -function, ideal theory)
1874 Neuer Beweis des Abelschen Theorems, Math. Ann. 8 (1874)	
1875 *Königsberg*	
1876 Edition of Riemann's work together with Dedekind	
	1879 *Dirichlet-Dedekind* Zahlentheorie, 3. Aufl. [basis for Weber's book of 1891]
1882 (22.10.1880) Theorie der algebraischen Funktionen einer Veränderlichen, Crelle 92(1882) [together with Dedekind]	
	1882 *Kronecker*: Grundzüge
1883 TH *Charlottenburg Berlin*	

1884 *Marburg*
1886 (March, October) Theorie der Abelschen
      Zahlkörper I,II,III, Acta Math. 8 and 9

1891 Elliptische Funktionen und algebrai-
      sche Zahlen, Braunschweig 1891
      (second edition 1908)[Theory of Com-
      plex Multiplication]
1892 Über Abelsche Zahlkörper dritten und
      vierten Grades, Sitz.ber. Marburg 1892
1892 *Göttingen*
1893 Zahlentheoretische Untersuchungen aus
      dem Gebiete der elliptischen Funct-
      ionen, [3 communications], Ges. Wiss.
      Göttingen (18.1.1893)

1895 *Strassburg*

1897 (31.8.1896)*Über Zahlgruppen in algebrai-*
      schen Körpern, I, II, III, Math. Ann. 48,
      49 and 50

1900 Die partiellen Differentialgleichungen
      der mathematischen Physik (2 Vol.),
      Braunschweig 1900-1901 [building on
      Riemann's lectures prepared by Karl
      Hattendorf]
1908 Volume III of *Lehrbuch der Algebra* as
      2nd edition of *Elliptische Funktionen*
      und algebraische Zahlen [it contains
      also Webers articles *Über Zahlgruppen*
      in algebraischen Körpern]

1886 *Kronecker*: Zur Theorie der elliptischen
      Functionen [basis for Weber's book of
      1891]

1894 *Hilbert*: Theorie des Galois'schen Kör-
      pers
      — Über den Dirichletschen biquadrati-
      schen Körper
      Dirichlet-Dedekind Zahlentheorie, 4. Aufl.

1896 *Hilbert*: Beweis des Satzes von Kro-
      necker und Weber

1897 (10.4.1897) *Hilbert*: Zahlbericht
1898 *Hilbert*: Theorie der relativ-Abelschen
      Zahlkörper

BIBLIOGRAPHY

[Ay-1974] Ayoub, R., *Euler and the Zeta Function*, Am. Math. Monthly 81
           (1974), pp. 1067–1086.

[Cl-1968] Clagett M., ed., *Nicole Oresme and the Medieval Geometry of Qual-*
           ities, Madison 1968.

[De]   Dedekind, R., *Gesammelte mathematishe Werke*, Braunschweig
           1930–1932.

[De-1877] Dedekind, R., *Über die Anzahl der Ideal-Klassen in den verschiedenen Ordnungen eines endlichen Körpers*, Festschrift zur Säkularfeier des Geburtstages von C. F. Gauss, Braunschweig 1877, pp. 1–55; Werke I, XII, pp. 105–158.

[Di-1837] Dirichlet, G. L., *Beweis des Satzes, dass jede unbegrenzte arithmetische Progression, deren erstes Glied und Differenz ganze Zahlen ohne gemeinschaftlichen Factor sind, unendlich viele Primzahlen enthält*, Abh. Akad. Wiss zu Berlin, 27.7.1837, pp. 45–81; Werke I, pp. 313–342.

[Di-1838] Dirichlet, G. L., *Sur l'usage des séries infinies dans la théorie des nombres*, J. reine angew. Math. 18 (May 1838), pp. 259–274; Werke I, pp. 357–374.

[DD-1863] Dirichlet-Dedekind, *Vorlesungen Über Zahlentheorie*, Braunschweig 1863 (first edition).

[DD-1871] Dirichlet-Dedekind, *Vorlesungen Über Zahlentheorie*, Braunschweig 1871 (second edition).

[DD-1879] Dirichlet-Dedekind, *Vorlesungen Über Zahlentheorie*, Braunschweig 1879 (third edition).

[DD-1894] Dirichlet-Dedekind, *Vorlesungen Über Zahlentheorie*, Braunschweig 1894 (fourth edition).

[El-1978] Ellison, W. and F., *Théorie des nombres*, chapitre V in *Abrégé d'histoire des mathématiques 1700–1900* edited by J. Dieudonné, Paris 1978.

[Eu] Euler, L., *Opera Omnia*, 1915- (E. refers to the Eneström number).

[Fr-1979] Frei, G., *On the Development of the Genus of Quadratic Forms*, Ann. Sc. Math. Québec, Vol. III., No. 1 (1979), pp. 5–62.

[Fr-1983] Frei, G., *Les nombres convenables de Leonhard Euler*, Publ. Math. Fac. Sci. Besançon, Théorie des Nombres, 1983–1984 (1985), pp. 1–58.

[Fr-1985] Frei, G., *Der Briefwechsel David Hilbert - Felix Klein (1886–1918) mit Anmerkungen herausgegeben*, Göttingen 1985.

[Fu-1907] Furtwängler, Ph., *Allgemeiner Existenzbeweis für den Klassenkörper eines beliebigen Zahlkörpers*, Math. Ann. 63 (1907), pp. 1–37.

[Fu-1930] Furtwängler, Ph., *Beweis des Hauptidealsatzes für Klassenkörper algebraischer Zahlkörper*, Abh. Math. Sem. Univ. Hamburg 7 (1930), pp. 14–36.

[Ga-1801] Gauss, C. F., *Disquisitiones Arithmeticae*, Leipzig 1801; translated by A. Clarke, New Haven 1966.

[Ha-1951] Hasse, H., *Zur Geschlechtertheorie in quadratischen Zahlkörpern*, J. Math. Soc. Japan 3.1 (1951), pp. 45–51.

[Ha-1967] Hasse, H., *History of Class Field Theory*, Chapter XI in *Algebraic Number Theory*, edited by Cassels and Fröhlich, Washington 1967, pp. 266–279.

[He-1912] Hecke, E., *Höhere Modulfunktionen und ihre Anwendung auf die Zahlentheorie*, Math. Ann. 71 (1912), pp. 1–37.

[He-1913] Hecke, E., *Über die Konstruktion relativ-Abelscher Zahlkörper durch Modulfunktionen von zwei Variabeln*, Math. Ann. 74 (1913), pp.465 –510.

[He-1917] Hecke, E., *Über die Zetafunktionen beliebiger algebraischer Zahlenkörper*, Nachr. Ges. Wiss. Göttingen (1917), pp. 77–89.

[Hi-1897] Hilbert, D., *Bericht: Die Theorie der algebraischen Zahlkörper*, J.ber. Dt. Math. Ver. 4 (1897), pp. 175–546; Werke I, pp. 63–363.

[Hi-1898] Hilbert, D., *Über die Theorie der relativ-Abelschen Zahlkörper*, Nachr. Ges. Wiss. Göttingen (1898), 377–399, or Acta Math. 26 (1902), pp. 99–132; Werke I, pp. 483–509).

[Hi-1899a] Hilbert, D., *Über die Theorie der relativ-quadratischen Zahlkörper*, J.ber. Dt. Math.-Ver. 6 (1899), pp. 88–94; Werke I, pp. 364–369.

[Hi-1899b] Hilbert, D., *Über die Theorie des relativ-quadratischen Zahlkörpers*, Math. Ann. 51 (1899), pp. 1–127; Werke I, pp. 370–482.

[Hi-1900] Hilbert, D., *Mathematische Probleme*. Vortrag auf dem internationalen Mathematiker Kongresse in Paris 1900, Nachr. Ges. Wiss. Göttingen (1900), pp. 253–297; Werke III, pp. 290–329.

[Kr-1853] Kronecker, L., *Über die algebraisch auflösbaren Gleichungen I*, Sitz. ber. Preuss. Akad. Wiss. zu Berlin 1853, pp. 365–374; Werke IV, pp. 1–11.

[Kr-1857] Kronecker, L., *Über die elliptischen Functionen für welche complexe Multiplication stattfindet*, Monatsber. der Preuss. Akad. der Wiss. zu Berlin, 29. Oct. 1857 (Berlin 1858), pp. 455–460; Werke IV, pp. 177–183.

[Kr-1862] Kronecker, L., *Über die complexe Multiplication der elliptischen Functionen*, Monatsber. der Preuss. Akad. der Wiss. zu Berlin, 26. Juni 1862 (Berlin 1863), pp. 363–372; Werke IV, pp. 207–217.

[Kr-1877] Kronecker, L., *Über Abelsche Gleichungen*, Monatsber. der Preuss. Akad. der Wiss. zu Berlin, 16. April 1877 (Berlin 1878), pp. 845–851; Werke IV, pp. 63–71.

[Kr-1881] Kronecker, L., *Zur Theorie der elliptischen Functionen*, Monatsber. der Preuss. Akad. der Wiss. zu Berlin, (Berlin 1882), pp. 1165–1172; Werke IV, pp. 309–318.

[Kr-1882] Kronecker, L., *Grundzüge einer arithmetischen Theorie der algebraischen Grössen*, J. reine angew. Math. 92 (1882), pp. 1–122; Werke II, pp. 237–388.

[Kr-1883] Kronecker, L., *Zur Theorie der elliptischen Functionen* I-XXII, Sitz. ber. Preuss. Akad. Wiss. zu Berlin 1883-1890; Werke IV, pp. 345–496 and V, pp. 1–132.

[Kl-1972] Kline, M., *Mathematical Thought from Ancient to Modern Times*, New York 1972.

[La-1903] Landau E., *Über die zu einem algebraischen Zahlkörper gehörige Zetafunktion*, J. reine angew. Math. 125 (1903).

[Ta-1920] Takagi, T., *Über eine Theorie des relativ-Abelschen Zahlkörpers*, J. Col. Sci. Imp. Univ. Tokyo 41, Nr. 9 (1920), pp. 1–133.

[Vo-1914] Voss, A., *Heinrich Weber*, J.ber. Dt. Math.-Ver. 23 (1914), pp. 431–444.

[We-1891] Weber, H., *Elliptische Funktionen und algebraische Zahlen*, Braunschweig 1891.

[We-1897] Weber, H., *Über Zahlengruppen in algebraischen Zahlkörpern*, I, II, III, Math. Ann. 48 (1897), pp. 433–473; 49 (1897), pp. 83–100; 50 (1898), pp. 1–26.

[We-1908] Weber, H., *Lehrbuch der Algebra*, Vol. III, Braunschweig 1908 (second edition of *Elliptische Funktionen und algebraische Zahlen*).

[Wl-1984] Weil, A., *Number Theory: An approach through history; From Hammurapi to Legendre*, Boston 1984.

Notes on the Contributors

Karine Chemla is affiliated with the Centre d'Histoire des Sciences et des Doctrines at the Centre National de la Recherche Scientifique (C.N.R.S.), Paris. Her recent publications include: "Paysages d'algorithmes, algorithmes de paysages," *Cahiers d'Histoire et de Philosophie des Sciences* (1987); (with R. Morelan and A. Allard), "La tradition arabe de Diophante d'Alexandrie," *L'Antiquité Classique* (1986); (with S. Pahaut), "Préhistoires de la dualité: explorations algébriques en trigonométrie sphérique (1753–1825)," R. Rashed, ed., *Science à l'époque de la Révolution*, 1988.

Roger Cooke, Professor of Mathematics at the University of Vermont, received his Ph. D. from Princeton University in 1966. Until 1980 he specialized in classical harmonic analysis, and since then he has been working on the history of mathematics in the 19th century. His scientific biography, *The Mathematics of Sonya Kovalevskaya* (Springer, 1984), represents the first detailed study of Kovaleskaya's mathematical achievements.

Harold M. Edwards is Professor of Mathematics at the Courant Institute for Mathematical Sciences of New York University. A leading authority on 19th-century algebra and number theory, his publications include: *Riemann's Zeta Function* (Academic Press, 1974); *Fermat's Last Theorem. A Genetic Introduction to Number Theory* (Springer, 1977); *Galois Theory* (Springer, 1984); and "Dedekind's Invention of Ideals," *Bulletin of the London Mathematical Society*, 1983.

Günther Frei is Professor of Mathematics at the Université Laval in Quebec, Canada. Among his recent publications are: "Felix Klein (1849–1925), A Biographical Sketch," *Jahrbuch Überblicke Mathematik*, 1984; and his edition of *Der Briefwechsel David Hilbert–Felix Klein (1886–1918)* (Göttingen: Vandenhoeck & Ruprecht, 1985).

Jeremy Gray is Lecturer in Mathematics at the Open University, Milton Keynes, England. He is an Associate Editor of *Historia Mathematica* and *The Mathematical Intelligencer*, and a well-known authority on 19th-century geometry and analysis. His publications include: *Ideas of Space* (Oxford

Univ. Press, 1979); *Linear Differential Equations and Group Theory from Riemann to Poincaré* (Birkhäuser, 1986); "A Commentary on Gauss's Mathematical Diary, 1796–1814," *Expositiones Mathematicae*, 1984.

Thomas Hawkins is Professor of Mathematics at Boston University. Since the publication of his book, *Lebesgue's Theory of Integration: Its Origins and Development* (Univ. of Wisconsin Press, 1970), his research has centered on developments in modern algebra, particularly the work of Killing, Cartan, and others on Lie groups and Lie algebras. Among his many publications are: "Wilhelm Killing and the Structure of Lie Algebras," *Archive for History of Exact Sciences*, 1982; "Non-Euclidean Geometry and Weierstrassian Mathematics: The Background to Killing's Work on Lie Algebras," *Historia Mathematica*, 1980; "The *Erlanger Programm* of Felix Klein: Reflections on its Place in the History of Mathematics," *Historia Mathematica*, 1984; "Hesse's Principle of Transfer and the Representation of Lie Algebras," *Archive for History of Exact Sciences*, 1988.

John McCleary is Associate Professor of Mathematics at Vassar College. He has published papers in topology, differential geometry, and number theory, and his *User's Guide to Spectral Sequences* is the first textbook treatment of a theory that has considerable applications in algebraic topology. He is particularly interested in historiographical issues surrounding the history of modern mathematics. An outline of some of his views on this topic can be found in an article co-authored with Audrey McKinney, "What Mathematics Isn't," *Mathematical Intelligencer*, 1986.

Gregory H. Moore is Assistant Professor of Mathematics at McMaster University. He is co-editor of the *Collected Works of Kurt Gödel*, vols. I and II (Oxford Univ. Press, 1986, 1987). Among his other recent publications are: *Zermelo's Axiom of Choice: Its Origins, Development, and Influence* (Springer, 1982); and "A House Divided Against Itself: The Emergence of First-Order Logic as the Basis for Mathematics," E. Phillips, ed., *Studies in the History of Mathematics* (Mathematical Association of America, 1987). He is also currently editing volumes 3 and 5 of the *Collected Papers of Bertrand Russell*.

Gregory Nowak holds an M.A. in mathematics from Princeton University where he is presently completing doctoral studies in the history of science. He is interested in the philosophical background that informs mathematical ideas, and he is currently writing a history of American topology during the first half of the 20th century.

Karen V. H. Parshall is Assistant Professor of Mathematics and the History of Science at the University of Virginia. Her research centers on the history of American mathematics and the development of modern algebra. Among her recent publications are: "Joseph H. M. Wedderburn and the Structure Theory of Algebras," *Archive for History of Exact Sciences*, 1985; "America's First School of Mathematical Research: James Joseph Sylvester at the Johns Hopkins University," *Archive for History of Exact Sciences*, 1988; (with David E. Rowe), "American Mathematics Comes of Age: 1875–1900," in P. Duren, ed. *A Century of Mathematics in America*, vol. 3, 1989.

Walter Purkert is Director of the Mathematics Institute at Leipzig University. During 1988 he was Visiting Research Professor at Pace University. Among his recent publications are: (with H. J. Ilgauds), *Georg Cantor, 1845–1918*, Vita Mathematica, vol. 1, 1987; "Georg Cantor und die Antinomien der Mengenlehre," *Bulletin de la Société Mathématique de Belgique*, 1987; "Die Bedeutung von Albert Einsteins Arbeit über Brownsche Bewegung für die Entwicklung der modernen Wahrscheinlichkeitstheorie," *Mitteilungen der Mathematischen Gesellschaft der DDR*, 1983.

Helena Pycior is Professor of History at the University of Wisconsin, Milwaukee, and managing editor of *Historia Mathematica*. Her principal fields of interest are 19th-century British algebra and its influence on mathematics in the United States. Among her recent publications are: "Internalism, Externalism, and Beyond: 19th-Century British Algebra," *Historia Mathematica*, 1984; "George Peacock and the British Origins of Symbolical Algebra," *Historia Mathematica*, 1981; "Benjamin Peirce's *Linear Associative Algebra*," *Isis*, 1979.

David E. Rowe is Associate Professor of Mathematics at Pace University, Pleasantville, New York, and book review editor of *Historia Mathematica*. His recent publications include: (with Karen V. H. Parshall), "American Mathematics Comes of Age: 1875–1900," in P. Duren, ed., A Century of Mathematics in America, vol. 3, 1989; "Klein, Hilbert, and the Göttingen Mathematical Tradition," *Osiris*, 1989; "Der Briefwechsel Sophus Lie – Felix Klein, eine Einsicht in ihre persönlichen und wissenschaftlichen Beziehungen," *NTM*, 1988; "'Jewish Mathematics' at Göttingen in the Era of Felix Klein," *Isis*, 1986.